AF411662

Oxidative Neural Injury

For other titles published in this series, go to
http://www.springer.com/series/7678

Sigrid C. Veasey

Editor

Oxidative Neural Injury

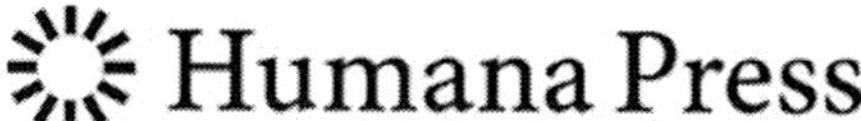 Humana Press

Editor
Sigrid C. Veasey
University of Pennsylvania
School of Medicine
Translational Research Laboratories
125 South 31st Street
Philadelphia, PA 19104
USA
veasey@mail.med.upenn.edu

ISBN 978-1-60327-341-1   e-ISBN 978-1-60327-342-8
DOI 10.1007/978-1-60327-342-8
Springer Dordrecht Heidelberg London New York

Library of Congress Control Number: 2009921900

Preface

Twenty-five years ago, Earl R. Stadtman, PhD discovered that specific enzymes regulating metabolism can be inactivated by oxidation [1]. He later showed that age-related oxidative modification contributes, at least in part, to age-related loss of function of the enzymes [2, 3]. Dr. Stadtman broke the ground for a new field of study to discover how oxidative stress contributes in significant ways to age-related cellular dysfunction and protein accumulation and that oxidation in the aging brain influences Alzheimer's disease, ischemia-reperfusion injury, amyotrophic lateral sclerosis, and lifespan [4–6]. Today, his research and mentorship have positively influenced the work of hundreds of scientists in this field. We dedicate this book to Dr. Earl R. Stadtman (1912–2008), in celebration of his passion for science and his superior collaborative and mentorship skills.

This book is comprised of three sections. The first describes the valuable roles reactive oxygen species (ROS) and reactive nitrogen species (RNS) play in cellular biology. The second section provides an overview of redox imbalance injury with effects on mitochondria, signaling, endoplasmic reticular function, and on aging in general. The third section takes these mechanisms to neurodegenerative disorders and provides a state-of-the-art look at the roles redox imbalances play in age-related susceptibility to disease and in the disease processes.

In the first section we attempt to answer a question posed by Dr. Stadtman, "Why have cells selected reactive oxygen species to regulate cell signaling events" [7]. It is imperative that we understand the biochemistry and the physiological significance of these molecules, in an effort to protect vital ROS/RNS functions in the cell and organism, as we attempt to minimize redox injury. As highly reactive molecules, ROS/RNS are expected to provide very rapid and focal signaling. In Chapter "Reactive Oxygen Species, Synaptic Plasticity and Memory", this concept and the newly identified mechanisms of ROS involvement in synaptic are discussed in detail. There are indeed significant recent breakthroughs to better address Dr. Stadtman's question, including discovery of NADPH oxidase as a major source of ROS in long-term plasticity. This identification has led to substantiation of the roles reactive oxygen species play in memory and delineates neural groups involved in NADPH oxidase-dependent plasticity. As a new discovery, future directions to further delineate the specific redox modifications resulting in long-term plasticity are described. Reactive nitrogen species also play important roles in cellular homeostasis and plasticity [8]. Chapter "Nitric Oxide Biochemistry: Pathophysiology of Nitric Oxide-Mediated Protein Modifications" makes sense of the elaborate biochemistry of nitric oxide species, describing the roles of specific reactive nitrogen species in learning and signaling, their sources and the relevance of their interactions with metals, thiols, and oxides in health and disease. In light of the pathological roles reactive oxygen and nitrogen species play in aging and many age-related neurodegenerative processes, the molecules must serve, in turn, extremely vital processes.

The following three chapters focus on redox pathophysiology in neural tissue. The common theme in these chapters is that while redox imbalance contributes to neural injury, there are important crosstalk mechanisms with the endoplasmic reticulum and mitochondria, and it is the combined interactions between endoplasmic reticulum, mitochondria and redox balance that determine susceptibility to injury. Thus, we describe how redox imbalance can perturb specific cellular functions, focusing on organelles implicated in neurodegenerative processes: the endoplasmic reticulum and mitochondria. Chapter "Redox Imbalance in the Endoplasmic Reticulum" describes a bidirectional influence between redox imbalance and endoplasmic reticulum function. As the protein-folding center, the endoplasmic reticulum lumen is a highly oxidative environment. Despite this, the endoplasmic reticulum lumen cannot handle further increases in oxidative stress. With meager reducing capacity, redox imbalance in the lumen can lead to significant oxidative folding modifications in lumen proteins sufficient to trigger an unfolded protein response. Small redox alterations can be addressed effectively by the endoplasmic reticulum by slowing protein synthesis, increasing degradation of poorly folded proteins, and increasing anti-oxidant enzymes and glutathione. Significant redox alterations initiate an executioner response with transcription and translation of pro-apoptotic factors. Endoplasmic reticulum stress and protein misfolding have been shown for most neurodegenerative processes. Thus, an understanding of the molecular mechanisms underlying redox homeostasis and dyshomeostasis in the endoplasmic reticulum is essential to understand toward the development of therapies for neurodegenerative processes. Mitochondrial function is impaired early on in many neurodegenerative processes. Chapter "Exocytosis, Mitochondrial Injury and Oxidative Stress in Neurodegenerative Diseases" describes specifically how mitochondrial injury and dysfunction can lead to neurobehavioral impairments. Here a major point is made that neurobehavioral impairments are the consequence not only of neuronal loss but also of impaired signaling in remaining neurons. This impaired signaling occurs when energy stores decline with early mitochondrial dysfunction. A model of how mitochondrial dysfunction would impact upon neuronal signaling in each of the major neurodegenerative diseases is provided. Understanding the early dysfunction of neurons in addition to mechanisms of cell loss is critical to identifying the most effective preventative therapies for neurodegenerative processes. Chapter "Neuronal Vulnerability to Oxidative Damage in Aging" adds consideration of calcium dysregulation in oxidative neural injury and places these interactions into the context of aging: a vicious cycle of cellular dyshomeostasis.

The third section begins to explore in greater depth specific conditions of neural injury. We begin in Chapter "Ischemia-Reperfusion Induces ROS Production from Three Distinct Sources" with ischemia-reperfusion injury. Neurodegenerative processes are typically insidious and redox imbalance is examined late in the course of injury. Ischemia-reperfusion affords a unique opportunity to examine the acute pattern of oxidative stress with high temporal resolution. Sources of ROS differ across this temporal course, and thus addressing each source will be important to identify optimal therapies for ischemia-reperfusion redox injury. The next three chapters detail oxidative injury in Alzheimer's disease. Chapter "Alzheimer Disease: Oxidative Stress and Compensatory Responses" describes anti-oxidant responses of β-amyloid and hyperphosphorylated tau to counter mitochondrial and metal abnormalities early on in Alzheimer's pathophysiology. Chapter "Oxidative Stress Associated Signal Transduction Cascades in Alzheimer Disease" delves further into specific enzyme abnormalities in mitochondrial dysfunction early on in Alzheimer's and how the neurons respond to the oxidative challenge. Here enters a second vicious cycle through the stress-activated protein kinase pathway upregulating β-amyloid and an initiation pathway for apoptosis. Chapter "Nitrated Proteins in the Progression of Alzheimer's Disease: A Proteomics Comparison of Mild Cognitive Impairment and Alzheimer's Disease Brain" complements the previous two chapters with molecules involved in nitrative injury. Nitrosative stress occurs early in Alzheimer's and is present in mild cognitive impairment. Proteomics has identified nitrated proteins in mid cognitive impairment that may be classified by biological function to provide insight into pathophysiology of nitrosative stress in mild cognitive impairment and in Alzheimer's disease, with involvement of energy, dendritic, signaling, and detoxification proteins. Differences between proteins nitrated in mild cognitive impairment versus Alzheimer's suggest early and late nitrosative stress effects.

The last three chapters focus on oxidative and nitrosative stress in neuronal injury and loss in Parkinson's disease. Chapter "Parkinson Disease: An Overview of Pathogenesis" provides an overview of genetic and environmental modifiers of Parkinson's disease, describes the strengths and weaknesses of each animal model of Parkinson's disease, and then places oxidative stress in the context of pathophysiology. The latter concept is elaborated upon in Chapter "Protein Oxidation Triggers the Unfolded Protein Response and Neuronal Injury in Chemically Induced Parkinson Disease", where protein oxidation is linked to endoplasmic reticulum stress and the unfolded protein response. This book ends on a positive note with Chapter "Treating Oxidative Neural Injury: Methionine Sulfoxide Reductase Therapy for Parkinson Disease" describing an effective approach for reducing oxidative stress and slowing Parkinson's symptoms in both the mouse and fly models.

It is anticipated that continued comprehensive exploration of early and late interactions between mitochondrial and endoplasmic reticulum function and redox imbalance are required to identify the most promising targets for preventing age-related neural injury and neurodegenerative processes. The true test for effectiveness should come back to Dr. Stadtman's earlier work with testing the functionality of perturbed metabolic enzymes.

We thank Jennifer Montoya, for without her tireless assistance this book would not have been completed.

Philadelphia, Pennsylvania

S.C. Veassey

References

1. Fucci L, Oliver CN, Coon MJ, Stadtman ER. Inactivation of key metabolic enzymes by mixed-function oxidation reactions: possible implication in protein turnover and ageing. Proc Natl Acad Sci U S A. 1983 Mar;80(6):1521–1525.
2. Oliver CN, Ahn BW, Moerman EJ, Goldstein S, Stadtman ER. Age-related changes in oxidized proteins. J Biol Chem. 1987 Apr 25;262(12):5488–5491.
3. Stadtman ER, Starke-Reed PE, Oliver CN, Carney JM, Floyd RA. Protein modification in aging. EXS. 1992;62:64–72.
4. Smith CD, Carney JM, Starke-Reed PE, Oliver CN, Stadtman ER, Floyd RA, Markesbery WR. Excess brain protein oxidation and enzyme dysfunction in normal aging and in Alzheimer disease. Proc Natl Acad Sci U S A. 1991 Dec 1;88(23):10540–10543.
5. Oliver CN, Starke-Reed PE, Stadtman ER, Liu GJ, Carney JM, Floyd RA. Oxidative damage to brain proteins, loss of glutamine synthetase activity, and production of free radicals during ischemia/reperfusion-induced injury to gerbil brain. Proc Natl Acad Sci U S A. 1990 Jul;87(13):5144–5147.
6. Moskovitz J, Bar-Noy S, Williams WM, Requena J, Berlett BS, Stadtman ER. Methionine sulfoxide reductase (MsrA) is a regulator of antioxidant defense and lifespan in mammals. Proc Natl Acad Sci U S A. 2001 Nov 6;98(23):12920–12925.
7. Stadtman ER, Levine RL. Why have cells selected reactive oxygen species to regulate cell signaling events? Hum Exp Toxicol. 2002 Feb;21(2):83.
8. Gow AJ, Ischiropoulos H. Nitric oxide chemistry and cellular signaling. J Cell Physiol. 2001 Jun;187(3):277–282.

Contents

Contributors

Andrey Y Abramov Department of Physiology, University College London, London, UK, a.abramov@ucl.ac.uk

Gjumrakch Aliev Department of Biology, University of Texas at San Antonio, San Antonio, TX, USA

Gábor Bánhegyi Department of Medical Chemistry, Molecular Biology and Pathobiochemistry, Semmelweis University, Budapest, Hungary; Pathobiochemistry Research Group of the Hungarian Academy of Sciences and Semmelweis University, Budapest, Hungary; Department of Pathophysiology, Experimental Medicine and Public Health, University of Siena, Via Aldo Moro, Siena, Italy, banhegyi@puskin.sote.hu

Alison I. Bernstein Washington University in St. Louis, St. Louis, MO, USA

D. Allan Butterfield Department of Chemistry, University of Kentucky, Lexington, KY, USA; Sanders-Brown Center on Aging, University of Kentucky, Lexington, KY, USA; Center of Membrane Sciences, University of Kentucky, Lexington, KY, USA, dabcns@uky.edu

Gemma Casadesus Department of Pathology, Case Western Reserve University, Cleveland, OH, USA

Pratap Chand Movement Disorders, Department of Neurology, Saint Louis University, University of School of Medicine, St Louis, Mo, USA, pchand1@slu.edu

Miklós Csala Department of Medical Chemistry, Molecular Biology and Pathobiochemistry, Semmelweis University, Budapest, Hungary

Andrew Gow Department of Pharmacology and Toxicology, Rutgers University, Piscataway, NJ, USA

Stefan H. Heinemann Department of Biophysics, Center for Molecular Biomedicine, Friedrich Schiller University Jena, Jena, Germany

Toshinori Hoshi Department of Physiology, University of Pennsylvania, Philadelphia, PA, USA, hoshi@hoshi.org

Damien J. Keating Department of Human Physiology and Center for Neuroscience, Flinders University, Adelaide, Australia, damien.keating@flinders.edu.au

Kenneth T. Kishida Computational Psychiatry Unit, Department of Neuroscience, Baylor College of Medicine, Houston, TX, USA

Eric Klann Center for Neural Science, New York University, New York, USA, eklann@cns.nyu.edu

Hyoung-Gon Lee Department of Pathology, Case Western Reserve University, Cleveland, OH, USA

Irene Litvan Division of Movement Disorders, Department of Neurology, University of Louisville School of Medicine, Louisville, KY, USA; University of Louisville National Parkinson Foundation Center of Excellence at Frazier Rehab Institute, Louisville, KY, USA, litvanirene@insightbb.com

József Mandl Department of Medical Chemistry, Molecular Biology and Pathobiochemistry, Semmelweis University, Budapest, Hungary; Pathobiochemistry Research Group of the Hungarian Academy of Sciences and Semmelweis University, Budapest, Hungary

Éva Margittai Department of Medical Chemistry, Molecular Biology and Pathobiochemistry, Semmelweis University; Budapest, Hungary

Mark P. Mattson Laboratory of Neurosciences, National Institute on Aging Intramural Research Program, Baltimore, MD, USA, mattsonm@grc.nia.nih.gov

Rosemary H Milton Department of Physiology, University College London, London, UK

Paula I. Moreira Center for Neuroscience and Cell Biology of Coimbra, University of Coimbra, Coimbra, Portugal

Akihiko Nunomura Department of Psychiatry and Neurology, Asahikawa Medical College, Asahikawa, Japan; and Interdisciplinary Graduate School of Medicine and Engineering, University of Yamanashi, Yamanashi, Japan

Karen L. O'Malley Washington University in St. Louis, St. Louis, MO, USA

Eitan Okun Laboratory of Neurosciences, National Institute on Aging Intramural Research Program, Baltimore, MD, USA, omalleyk@pcg.wustl.edu

George Perry Department of Pathology, Case Western Reserve University, Cleveland, OH, USA; College of Sciences, University of Texas at San Antonio, San Antonio, TX, USA, george.perry@utsa.edu

Robert B. Petersen Department of Pathology, Case Western Reserve University, Cleveland, OH, USA, robert.petersen@case.edu

Alba Rossi-George Department of Pharmacology and Toxicology, Rutgers University, Piscataway, NJ, USA

Mark A. Smith Department of Pathology, Case Western Reserve University, Cleveland, OH, USA

Renã A. Sowell Department of Chemistry, University of Kentucky, Lexington, KY, USA; Sanders-Brown Center on Aging, University of Kentucky, Lexington, KY, USA

Rukhsana Sultana Department of Chemistry, University of Kentucky, Lexington, KY, USA; Sanders-Brown Center on Aging, University of Kentucky, Lexington, KY, USA

Sigrid Veasey Center for Sleep and Respiratory Neurobiology, University of Pennsylvania, Philadelphia, PA, USA

Ramez Wassef Department of Physiology, University of Pennsylvania, Philadelphia, PA, USA

Mark P. Zanin Department of Human Physiology and Centre for Neuroscience, Flinders University, Adelaide, Australia

Xiongwei Zhu Department of Pathology, Case Western Reserve University, Cleveland, OH, USA

Reactive Oxygen Species, Synaptic Plasticity, and Memory

Kenneth T. Kishida and Eric Klann

Abstract Increasing evidence suggests that reactive oxygen species (ROS), such as superoxide and hydrogen peroxide, act as signaling molecules necessary for neuronal processes underlying cognition. Specifically, ROS have been shown to be necessary in molecular processes underlying signal transduction, synaptic plasticity, and memory formation. Research from several laboratories suggests that NADPH oxidase is an important source of superoxide in the brain. Herein we review evidence showing that ROS are important signaling molecules involved in synaptic plasticity and memory formation. Moreover, we will discuss the evidence that a neuronal NADPH oxidase complex is a key regulator of ROS generation in synaptic plasticity and memory formation. Understanding redox signaling in the brain, including the sources and molecular targets of ROS, is important for a full understanding of the signaling pathways that underlie synaptic plasticity and memory. Moreover, knowledge of ROS function in the brain is critical for understanding aging and neurodegenerative diseases such as Alzheimer's disease and Parkinson's disease that may be exacerbated by the unregulated generation of ROS.

Key words Synaptic plasticity · Learning · Memory · Cognition · NADPH oxidase · Long-term potentiation · Hippocampus · Redox signaling · NMDA receptor · High-frequency stimulation

1 Introduction

Over the last 10 years, multiple studies have indicated that reactive oxygen species (ROS), such as superoxide and hydrogen peroxide (H_2O_2), are important signaling molecules underlying mammalian learning and memory. Historically, ROS have been described as a class of destructive molecules that hinder neuronal function and as such have been implicated in degenerative processes

E. Klann (✉)
Center for Neural Science, New York University, New York, USA
e-mail: eklann@cns.nyu.edu

S.C. Veasey (ed.), *Oxidative Neural Injury*,
Contemporary Clinical Neuroscience, DOI 10.1007/978-1-60327-342-8_1,

underlying Alzheimer's and Parkinson's diseases [1–4], as well as processes thought to underlie the general aging-associated decline of cognitive function [5–7]. However, several studies utilizing cellular and behavioral levels of analyses have shown that ROS are required for normal cognitive function. Specifically, ROS have been shown to be required for learning and memory, a form of synaptic plasticity called long-term potentiation (LTP), and for biochemical signal transduction cascades that are believed to underlie both LTP and memory formation. Thus, the source of ROS responsible for these brain functions and how these small highly reactive molecules are regulated are important questions that we will discuss in this chapter.

Positive correlations have been made between LTP and learning. One such correlation is that signal transduction cascades that are activated during LTP have been shown to be required for memory formation (see Fig. 1 for signal transduction cascades implicated in LTP and memory). For instance, inhibition of NMDA receptors can block LTP and can interfere with certain types of memory formation [8]. Furthermore, deficiencies in memory formation are observed in mice with gene-specific mutations in loci coding for signal transduction elements that are critical for LTP [9–12]. Many of these signal transduction elements, such as small messenger molecules and protein kinases, become active once NMDA receptors are activated, thereby permitting the influx of Ca^{2+} into the postsynaptic terminal. Thus, NMDA receptor-dependent LTP in acute hippocampal slices has been used as an in vitro model to study synaptic plasticity and the cellular mechanisms underlying memory formation.

In the hippocampus, high-frequency stimulation of CA3 Schaffer collaterals can induce LTP at synapses in area CA1. High-frequency stimulation induces the activation of postsynaptic NMDA receptors, resulting in the influx of Ca^{2+} into the postsynaptic terminal. An example of an LTP experimental setup to measure postsynaptic potentials and a typical LTP response to a high-frequency stimulus are provided in Fig. 2. The transient rise in Ca^{2+} results in the production of small messenger molecules, such as cyclic adenosine monophosphate (cAMP), nitric oxide, arachidonic acid, and ROS [13]. These small messenger molecules in turn activate kinases such as protein kinase C (PKC) and extracellular signal-regulated kinase (ERK) and typically inhibit phosphatases such as calcineurin (protein phosphatase 2B or PP2B) [14, 15]. Many of these signaling molecules and enzymes have been shown to be involved in the induction and expression of LTP, as well as memory formation [14–16]. In addition to these well-studied signaling pathways, NMDA receptor activation has been shown to produce ROS [17], which are also required for LTP and hippocampus-dependent memory [18–21]. Until very recently, the source of ROS required for LTP and memory formation was unclear.

NADPH oxidase has been shown in multiple cell types to generate superoxide as a response to specific extracellular and intracellular stimuli [22, 23]. The best known of these responses is the phagocytic oxidative burst [23]. This burst generates large amounts of superoxide rapidly, transiently, and in a well-controlled manner. The pool of superoxide produced by NADPH oxidase is

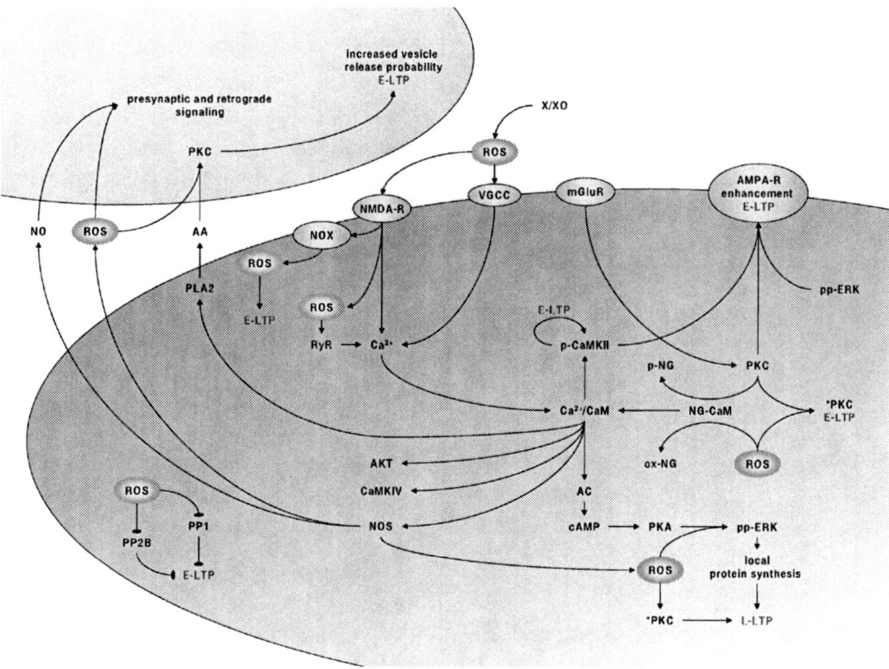

Fig. 1 Schematic summarizing signaling events underlying early and late phases of LTP. NMDA receptors (NMDA-R), voltage-gated calcium channels (VGCC), metabotropic-glutamate receptors (mGluR), and AMPA receptors (AMPA-R) become active in the postsynaptic membrane following the appropriate stimulation from presynaptic terminals. Activation of these receptors, individually or in combination, leads to the activation of multiple biochemical signaling cascades, many of which lead to the production of reactive oxygen species (ROS) and the activation of protein kinases, such as protein kinase C (PKC), protein kinase A (PKA), extracellular signal-regulated kinase (ERK), and calcium/calmodulin-dependent kinase II (CaMKII). Stimulation of pathways leading to LTP also includes the inhibition of protein phosphatases, such as protein phosphatase 1 and calcineurin (PP1 and CnA, respectively), which can be mediated by ROS directly and indirectly. Interestingly, postsynaptic signaling cascades are not the sole targets of biochemical signals that regulate synaptic plasticity. Presynaptic molecules also become altered by specific signaling cascades via presynaptic membrane receptors and retrograde signaling molecules, which include nitric oxide (NO), ROS, and arachidonic acid (AA)

used to break down phagocytosed material and as a signal to initiate signal transduction cascades underlying the bactericidal response [23]. Furthermore, in several cell types, production of superoxide by NADPH oxidase has been implicated in signaling required for transcriptional activation and cell proliferation [24–28].

In this chapter we will briefly discuss the evidence that ROS are important signaling molecules underlying LTP and memory, including known biochemical targets of ROS signaling in the brain. We will also discuss some of the hypothesized sources of ROS focusing on recent evidence pointing to the role

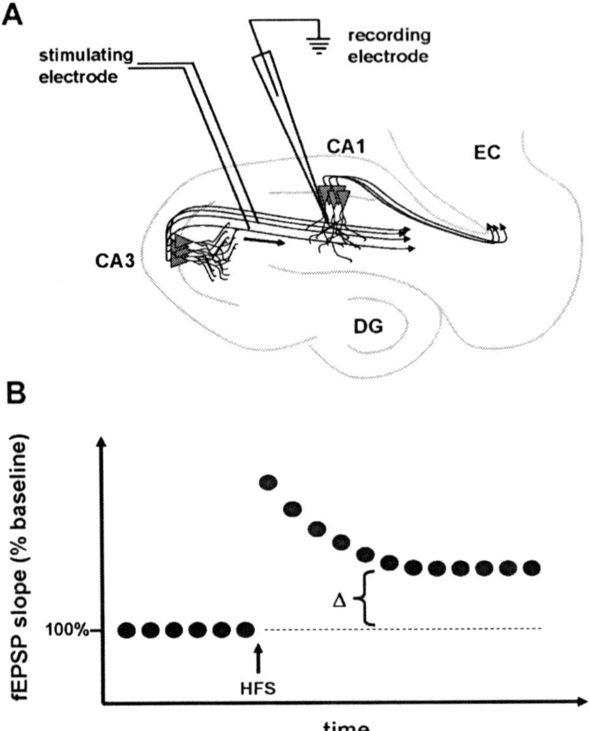

Fig. 2 Depiction of a hippocampal slice with stimulating and recording electrodes placed for measuring field excitatory postsynaptic potentials (fEPSPs) during a typical LTP experiment. (**A**) A drawing of a hippocampal slice indicating the dentate gyrus (DG), area CA3, and area CA1. Schaffer collaterals emanating from pyramidal neurons in area CA3 are stimulated using a bipolar stimulating electrode. Schaffer collaterals make synaptic contact with the dendritic fields of pyramidal neurons in area CA1. Synaptic communication in area CA1 is monitored using a glass electrode placed in *stratum radiatum* (the dendritic region) of area CA1. (**B**) Depiction of fEPSP measurements from a typical LTP experiment. The *x*-axis is time; the *y*-axis is the measured response, which is quantified as the slope of the fEPSP normalized to baseline responses. A high-frequency stimulus (HFS) can be used to induce long-term potentiation, which can be measured as the long-lasting increase in the slope of the fEPSP (Δ)

that NADPH oxidase plays in the generation of ROS that are required for NMDA receptor-dependent signal transduction, synaptic plasticity, and memory formation.

2 ROS Are Critical Signaling Molecules in Synaptic Plasticity

Synaptic plasticity is the physiological process that is thought to underlie learning and memory at the cellular level. One form of plasticity that has been commonly studied is LTP. Many of the molecular processes underlying LTP are also required for learning and memory [18, 29]. Thus, LTP has become a popular model to study the mechanisms that underlie learning and memory behavior.

Investigations into the role of ROS in LTP have revealed an interesting, yet complex role for oxidative species such as superoxide and H_2O_2 in the molecular processes that lead to changes in synaptic strength. The role of ROS is specific to the identity of the oxidative molecule involved (i.e., superoxide or H_2O_2) [19, 30] and the concentration of ROS during the process [19, 31, 32].

The use of pharmacological approaches and genetically modified mice that overexpress antioxidant enzymes to block the activity of ROS has been useful in identifying the necessity of these signaling molecules in synaptic plasticity and memory formation. LTP studies in the rodent hippocampus have revealed that scavenging superoxide blocks LTP induced using high-frequency stimulation (HFS-LTP) [33, 34], suggesting that superoxide is required for HFS-LTP. On the other hand, overproduction of H_2O_2 as a byproduct of superoxide dismutase (SOD) activity was also shown to inhibit LTP [35]. In order to better understand the specific role of ROS, investigators have also applied exogenous sources of either superoxide or H_2O_2 to various in vitro systems and have observed interesting effects on neuronal plasticity.

Xanthine–xanthine oxidase (X/XO) is often used to generate superoxide in vitro [36]. Exogenous application of X/XO to hippocampal slices resulted in a short-term depression of synaptic transmission that eventually potentiated in an LTP-like manner that was inhibited by SOD [36]. The role of H_2O_2 has also been investigated using the application of exogenous H_2O_2 to study the effect of this ROS on the molecular mechanisms underlying cellular plasticity. Kamsler and colleagues have shown that H_2O_2 could either potentiate or depress HFS-LTP in a concentration-dependent manner [32]. Also, low concentrations of H_2O_2 have been shown to potentiate or depress intracellular Ca^{2+} levels [37, 38].

Many of the studies implicating ROS involvement in the molecular mechanisms underlying synaptic plasticity and memory investigate a narrow range of superoxide or H_2O_2 concentrations. However, very interesting results were obtained when a range of ROS concentrations were tested for their effects on synaptic plasticity [19, 31, 32, 36]. For example, high concentrations of superoxide or H_2O_2 resulted in the depression of excitatory postsynaptic field potentials (fEPSPs) measured in the hippocampal area CA1, whereas lower concentrations resulted in a potentiation of the fEPSP [32, 36].

The role of ROS in aging-related changes in synaptic plasticity and cognitive performance is an interesting and thriving area of research. During aging, there appears to be a shift in the role of ROS in synaptic plasticity and memory that is dependent on either the accumulation of oxidative damage or changes in oxidant regulation via enzymatic antioxidants [6, 39]. Although not a focus of this chapter, understanding the aging-dependent changes in ROS function in the context of synaptic plasticity and memory will be important for future consideration.

3 ROS Are Required for Learning and Memory

SOD mutant mice have been critical models for determining the role of ROS in learning and memory. SOD scavenges superoxide and dismutates it to H_2O_2, which can then be rapidly degraded by catalase. There are three SOD isozymes: cytoplasmic-Cu/Zn-SOD (SOD-1), manganese-containing mitochondrial SOD (SOD-2 or MnSOD), and extracellular-SOD (EC-SOD), which is also a Cu/Zn-containing SOD. Mice that either overexpress SOD or express a dysfunctional SOD have been analyzed using various behavioral paradigms with surprising results.

The earliest evidence that ROS are required for learning and memory came from mice that overexpress SOD-1. Behavioral analyses of the SOD-1 over-expressing mice indicated that the mice displayed altered behavior in open field analyses and more interestingly showed decreased escape latencies in the hippocampus- and NMDA receptor-dependent Morris water maze paradigm as compared to wild-type mice [8, 35, 40]. These data were the first to suggest that scavenging of superoxide disrupted learning and memory. Consistent with a requirement of superoxide in learning and memory, Levin et al. showed that mice that either overexpress EC-SOD or are genetically deficient for the EC-SOD gene had impaired performance in the win-shift eight-arm radial maze [41], also a hippocampus-dependent task [42]. Further analysis of the EC-SOD overexpressing mice revealed that these mice were impaired in hippocampus-dependent contextual fear conditioning [43, 44] but were normal for cue-dependent fear conditioning [5, 20, 44].

Hippocampus-dependent learning and memory formation has been a useful model in determining the role of ROS in cognition; however, behavioral analysis of SOD overexpressing mice has revealed potential roles for ROS in cognitive performance associated with other brain regions as well. EC-SOD overexpressing mice displayed alterations in performance on the eight-arm radial maze task that was dependent on motivational state induced by food restriction [41, 45]. In a low motivational state EC-SOD overexpressing mice showed impairments in learning; however, when highly motivated EC-SOD overexpressing mice were able to learn, albeit at a slower rate, and express long-term memory [45].

Taken as a whole, the data generated from the analysis of SOD mutant mice suggest that ROS are critically involved in the molecular processes involved in cognition, particularly processes underlying learning and memory formation.

4 ROS Are Critical Signaling Molecules that Modulate Signaling Pathways Involved in LTP and Memory

The molecular mechanisms underlying synaptic plasticity and memory have been intensely studied and remain a thriving area of research (see Fig. 1) [15]. Investigations of signal transduction cascades involved in synaptic plasticity have revealed a wide range of molecular players, including protein kinases [46], phosphatases [46, 47], transcription factors [48], translation factors [49], GTPases [50], and Ca^{2+}-dependent enzymes [16, 51]. Interestingly, ROS have been shown to be important modulators of many of these pathways [30], including some evidence of direct modification of signaling enzymes. As mentioned earlier, NMDA receptor-dependent signaling is one critical pathway that is thought to underlie synaptic plasticity and long-term memory formation that has been shown to contain several signaling molecules that are directly affected by ROS.

Glutamatergic synaptic transmission results in the activation of AMPA receptors and voltage-dependent calcium channels during fast synaptic transmission. Given the appropriate pattern of stimulation, NMDA receptors can become activated, which results in a transient spike in postsynaptic calcium levels that leads to the activation of numerous signal transduction cascades. Among the signaling events that occur following NMDA receptor activation are the activation of various protein kinases, including calcium/calmodulin-dependent protein kinase II (CaMKII) [16], protein kinase C (PKC) [34, 52], and extracellular signal-regulated kinase (ERK) [10, 53, 54], and the inhibition of protein phosphatases such as calcineurin and protein phosphatase 1 [55–58]. NMDA receptor stimulation has also been shown to result in the production of ROS [59]. Following ROS generation, several target proteins have been shown to become oxidatively modulated. For example, NMDA receptor activation was shown to result in the oxidation of neurogranin (NG) in a time- and dose-dependent manner that is sensitive to NMDA receptor antagonists [60]. Neurogranin is a postsynaptic PKC substrate that binds to and sequesters calmodulin (CaM) [60]; either oxidation or phosphorylation [52, 61] of neurogranin promotes the release of CaM, thereby promoting Ca^{2+}/CaM signaling via the activation of CaMKII. This signaling module is likely to trigger the induction of LTP [16]. NMDA receptor-mediated oxidation of neurogranin occurs within 3–5 minutes of stimulation and quickly returns to basal oxidation levels [60] and it has been shown that H_2O_2 can directly cause the oxidation of NG [60]. In addition, the generation of superoxide by X/XO increases the phosphorylation of neurogranin by the activation of autonomous PKC [52]. ROS also can directly oxidize and activate PKC (will be discussed below). Taken together, these reports suggest that neurogranin is an important target of NMDA receptor-induced ROS production, and when either oxidized or phosphorylated by PKC, promotes LTP and possibly long-term memory formation [11, 60, 62, 63].

Another signaling enzyme that is an important modulator of LTP and memory formation is PKC. LTP-inducing stimulation has been shown to result in the activation of PKC that is NMDA receptor-dependent [34, 36, 64]. Interestingly, the exogenous application of SOD and catalase was shown to inhibit not only LTP but also the LTP-induced activation of PKC [34, 36], suggesting that superoxide and H_2O_2 are necessary for PKC activation. In addition, direct application of the superoxide-generating system X/XO to hippocampal slices induced not only an LTP-like potentiation but also the persistent activation of PKC [65]. SOD, but not catalase, was shown to block the X/XO-induced activation of PKC [36, 65]. Furthermore, the PKC inhibitor bisindomaleimide blocked X/XO-induced LTP and HFS-induced LTP [36], suggesting a common pathway involving PKC for both HFS-LTP and X/XO-induced LTP. The mechanism of PKC activation via oxidation seems to be mediated via direct modification of the zinc finger region of the kinase; $ZnCl_2$ blocked the X/XO-induced activation of PKC, whereas the zinc chelator (TPEN) activated PKC [65]. Furthermore, X/XO stimulated the release of zinc from PKC [65]. Interestingly, peroxynitrite, a strong oxidant that is generated via the reaction of superoxide and NO [66], also can modulate PKC activity in a concentration-dependent manner. Low concentrations of peroxynitrite were shown to activate PKC, whereas high concentrations of peroxynitrite were shown to inhibit PKC [67]. At all concentrations tested, peroxynitrite increased the nitration of PKC [67]. Interestingly, although the activation of PKC by peroxynitrite was inhibited by reducing agents, peroxynitrite-induced inhibition of PKC activity was resistant to reducing agents [67]. It remains to be determined whether peroxynitrite-induced modulation of PKC occurs during LTP.

Activation of NMDA receptors, either pharmacologically or electrically with LTP-inducing HFS, has been shown to activate ERK [68–70]. Consistent with the importance of ROS involvement in NMDA receptor-dependent signaling, superoxide and to a lesser extent H_2O_2 were shown to be required for NMDA receptor-dependent activation of ERK [68]. In this study, NMDA receptor-dependent activation of ERK in acute hippocampal slices was blocked by exogenously added SOD, SOD mimetics, and catalase. Furthermore, it was shown that application of either X/XO or H_2O_2 to hippocampal slices resulted in the activation of ERK, which was inhibited by the general antioxidant N-acetyl-cysteine [71]. Thus, the activation of ERK during NMDA receptor-dependent LTP and long-term memory may require ROS.

Protein phosphatases have also been shown to be important modulators of synaptic plasticity [14] that are regulated by ROS. One such phosphatase is calcineurin (CnA), which is thought to suppress LTP and long-term memory formation by opposing the effect of LTP-inducing kinases such as CaMKII and PKC [14]. Calcineurin has been shown to be highly sensitive to redox modification by ROS [72]. For example, basal calcineurin activity was reduced by either X/XO or H_2O_2 but was enhanced by the addition of SOD. Furthermore, strong oxidizing agents inhibit calcineurin, whereas reducing agents were shown to enhance calcineurin activity possibly through direct oxidative modification of calcineurin protein [72]. Consistent with the idea that a redox-sensitive calcineurin might play a role in LTP, FK506 (a calcineurin inhibitor) blocked H_2O_2-mediated enhancement of LTP and restored LTP in slices treated with concentrations of H_2O_2 that normally inhibited LTP [32]. Interestingly, aged wild-type mice showed increased phosphatase activity that was similar to the levels of phosphatase activity observed in young wild-type mice that had been treated with exogenously applied H_2O_2. Thus, modulation of calcineurin by ROS may promote the expression of LTP following NMDA receptor activation.

Another potential target of ROS-mediated signal transduction during hippocampal synaptic plasticity and memory is the redox-sensitive transcription factor NF-κB. Several lines of evidence suggest that LTP and long-term memory may involve redox-dependent activation of NF-κB. It has been observed that NF-κB becomes activated following LTP induction [73, 74] and that NF-κB can be activated by redox-mediated signaling [48, 75, 76]. Furthermore, NF-κB has been implicated in the formation of long-term memory in nonmammalian organisms [77, 78] as well as in the consolidation of fear memories in rodents [77]. Taken together, these findings suggest that NF-κB may be an important target of redox-mediated signaling during synaptic plasticity and normal cognitive function. However, further studies are required to directly demonstrate that ROS-dependent activation of NF-κB occurs during synaptic plasticity and memory.

Regulation of intracellular Ca^{2+} is an important function that controls synaptic plasticity and memory [51]. ROS-mediated signaling could modulate intracellular Ca^{2+} via the oxidative modification of Ca^{2+} channels, thereby altering the Ca^{2+} response during plasticity-inducing stimuli. As mentioned previously, ROS can directly modulate voltage-dependent Ca^{2+} channels [37]; however, there is also evidence that oxidative regulation of ryanodine receptors could be involved in redox-mediated modulation of intracellular Ca^{2+}. Ryanodine receptors are very sensitive to redox modulation, which results in alterations in channel function [79]. This is intriguing because mutant mice that lack one form of the ryanodine receptor, RyR3, have been shown to express alterations in HFS-LTP and diminished spatial memory [80, 81]. Thus, ROS-mediated alteration of ryanodine receptor function may be an important step during the molecular events involved in the modulation of intracellular Ca^{2+} that underlies LTP and memory. To test this mechanism directly, recent experiments have revealed that application of X/XO directly to hippocampal slices stimulated the activation of ryanodine receptors [82]. Also, potentiation of fEPSPs recorded in hippocampal area CA1 via application of X/XO could be blocked by inhibiting ryanodine receptor function. Furthermore, hippocampal slices prepared from RyR3 knockout mice failed to exhibit X/XO-induced potentiation. Interestingly, application of a general ryanodine receptor inhibitor had no further effect on X/XO-treated hippocampal slices prepared from RyR3 knockout mice, suggesting that the effects of pharmacological ryanodine receptor inhibition on ROS-induced potentiation occur specifically on the RyR3 isoform. Huddleston and colleagues went on to show in hippocampal slices that ryanodine receptor inhibition (either pharmacological or genetic via RyR3 knockout mice) during X/XO treatment also inhibited ERK2 phosphorylation, which is also a downstream effector of Ca^{2+}-dependent signaling

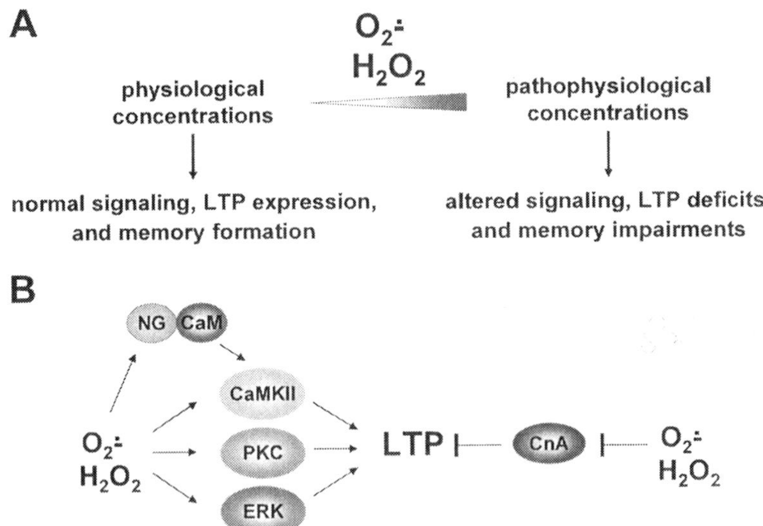

Fig. 3 Schematic depicting the role of ROS in the molecular mechanisms underlying cognition. (**A**) Low concentrations of ROS are required for signaling, synaptic plasticity, and memory formation; however, as the concentration of ROS increases, their function switches from a signaling molecule to an inhibitory or even toxic molecule. (**B**) On the *left*, superoxide and H_2O_2 have been shown to activate protein kinase C (PKC) and extracellular signal-regulated kinase (ERK) and to oxidize neurogranin (NG), which then releases calmodulin, resulting in the activation of calcium/calmodulin-dependent protein kinase II (CaMKII); whereas on the *right*, superoxide and H_2O_2 have also been shown to inhibit calcineurin (CnA). Activation of PKC, ERK, and CaMKII promotes LTP, whereas the activity of PP2B tends to block LTP; thus the activation of PKC, ERK, and CaMKII, along with the inhibition of PP2B, are all plausible, redox-sensitive, mechanisms by which ROS could promote synaptic plasticity in a concentration-dependent and cellular signaling state-dependent manner

during LTP (see Fig. 1, and 3B). Consistent with the idea that superoxide-dependent activation of ERK is involved in LTP, X/XO-induced potentiation was blocked by inhibiting ERK signaling [82]. Importantly, RyR3 is not the only redox-sensitive calcium channel that is required for X/XO-induced potentiation in hippocampal area CA1; inhibition of L-type calcium channels is also sufficient to block superoxide-induced potentiation [83].

It seems clear that ROS are important modulators of signal transduction cascades that underlie synaptic plasticity and memory formation. Further work to identify other targets of ROS signaling will be important to better understand the significance of redox signaling in the brain. Another important goal will be to identify the sources of ROS that are involved in modulating redox-sensitive signaling pathways during LTP and memory formation. Uncontrolled ROS production would quickly become unhealthy for the neural tissue producing it; thus any mechanism involving ROS must contain a potent, yet controlled source of ROS. There are several candidates for sources of ROS that meet these criteria, which will be discussed in the next section.

5 Potential Sources of ROS Involved in LTP and Memory

Although ROS have been shown to be critical signaling molecules that are required in the molecular events that underlie synaptic plasticity and memory formation, the source of ROS during these events has yet to be determined. Several sources have been hypothesized and evidence showing the plausibility of ROS generation from these sources has been provided; however, experiments determining the distinct physiological significance of these potential sources of ROS have been elusive. Among the hypothesized sources of ROS are mitochondria, monoamine oxidase, cyclooxygenase, nitric oxide synthase, and NADPH oxidase. Each of these potential sources is known to generate ROS under pathophysiological conditions, but the physiological role of the ROS produced by these sources has not been determined. Experiments designed to determine the physiological significance of each of these sources of ROS, especially with respect to a role in synaptic plasticity and memory formation will be an important goal for understanding the role of ROS signaling in the brain.

5.1 Mitochondria

Mitochondria produce superoxide as a metabolic byproduct of the electron transport chain and oxidative phosphorylation. Typically, mitochondrial production of superoxide is studied in models of oxidative stress, apoptosis, and neurodegeneration; however, there is evidence that suggests that mitochondria may be a source of ROS that is stimulated by appropriate physiological stimuli. For instance, elevating Ca^{2+} and Na^+ is sufficient to produce free radicals from isolated rat mitochondria [84]. Also, NMDA receptor activation via glutamate application to cultured forebrain neurons induced a localized generation of ROS that was blocked by MK-801 [85]. Interestingly, glutamate application caused intracellular pH to decrease in a Ca^{2+}-dependent manner [85], which suggests that NOS could play a role in superoxide generation as well [86]. NMDA receptor stimulation also was shown to result in the production of superoxide that occluded superoxide production induced by the uncoupling of proton transport with FCCP [59]. The authors of this study suggested that this was evidence for mitochondrial production of superoxide; however, these data do not rule out the possibility that FCCP affected proton uncoupling on NADPH oxidase [87]. Dugan et al. showed that NMDA-induced dihydrorhodamine (DHR) oxidation may be caused by ROS generated by mitochondria because the oxidation was inhibited by inhibitors of mitochondrial complex I (rotenone) and III (antimycin A) [88]. Interestingly, activation of NMDA receptors through the application of glutamate seemed to be necessary; however, FCCP treatment in the presence of MK-801 and NBQX was sufficient to cause the oxidation of DHR [88]. Although these reports suggest that the mitochondrial electron transport chain may be a potent source of ROS during increased Ca^{2+} signaling, none of these reports distinguish the effects of high Ca^{2+}

and ROS generation during the induction of synaptic plasticity from those underlying neurotoxicity and apoptotic signaling. Moreover, mice that overexpress SOD-2, which should scavenge superoxide produced by mitochondria, exhibit normal hippocampal LTP and memory [39]. Finally, there is much evidence suggesting that mitochondrial production of ROS is tightly regulated and that increased ROS production by the mitochondria quickly produces oxidative disease states, including the induction of apoptosis and neurodegeneration [89]. Thus, at this time the evidence suggest that mitochondria are not a likely source of ROS signaling during synaptic plasticity and memory.

5.2 Monoamine Oxidase and Cyclooxygenase

Monoamine oxidase and cyclooxygenase are also potential sources, albeit indirectly, of ROS that could be involved in synaptic plasticity and cognition. In cerebellar granule cells a cyclooxygenase inhibitor (indomethican) and a monoamine oxidase inhibitor (nialamide) blocked NMDA- and kainic acid (KA)-induced ROS production, which suggests that the oxidative-metabolic activity of each of these enzymes may be responsible for the generation of ROS [90]. Rotenone did not block either NMDA- or KA-induced ROS production, which suggests that mitochondria is not the source of ROS as assessed in this system [90]. In addition, the phorbol ester phorbol-12-myristate-13-acetate (PMA), which is a potent stimulator of NADPH oxidase, stimulated ROS production in a mechanism that was different than NMDA- or KA-induced ROS production because indomethican, nialamide, and rotenone were unable to block PMA-stimulated ROS production [90]. These data indicate that there may be as many as three different sources of ROS in cerebellar granule cells. Further investigation into the potential role of monoamine oxidase in ROS generation during synaptic plasticity may lead to important insights regarding psychiatric disease, given the known effects of monoamine oxidase inhibitors on psychiatric disorders such as major depression [91].

5.3 Nitric Oxide Synthase

Nitric oxide synthase is well known for its role in generating nitric oxide (NO) gas, which has been observed to be a critical signaling molecule involved in synaptic plasticity and memory [92, 93]. However, NOS may also be an important source of ROS for the molecular events required for LTP and memory formation. For example, purified NOS has been shown to generate superoxide that was dependent on Ca^{2+} and calmodulin [94]. Interestingly, L-NAME, but not L-NMMA, could block the superoxide generation by NOS [94]. In addition, it was shown that NOS can generate H_2O_2 under a variety of conditions [86]. Specifically, low L-arginine and low pH each were shown to independently promote the generation of H_2O_2 production by purified NOS [86]. On the other hand H_4-biopterin inhibited H_2O_2 and promoted NO production by NOS [86]. Importantly, only certain NOS inhibitors were shown to inhibit NO and H_2O_2 formation. For instance, L-NNA, L-NAME, and L-NMMA all inhibited NO formation; however, only L-NNA and L-NAME also blocked H_2O_2 formation by NOS, whereas L-NMMA failed to block H_2O_2 formation [86, 95]. Further evidence for NOS generation of superoxide has come from experiments using cultures of primary cerebellar granule neurons, where glutamate receptor stimulation induced NOS-dependent production of superoxide if the cultures were pretreated with arginase [96].

NOS contains two enzymatic domains, one that generates NO and another that contains NADPH oxidase activity. NOS expression was shown to result in increased ERK activation that was inhibited by cotransfection of SOD [97], which suggested a role for either superoxide or H_2O_2 in NOS-mediated ERK activation. Interestingly, a mutation in NOS that rendered the NO synthase portion of the enzyme incompetent, yet retained NADPH oxidase activity, had no effect on ERK activation. In contrast, deletion of the NADPH-binding region of NOS blocked NOS-dependent ERK activation [97]. These results suggest that NOS may generate superoxide that is critical for signal transduction cascades that is separable from NO generation.

Experiments that use various pharmacological agents to inhibit NOS activity often fail to discriminate between the specific roles that each of these domains play. For instance, L-NAME, which is often proposed to inhibit NO generation by NOS, is actually a very good inhibitor of NOS-mediated NADPH oxidation and superoxide generation [95]. In addition, L-NMMA was shown to have little to no effect on the detection of the superoxide-dependent formation of DMPO-OOH, a superoxide-dependent electron spin-adduct, via neuronal NOS [95], suggesting that this pharmacological agent is good at dissecting the dual enzymatic function of NOS. Furthermore, diphenylene iodonium (DPI) is often used as an NADPH oxidase inhibitor [98]; however, DPI also was shown to inhibit the effect of L-arginine inhibition of NOS-mediated superoxide formation [95], likely via interaction with the NADPH oxidase domain of the NOS enzyme. Unfortunately, many of the experiments that implicate generation of NO by NOS in synaptic plasticity use pharmacological inhibitors of NOS that fail to distinguish the enzymatic generation of NO from NOS from the generation of superoxide by NOS. There is substantial evidence that NO is an important signaling molecule during synaptic plasticity [93], but the potential role of superoxide generated by NOS is relatively unexplored.

Taken together these reports suggest the possibility that NADPH oxidation by NOS could be an important source of ROS that is independent of NO generation during synaptic plasticity and memory formation. However, there is another enzyme complex whose primary enzymatic activity is the oxidation of NADPH and concomitant production of superoxide. The role of NADPH oxidase in the generation of superoxide will be the focus of the remainder of this chapter.

5.4 NADPH Oxidase

The NADPH oxidase complex is a plausible source of ROS in synaptic plasticity and memory that previously has been considered primarily as a generator of superoxide in nonneuronal cells, including immune system cells [23, 99], endothelial cells [100], and glia [3, 99]. In these previous studies, NADPH oxidase has been shown to be tightly regulated, such that a burst of superoxide could be generated in response to particular stimuli, which could subsequently be turned off; thus, the generation of ROS via the NADPH oxidase system has been shown to be rapid, tightly controlled, and specific to particular signaling events [22]. Interestingly, several of the activators and effectors of the NADPH oxidase complex also have been implicated in signal transduction mechanisms that underlie synaptic plasticity and memory formation. Moreover, recent evidence has been provided that directly supports a role of NADPH oxidase-dependent superoxide generation during brain function, which may explain why human patients with mutations in genes encoding subunits in the NADPH oxidase complex may display mild cognitive deficits [101].

The structure and regulation of the NADPH oxidase has been well studied [102]. Briefly, the NADPH oxidase complex consists of five subunits, three cytosolic ($p67^{phox}$, $p47^{phox}$, and rac) and two membrane-spanning ($gp91^{phox}$ and $p22^{phox}$). The membrane-spanning components exist as a heterodimer; $gp91^{phox}$ is the catalytic subunit that is responsible for the transfer of electrons between NADPH and molecular oxygen, as well as the H^+ conductance that has been associated with this process. The regulation of NADPH oxidase activity is mediated through complex interactions between the cytosolic and membrane-associated components. Translocation of $p67^{phox}$ and GTP-bound active Rac to the membrane is essential for the activation of $gp91^{phox}$-mediated electron transfer. $p47^{phox}$, once phosphorylated, acts as an "organizer" of the complex and mediates correct positioning and association of the $p67^{phox}$ "activator" subunit with the $gp91^{phox}$ and $p22^{phox}$ heterodimer. Thus, translocation of all the cytosolic subunits in response to specific stimuli is required for the full activation of the NADPH oxidase complex. A model of NADPH oxidase is shown in Fig. 4.

Interestingly, many of the signaling agents involved in LTP and memory formation are also known to regulate NADPH oxidase activity. The transcription-dependent regulation of NADPH oxidase subunits implicates transcription factor components known to be important in the regulation of LTP and memory formation. Specifically, treatment of murine monocytic cell with lipopolysaccharide (LPS) and interferon-γ (IFN-γ) resulted in the increased expression of $gp91^{phox}$ mRNA and protein that was dependent on NF-κB [103], which has been shown to be an important transcription factor involved in LTP and memory [73, 74, 77]. More near term signaling events have also

Fig. 4 Subunit composition and activation of the NADPH oxidase complex. (**A**) The NADPH oxidase complex consists of two membrane-bound subunits (gp91^{phox} and p22^{phox}) and three cytosolic subunits (p67^{phox}, rac, and p47^{phox}), which upon activation translocate and associate with the membrane-bound subunits (**B**). Upon activation, the NADPH oxidase complex transfers electrons from NADPH substrate to molecular oxygen, thus producing superoxide. During this process NADPH oxidase also pumps protons across the membrane

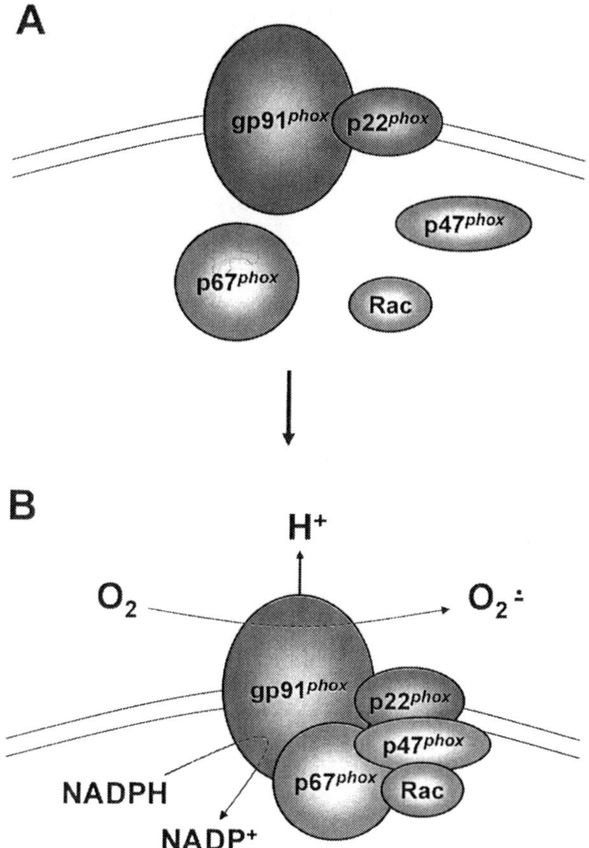

implicated a potential role for NADPH oxidase in the generation of ROS-mediated signaling in neurons.

The phorbol ester PMA is a widely used compound that induces NADPH oxidase activation via the phosphorylation of p47^{phox} through activated PKC [104]. In cerebellar granule cells, PMA was shown to stimulate ROS production [90] and in hippocampal slices, PMA treatment leads to an LTP-like potentiation that is dependent on PKC activation [36]. Furthermore, phospholipase A_2 (PLA$_2$)-dependent genesis of arachidonic acid was also shown to induce the activation of NADPH oxidase in intact neutrophils [105]. Consistent with the possibility that this type of signaling might be involved in synaptic plasticity, hippocampal slices treated with arachidonic acid during a brief train of tetanization resulted in an LTP-like potentiation [106]. Moreover, NMDA applied to cultured cerebellar granule cells led to the generation of superoxide that was mimicked by the application of arachidonic acid and inhibited by mepacrine, a PLA$_2$ inhibitor [107]. In addition, NMDA- and glutamate-induced oxidation of dichlorofluorescein (DCF) in cerebellar granule cells could be blocked via PLA_2 inhibition [17]. The PI3 kinase-Akt pathway is another kinase signaling pathway that plays a critical role in synaptic plasticity [108–110] and has been shown to be an important activator of NADPH oxidase in nonneuronal cells [27, 111].

Not only have upstream activators of NADPH oxidase been shown to be important regulators of plasticity and memory formation, but downstream effectors of NADPH oxidase-generated ROS in nonneuronal cells also parallel important signaling pathways involved in synaptic plasticity and memory. For instance, NADPH oxidase-generated ROS have been shown in T cells to regulate phosphorylation and activation of the ERK signal transduction pathway [24], which has been shown to be a critical signaling pathway in LTP and memory formation in neurons. The evidence above is consistent with the hypothesis that NAPDPH oxidase is an important source of ROS in signal transduction pathways in the brain. More direct evidence recently has been provided that implicates NADPH oxidase function in the brain.

6 NADPH Oxidase Expression in the Brain

Consistent with an important function for the NADPH oxidase complex in the central nervous system, all components of the complex, including the various homologs of specific subunits, have been shown to be expressed in various regions throughout the brain [112–118]. Serrano et al. have shown that mouse hippocampi are immunoreactive for gp91, p47, p67, p40, and $p22^{phox}$ proteins [112, 113] and that $p47^{phox}$ and $gp91^{phox}$ immunoreactivities were observed in pyramidal neurons in area CA1 [113]. Furthermore, the NADPH oxidase subunits $gp91^{phox}$ and $p67^{phox}$ have been found in synaptosomal fractions prepared from the whole brain and the hippocampus [114], suggesting a localized distribution that is consistent with a role for NADPH oxidase in synaptic plasticity. Interestingly, $gp91^{phox}$ and $p47^{phox}$ proteins were also found in several other areas of the brain including the cortex, habenula, paraventricular thalamic nucleus, anterior and posterior basolateral nucleus, basomedial nucleus of the amygdala, and striatum [113].

As discussed earlier, $gp91^{phox}$ is the catalytic subunit of NADPH oxidase that is responsible for the transfer of electrons between NADPH and molecular oxygen. Recently, homologs of $gp91^{phox}$ have been described, several of which are also expressed in the brain. In addition to $gp91^{phox}$ (also referred to as NOX-2), NOX-4 [116, 117] and NOX-5 [116] have been shown by rt-PCR to be expressed in the adult brain and NOX-4 was detected using in situ hybridization in the mouse cortex, cerebellum, and pyramidal cells of the hippocampus [117]. Furthermore, NOX-3 was shown to be highly expressed in the inner ear by rt-PCR and by in situ hybridization [115]. $p47^{phox}$ and $p67^{phox}$ and their respective homologs were also detected in the brain using rt-PCR analysis [115] and by Northern blot analysis [112]. Rao et al. also showed that NADPH oxidase is expressed and functional in lens epithelium [118]. Thus, there are likely to be multiple homologs of gp91^{phox} that are important for normal cognitive function. However, whether the other NOX proteins are regulated in a similar manner to gp91^{phox} and whether they are critical for ROS signaling in the brain remain to be determined.

The expression pattern of NADPH oxidase suggests that it may be involved in ROS-dependent signaling throughout the brain. Consistent with this notion, it was shown in cultured hippocampal neurons that PMA could stimulate the redistribution of the cytosolic subunits of the NADPH oxidase complex to the membrane [114]. Also in hippocampal slices, PMA induced the generation of superoxide that was inhibited by either DPI or AEBSF [114], two pharmacological inhibitors of the NADPH oxidase complex [98, 119]. Furthermore, stimulation of the cellular prion protein (PrPc) was shown in a number of neuronal and nonneuronal cell lines to lead to the activation of NADPH oxidase, which induced the activation of the MEK-ERK pathway in an NADPH oxidase- and ROS-dependent manner [28]. Thus, a functional NADPH oxidase is expressed in the brain, suggesting that this superoxide-generating complex may be involved in signaling, synaptic plasticity, and memory.

7 NADPH Oxidase-Mediated Signaling in the Brain

NMDA receptor-dependent activation of ERK is well known to be involved in various forms of synaptic plasticity and memory [13, 15, 51]. Consistent with a role for NADPH oxidase in mediating this type of signaling, DPI, an NADPH oxidase inhibitor, was shown to inhibit NMDA receptor-mediated ERK activation [68]. Moreover, mice that lacked the p47^{phox} subunit [120] also lacked the NMDA receptor-dependent activation of ERK [68]. These findings are consistent with previous reports that have implicated NADPH oxidase activity with MEK-ERK signal transduction in nonneuronal cells [24, 28]. However, it was unclear from these studies whether the response to NMDA receptor activation was one typical of synaptic plasticity or one typical of neurotoxicity. However, a recent series of studies with NADPH oxidase mutant mice indicate that this enzyme is indeed critical for synaptic plasticity.

8 A Role for NADPH Oxidase in Hippocampal Synaptic Plasticity and Memory

Recent studies with pharmacological inhibitors of NADPH oxidase as well as studies with mutant mice that are genetically deficient for either gp91^{phox} [121] or p47^{phox} [120] indicate that NADPH oxidase is involved in LTP. Two pharmacological inhibitors of the NADPH oxidase complex, DPI and apocynin [98, 122], blocked early phase LTP (E-LTP) and mutant mice that lacked either the gp91^{phox} or p47^{phox} subunits also expressed deficient E-LTP. Interestingly, slices from gp91^{phox} knockout mice and slices from wild-type mice treated with DPI also expressed deficient post-tetanic potentiation (PTP) which is a form of NMDA receptor-independent short-term plasticity [123]. Other forms of presynaptic plasticity were normal in both p47^{phox} and gp91^{phox} knockout mice [123]. Thus, NADPH oxidase appears to be required for E-LTP.

Behavioral studies with the NADPH oxidase mutant mice indicate that superoxide produced by this enzyme may have an important role in hippocampus-dependent learning and memory. Consistent with the idea that NADPH oxidase-generated superoxide is necessary for learning and memory, it was shown that gp91^{phox} knockout mice displayed mild deficits in the Morris water maze and that p47^{phox} knockout mice displayed deficits in the contextual fear conditioning [123]. Interestingly, these mice also showed differences in the accelerating rotating rod apparatus and in the open field analysis suggesting that areas other than the hippocampus may be affected in these mutant animals [123]. Thus, in addition to its role in hippocampal LTP, NADPH oxidase appears to play a role in several types of hippocampus-dependent memory.

9 Conclusions and Future Directions

Here we have discussed evidence that ROS, including superoxide and H_2O_2, are important signaling molecules in a variety of neuronal and nonneuronal systems. Importantly, ROS have been shown to be critical signaling molecules underlying fundamental cognitive functions including learning and memory. This is atypical of previous views that placed ROS in a class of oxidatively destructive molecules that when produced lead to toxic processes underlying cellular degeneration and apoptosis [124]. We have also presented evidence that NADPH oxidase is likely an important source of ROS in the brain. This is evidenced by the fact that NADPH oxidase has been shown to be required for biochemical signal transduction cascades, synaptic plasticity, and cognitive behaviors involved in the formation and expression of memory. The relatively small amount of research aimed at determining the role of NADPH oxidase in brain function already has uncovered interesting results that warrant further investigation.

NADPH oxidase was shown to be required for hippocampus-dependent learning and memory, as well as for normal performance in behavioral paradigms that require other brain regions. This is consistent with the observation that ROS-dependent learning and memory has a motivational component [125] and that subunits of the NADPH oxidase complex have been shown to be expressed in brain regions other than the hippocampus, including the cortex, cerebellum, striatum, and amygdala [113]. Future research into the role that NADPH oxidase plays in cognitive function should address the role of ROS signaling in these other brain regions as well.

Interestingly, there are several homologs of the main catalytic subunit gp91^{phox} that have been shown to be expressed in various regions of the brain including the cortex and the hippocampus. These other NADPH oxidase subunit homologs may be important for plasticity and cognition. Interestingly, it is known that each of these homologs requires different upstream signals for oxidase activation. For example, NOX-5 contains EF-hand regions that respond directly to Ca^{2+} influx [126]. Determining the role that the various NADPH oxidase subunit homologs play in synaptic plasticity and memory should be an important goal for future investigations.

Not only is it important to determine the role of NADPH oxidase in generating ROS involved in plasticity and cognition, but also it will be equally important to determine the role of other sources of ROS in these processes. We have mentioned several other potential sources of ROS that have been implicated in signal transduction and synaptic plasticity. Future research should address the distinct roles each of these sources of ROS play in mediating the molecular signaling underlying synaptic plasticity and memory.

An important issue that has been addressed only sparingly is the identity of the relevant targets of ROS signaling during synaptic plasticity and memory. As discussed earlier, several studies point directly to activation and inactivation of various kinases and phosphatases [36, 46, 64, 65, 71, 127]. Although there have been few reports in neuronal systems, direct oxidative modification of ion channels, including voltage-gated Ca^{2+} channels and the ryanodine receptor, has been shown to regulate the levels of intracellular Ca^{2+} [37, 80]. Also shown to be an important signaling target of ROS are redox-sensitive transcription factors such as NF-κB [48]. The elucidation of all of the targets of ROS signaling, in addition to the sources of ROS responsible for redox regulation of these targets, will be critical in understanding the roles that these highly reactive molecules play in normal cognitive function, and importantly, how perturbation of these signaling systems could lead to alterations in cognition, including neurodegenerative conditions mediated by oxidative stress.

We have argued that ROS signaling plays an important and necessary role in synaptic plasticity and memory formation. Moreover, NADPH oxidase is likely to be one of the important sources of ROS mediating these effects. Two key features make NADPH oxidase an attractive candidate for an ROS source in these physiological processes. First, the enzymatic complex generates large amounts of superoxide quickly and second, it can do so in a well-regulated manner. Misregulation in either of these aspects could lead to neuronal dysfunction, as well as potential cognitive problems. Recent work indicates that not generating enough superoxide via NADPH oxidase leads to deficient synaptic plasticity and cognitive function in mice [123]. Interestingly, human patients with mutations that render NADPH oxidase inactive may also express mild cognitive deficits [101]. One can imagine that if NADPH oxidase regulatory mechanisms were altered, especially the mechanisms responsible for shutting down superoxide production, the well-known destructive role that ROS are known for in the brain could be fulfilled. Exuberant ROS generation could quickly lead to oxidative stress and subsequent neuronal dysfunction and damage. A fuller understanding of ROS signaling may lead to a better understanding of the degenerative mechanisms that underlie disorders such as Parkinson's disease [128] and Alzheimer's disease [4] that may be in part caused by aberrant ROS generation. Thus, understanding the regulatory mechanisms that underlie NADPH oxidase-mediated signaling in the brain, as well as the regulation of other potential sources of ROS, should be an imperative for future research.

10 List of Abbreviations

CaM – calmodulin
CaMKII – calmodulin dependent kinase II
CnA – calcineurin
DCF – dichlorofluorescein
DHR – dihydrorhodamine
DPI – diphenylene iodonium
EC-SOD – extracellular-superoxide dismutase
E-LTP – early long-term potentiation
ERK – extracellular signal-regulated kinase
FCCP – carbonylcyanide p-trifluoromethoxyphenylhydrazone
fEPSPs – field excitatory postsynaptic potentials
H_2O_2 – hydrogen peroxide
HFS – high-frequency stimulation
HFS-LTP – high-frequency stimulation induced long-term potentiation
IFN-γ – interferon gamma
KA – kainic acid
L-NAME – N-nitro-L-arginine methyl ester
L-NMMA – N-methyl-L-arginine acetate
L-NNA – nitro-L-arginine
LPS – lipopolysacharide
LTP – long-term potentiation
NG – neurogranin
NMDA – N-methyl-D-aspartate
NO – nitric oxide
NOS – nitric oxide synthatse
PKC – protein kinase C
PLA2 – phospholipase A2
PMA – phorbol-12-myristate-13-acetate
PP2B – protein phosphase 2B
PrPc – prion protein
PTP – post-tetanic potentiation
ROS – reactive oxygen species
SOD – superoxide dismutase
TPEN – N,N,N′,N′-Tetrakis(2-pyridylmethyl)ethylenediamine
X/XO – xanthine/xanthine oxidase

References

1. Smythies J. Redox aspects of signaling by catecholamines and their metabolites. Antioxid Redox Signal. 2000 Fall;2(3):575–583.
2. Mattson MP, Liu D. Energetics and oxidative stress in synaptic plasticity and neurodegenerative disorders. Neuromolecular Med. 2002;2(2):215–231.
3. Zekry D, Epperson TK, Krause KH. A role for NOX NADPH oxidases in Alzheimer's disease and other types of dementia? IUBMB Life. 2003 Jun;55(6):307–313.
4. Abramov AY, Canevari L, Duchen MR. Calcium signals induced by amyloid beta peptide and their consequences in neurons and astrocytes in culture. Biochim Biophys Acta. 2004 Dec 6;1742(1–3):81–87.
5. Hu D, Serrano F, Oury TD, Klann E. Aging-dependent alterations in synaptic plasticity and memory in mice that overexpress extracellular superoxide dismutase. J Neurosci. 2006 Apr 12;26(15):3933–3941.
6. Serrano F, Klann E. Reactive oxygen species and synaptic plasticity in the aging hippocampus. Ageing Res Rev. 2004 Nov;3(4):431–443.
7. Brewer GJ. Neuronal plasticity and stressor toxicity during aging. Exp Gerontol. 2000 Dec;35(9–10):1165–1183.
8. Tsien JZ, Huerta PT, Tonegawa S. The essential role of hippocampal CA1 NMDA receptor-dependent synaptic plasticity in spatial memory. Cell. 1996 Dec 27;87(7): 1327–1338.
9. Banko JL, Poulin F, Hou L, DeMaria CT, Sonenberg N, Klann E. The translation repressor 4E-BP2 is critical for eIF4F complex formation, synaptic plasticity, and memory in the hippocampus. J Neurosci. 2005 Oct 19;25(42):9581–9590.
10. Atkins CM, Selcher JC, Petraitis JJ, Trzaskos JM, Sweatt JD. The MAPK cascade is required for mammalian associative learning. Nat Neurosci. 1998 Nov;1(7):602–609.
11. Miyakawa T, Yared E, Pak JH, Huang FL, Huang KP, Crawley JN. Neurogranin null mutant mice display performance deficits on spatial learning tasks with anxiety related components. Hippocampus. 2001;11(6):763–775.
12. Pak JH, Huang FL, Li J, Balschun D, Reymann KG, Chiang C, Westphal H, Huang KP. Involvement of neurogranin in the modulation of calcium/calmodulin-dependent protein kinase II, synaptic plasticity, and spatial learning: a study with knockout mice. Proc Natl Acad Sci U S A. 2000 Oct 10;97(21):11232–11237.
13. Sweatt JD. The neuronal MAP kinase cascade: a biochemical signal integration system subserving synaptic plasticity and memory. J Neurochem. 2001 Jan;76(1):1–10.
14. Winder DG, Sweatt JD. Roles of serine/threonine phosphatases in hippocampal synaptic plasticity. Nat Rev Neurosci. 2001 Jul;2(7):461–474.
15. Sweatt JD. Mitogen-activated protein kinases in synaptic plasticity and memory. Curr Opin Neurobiol. 2004 Jun;14(3):311–317.
16. Xia Z, Storm DR. The role of calmodulin as a signal integrator for synaptic plasticity. Nat Rev Neurosci. 2005 Apr;6(4):267–276.
17. Gunasekar PG, Kanthasamy AG, Borowitz JL, Isom GE. NMDA receptor activation produces concurrent generation of nitric oxide and reactive oxygen species: implication for cell death. J Neurochem. 1995 Nov;65(5):2016–2021.
18. Lynch MA. Long-term potentiation and memory. Physiol Rev. 2004 Jan;84(1):87–136.
19. Knapp LT, Klann E. Role of reactive oxygen species in hippocampal long-term potentiation: contributory or inhibitory? J Neurosci Res. 2002 Oct 1;70(1):1–7.
20. Thiels E, Urban NN, Gonzalez-Burgos GR, Kanterewicz BI, Barrionuevo G, Chu CT, Oury TD, Klann E. Impairment of long-term potentiation and associative memory in mice that overexpress extracellular superoxide dismutase. J Neurosci. 2000 Oct 15;20(20): 7631–7639.
21. Medina JH, Izquierdo I. Retrograde messengers, long-term potentiation and memory. Brain Res Brain Res Rev. 1995 Sep;21(2):185–194.

22. Quinn MT, Gauss KA. Structure and regulation of the neutrophil respiratory burst oxidase: comparison with nonphagocyte oxidases. J Leukoc Biol. 2004 Oct;76(4): 760–781.
23. Nauseef WM. Assembly of the phagocyte NADPH oxidase. Histochem Cell Biol. 2004 Oct;122(4):277–291.
24. Jackson SH, Devadas S, Kwon J, Pinto LA, Williams MS. T cells express a phagocyte-type NADPH oxidase that is activated after T cell receptor stimulation. Nat Immunol. 2004 Aug;5(8):818–827.
25. Segal BH, Kuhns DB, Ding L, Gallin JI, Holland SM. Thioglycollate peritonitis in mice lacking C5, 5-lipoxygenase, or p47(phox): complement, leukotrienes, and reactive oxidants in acute inflammation. J Leukoc Biol. 2002 Mar;71(3):410–416.
26. Lambeth JD. Nox/Duox family of nicotinamide adenine dinucleotide (phosphate) oxidases. Curr Opin Hematol. 2002 Jan;9(1):11–17.
27. Seshiah PN, Weber DS, Rocic P, Valppu L, Taniyama Y, Griendling KK. Angiotensin II stimulation of NAD(P)H oxidase activity: upstream mediators. Circ Res. 2002 Sep 6;91(5):406–413.
28. Schneider B, Mutel V, Pietri M, Ermonval M, Mouillet-Richard S, Kellermann O. NADPH oxidase and extracellular regulated kinases 1/2 are targets of prion protein signaling in neuronal and nonneuronal cells. Proc Natl Acad Sci U S A. 2003 Nov 11;100(23):13326–13331.
29. Shapiro M. Plasticity, hippocampal place cells, and cognitive maps. Arch Neurol. 2001 Jun;58(6):874–881.
30. Kamsler A, Segal M. Hydrogen peroxide as a diffusible signal molecule in synaptic plasticity. Mol Neurobiol. 2004 Apr;29(2):167–178.
31. Kamsler A, Segal M. Paradoxical actions of hydrogen peroxide on long-term potentiation in transgenic superoxide dismutase-1 mice. J Neurosci. 2003 Nov 12;23(32): 10359–10367.
32. Kamsler A, Segal M. Hydrogen peroxide modulation of synaptic plasticity. J Neurosci. 2003 Jan 1;23(1):269–276.
33. Klann E. Cell-permeable scavengers of superoxide prevent long-term potentiation in hippocampal area CA1. J Neurophysiol. 1998 Jul;80(1):452–457.
34. Klann E, Roberson ED, Knapp LT, Sweatt JD. A role for superoxide in protein kinase C activation and induction of long-term potentiation. J Biol Chem. 1998 Feb 20;273(8): 4516–4522.
35. Gahtan E, Auerbach JM, Groner Y, Segal M. Reversible impairment of long-term potentiation in transgenic Cu/Zn-SOD mice. Eur J Neurosci. 1998 Feb;10(2):538–544.
36. Knapp LT, Klann E. Potentiation of hippocampal synaptic transmission by superoxide requires the oxidative activation of protein kinase C. J Neurosci. 2002 Feb 1;22(3): 674–683.
37. Li A, Segui J, Heinemann SH, Hoshi T. Oxidation regulates cloned neuronal voltage-dependent Ca2+ channels expressed in Xenopus oocytes. J Neurosci. 1998 Sep 1;18(17): 6740–6747.
38. Yermolaieva O, Brot N, Weissbach H, Heinemann SH, Hoshi T. Reactive oxygen species and nitric oxide mediate plasticity of neuronal calcium signaling. Proc Natl Acad Sci U S A. 2000 Jan 4;97(1):448–453.
39. Hu D, Klann E, Thiels E. Superoxide dismutase and hippocampal function: age and isozyme matter. Antioxid Redox Signal. 2007 Feb;9(2):201–210.
40. Logue SF, Paylor R, Wehner JM. Hippocampal lesions cause learning deficits in inbred mice in the Morris water maze and conditioned-fear task. Behav Neurosci. 1997 Feb;111(1):104–113.
41. Levin ED, Brady TC, Hochrein EC, Oury TD, Jonsson LM, Marklund SL, Crapo JD. Molecular manipulations of extracellular superoxide dismutase: functional importance for learning. Behav Genet. 1998 Sep;28(5):381–390.

42. Packard MG, Hirsh R, White NM. Differential effects of fornix and caudate nucleus lesions on two radial maze tasks: evidence for multiple memory systems. J Neurosci. 1989 May;9(5):1465–1472.

43. Phillips RG, LeDoux JE. Lesions of the fornix but not the entorhinal or perirhinal cortex interfere with contextual fear conditioning. J Neurosci. 1995 Jul;15(7 Pt 2): 5308–5315.

44. Phillips RG, LeDoux JE. Differential contribution of amygdala and hippocampus to cued and contextual fear conditioning. Behav Neurosci. 1992 Apr;106(2):274–285.

45. Levin ED, Christopher NC, Lateef S, Elamir BM, Patel M, Liang LP, Crapo JD. Extracellular superoxide dismutase overexpression protects against aging-induced cognitive impairment in mice. Behav Genet. 2002 Mar;32(2):119–125.

46. Klann E, Thiels E. Modulation of protein kinases and protein phosphatases by reactive oxygen species: implications for hippocampal synaptic plasticity. Prog Neuropsychopharmacol Biol Psychiatry. 1999 Apr;23(3):359–376.

47. Rusnak F, Reiter T. Sensing electrons: protein phosphatase redox regulation. Trends Biochem Sci. 2000 Nov;25(11):527–529.

48. Kabe Y, Ando K, Hirao S, Yoshida M, Handa H. Redox regulation of NF-kappaB activation: distinct redox regulation between the cytoplasm and the nucleus. Antioxid Redox Signal. 2005 Mar–Apr;7(3–4):395–403.

49. Klann E, Dever TE. Biochemical mechanisms for translational regulation in synaptic plasticity. Nat Rev Neurosci. 2004 Dec;5(12):931–942.

50. Werner E. GTPases and reactive oxygen species: switches for killing and signaling. J Cell Sci. 2004 Jan 15;117(Pt 2):143–153.

51. Poser S, Storm DR. Role of Ca2+-stimulated adenylyl cyclases in LTP and memory formation. Int J Dev Neurosci. 2001 Jul;19(4):387–394.

52. Klann E, Chen SJ, Sweatt JD. Mechanism of protein kinase C activation during the induction and maintenance of long-term potentiation probed using a selective peptide substrate. Proc Natl Acad Sci U S A. 1993 Sep 15;90(18):8337–8341.

53. English JD, Sweatt JD. A requirement for the mitogen-activated protein kinase cascade in hippocampal long term potentiation. J Biol Chem. 1997 Aug 1;272(31):19103–19106.

54. English JD, Sweatt JD. Activation of p42 mitogen-activated protein kinase in hippocampal long term potentiation. J Biol Chem. 1996 Oct 4;271(40):24329–24332.

55. Malleret G, Haditsch U, Genoux D, Jones MW, Bliss TV, Vanhoose AM, Weitlauf C, Kandel ER, Winder DG, Mansuy IM. Inducible and reversible enhancement of learning, memory, and long-term potentiation by genetic inhibition of calcineurin. Cell. 2001 Mar 9;104(5):675–686.

56. Onuma H, Lu YF, Tomizawa K, Moriwaki A, Tokuda M, Hatase O, Matsui H. A calcineurin inhibitor, FK506, blocks voltage-gated calcium channel-dependent LTP in the hippocampus. Neurosci Res. 1998 Apr;30(4):313–319.

57. Atkins CM, Davare MA, Oh MC, Derkach V, Soderling TR. Bidirectional regulation of cytoplasmic polyadenylation element-binding protein phosphorylation by Ca2+/calmodulin-dependent protein kinase II and protein phosphatase 1 during hippocampal long-term potentiation. J Neurosci. 2005 Jun 8;25(23):5604–5610.

58. Woo NH, Abel T, Nguyen PV. Genetic and pharmacological demonstration of a role for cyclic AMP-dependent protein kinase-mediated suppression of protein phosphatases in gating the expression of late LTP. Eur J Neurosci. 2002 Nov;16(10):1871–1876.

59. Bindokas VP, Jordan J, Lee CC, Miller RJ. Superoxide production in rat hippocampal neurons: selective imaging with hydroethidine. J Neurosci. 1996 Feb 15;16(4):1324–1336.

60. Li J, Pak JH, Huang FL, Huang KP. N-methyl-D-aspartate induces neurogranin/RC3 oxidation in rat brain slices. J Biol Chem. 1999 Jan 15;274(3):1294–1300.

61. Huang KP, Huang FL, Li J, Schuck P, McPhie P. Calcium-sensitive interaction between calmodulin and modified forms of rat brain neurogranin/RC3. Biochemistry. 2000 Jun 20;39(24):7291–7299.

62. Wu J, Huang KP, Huang FL. Participation of NMDA-mediated phosphorylation and oxidation of neurogranin in the regulation of Ca2+- and Ca2+/calmodulin-dependent neuronal signaling in the hippocampus. J Neurochem. 2003 Sep;86(6):1524–1533.

63. Wu J, Li J, Huang KP, Huang FL. Attenuation of protein kinase C and cAMP-dependent protein kinase signal transduction in the neurogranin knockout mouse. J Biol Chem. 2002 May 31;277(22):19498–19505.

64. Thiels E, Kanterewicz BI, Knapp LT, Barrionuevo G, Klann E. Protein phosphatase-mediated regulation of protein kinase C during long-term depression in the adult hippocampus in vivo. J Neurosci. 2000 Oct 1;20(19):7199–7207.

65. Knapp LT, Klann E. Superoxide-induced stimulation of protein kinase C via thiol modification and modulation of zinc content. J Biol Chem. 2000 Aug 4;275(31): 24136–24145.

66. Radi R, Peluffo G, Alvarez MN, Naviliat M, Cayota A. Unraveling peroxynitrite formation in biological systems. Free Radic Biol Med. 2001 Mar 1;30(5):463–488.

67. Knapp LT, Kanterewicz BI, Hayes EL, Klann E. Peroxynitrite-induced tyrosine nitration and inhibition of protein kinase C. Biochem Biophys Res Commun. 2001 Aug 31; 286(4):764–770.

68. Kishida KT, Pao M, Holland SM, Klann E. NADPH oxidase is required for NMDA receptor-dependent activation of ERK in hippocampal area CA1. J Neurochem. 2005 Jul;94(2):299–306.

69. Selcher JC, Weeber EJ, Christian J, Nekrasova T, Landreth GE, Sweatt JD. A role for ERK MAP kinase in physiologic temporal integration in hippocampal area CA1. Learn Mem. 2003 Jan–Feb;10(1):26–39.

70. Llansola M, Saez R, Felipo V. NMDA-induced phosphorylation of the microtubule-associated protein MAP-2 is mediated by activation of nitric oxide synthase and MAP kinase. Eur J Neurosci. 2001 Apr;13(7):1283–1291.

71. Kanterewicz BI, Knapp LT, Klann E. Stimulation of p42 and p44 mitogen-activated protein kinases by reactive oxygen species and nitric oxide in hippocampus. J Neurochem. 1998 Mar;70(3):1009–1016.

72. Namgaladze D, Hofer HW, Ullrich V. Redox control of calcineurin by targeting the binuclear Fe(2+)-Zn(2+) center at the enzyme active site. J Biol Chem. 2002 Feb 22; 277(8):5962–5969.

73. Freudenthal R, Romano A, Routtenberg A. Transcription factor NF-kappaB activation after in vivo perforant path LTP in mouse hippocampus. Hippocampus. 2004;14(6):677–683.

74. Albensi BC, Mattson MP. Evidence for the involvement of TNF and NF-kappaB in hippocampal synaptic plasticity. Synapse. 2000 Feb;35(2):151–159.

75. Haddad JJ. Antioxidant and prooxidant mechanisms in the regulation of redox(y)-sensitive transcription factors. Cell Signal. 2002 Nov;14(11):879–897.

76. Ginn-Pease ME, Whisler RL. Redox signals and NF-kappaB activation in T cells. Free Radic Biol Med. 1998 Aug;25(3):346–361.

77. Yeh SH, Lin CH, Lee CF, Gean PW. A requirement of nuclear factor-kappaB activation in fear-potentiated startle. J Biol Chem. 2002 Nov 29;277(48):46720–46729.

78. Freudenthal R, Romano A. Participation of Rel/NF-kappaB transcription factors in long-term memory in the crab Chasmagnathus. Brain Res. 2000 Feb 14;855(2):274–281.

79. Hidalgo C, Donoso P, Carrasco MA. The ryanodine receptors Ca2+ release channels: cellular redox sensors? IUBMB Life. 2005 Apr–May;57(4–5):315–322.

80. Futatsugi A, Kato K, Ogura H, Li ST, Nagata E, Kuwajima G, Tanaka K, Itohara S, Mikoshiba K. Facilitation of NMDAR-independent LTP and spatial learning in mutant mice lacking ryanodine receptor type 3. Neuron. 1999 Nov;24(3):701–713.

81. Balschun D, Wolfer DP, Bertocchini F, Barone V, Conti A, Zuschratter W, Missiaen L, Lipp HP, Frey JU, Sorrentino V. Deletion of the ryanodine receptor type 3 (RyR3) impairs forms of synaptic plasticity and spatial learning. Embo J. 1999 Oct 1;18(19): 5264–5273.

82. Huddleston AT, Tang W, Takeshima H, Hamilton SL, Klann E. Superoxide-induced potentiation in the hippocampus requires activation of ryanodine receptor type 3 and ERK. J Neurophysiol. 2008 Mar;99(3):1565–1571.

83. Huddleston AT, Tang W, Takeshima H, Hamilton SL, Klann E. Superoxide-induced potentiation in the hippocampus requires activation of ryanodine receptor type 3 and ERK. J Neurophysiol. 2008 Mar;99(3):1565–1571.

84. Dykens JA. Isolated cerebral and cerebellar mitochondria produce free radicals when exposed to elevated CA2+ and Na+: implications for neurodegeneration. J Neurochem. 1994 Aug;63(2):584–591.

85. Reynolds IJ, Hastings TG. Glutamate induces the production of reactive oxygen species in cultured forebrain neurons following NMDA receptor activation. J Neurosci. 1995 May;15(5 Pt 1):3318–3327.

86. Heinzel B, John M, Klatt P, Bohme E, Mayer B. Ca2+/calmodulin-dependent formation of hydrogen peroxide by brain nitric oxide synthase. Biochem J. 1992 Feb 1;281 (Pt 3):627–630.

87. Abramov AY, Jacobson J, Wientjes F, Hothersall J, Canevari L, Duchen MR. Expression and modulation of an NADPH oxidase in mammalian astrocytes. J Neurosci. 2005 Oct 5;25(40):9176–9184.

88. Dugan LL, Sensi SL, Canzoniero LM, Handran SD, Rothman SM, Lin TS, Goldberg MP, Choi DW. Mitochondrial production of reactive oxygen species in cortical neurons following exposure to N-methyl-D-aspartate. J Neurosci. 1995 Oct;15(10): 6377–6388.

89. Swerdlow RH. Treating neurodegeneration by modifying mitochondria: potential solutions to a "complex" problem. Antioxid Redox Signal. 2007 Oct;9(10):1591–1603.

90. Boldyrev AA, Carpenter DO, Huentelman MJ, Peters CM, Johnson P. Sources of reactive oxygen species production in excitotoxin-stimulated cerebellar granule cells. Biochem Biophys Res Commun. 1999 Mar 16;256(2):320–324.

91. Pare CM. New pharmacological developments in antidepressants. Psychopathology. 1986;19 Suppl 2:103–107.

92. Schuman EM, Madison DV. A requirement for the intercellular messenger nitric oxide in long-term potentiation. Science. 1991 Dec 6;254(5037):1503–1506.

93. Zorumski CF, Izumi Y. Nitric oxide and hippocampal synaptic plasticity. Biochem Pharmacol. 1993 Sep 1;46(5):777–785.

94. Pou S, Pou WS, Bredt DS, Snyder SH, Rosen GM. Generation of superoxide by purified brain nitric oxide synthase. J Biol Chem. 1992 Dec 5;267(34):24173–24176.

95. Pou S, Keaton L, Surichamorn W, Rosen GM. Mechanism of superoxide generation by neuronal nitric-oxide synthase. J Biol Chem. 1999 Apr 2;274(14):9573–9580.

96. Culcasi M, Lafon-Cazal M, Pietri S, Bockaert J. Glutamate receptors induce a burst of superoxide via activation of nitric oxide synthase in arginine-depleted neurons. J Biol Chem. 1994 Apr 29;269(17):12589–12593.

97. Wang W, Wang S, Nishanian EV, Del Pilar Cintron A, Wesley RA, Danner RL. Signaling by eNOS through a superoxide-dependent p42/44 mitogen-activated protein kinase pathway. Am J Physiol Cell Physiol. 2001 Aug;281(2):C544–554.

98. O'Donnell BV, Tew DG, Jones OT, England PJ. Studies on the inhibitory mechanism of iodonium compounds with special reference to neutrophil NADPH oxidase. Biochem J. 1993 Feb 15;290 (Pt 1):41–49.

99. Dringen R. Oxidative and antioxidative potential of brain microglial cells. Antioxid Redox Signal. 2005 Sep–Oct;7(9–10):1223–1233.

100. Hwang J, Saha A, Boo YC, Sorescu GP, McNally JS, Holland SM, Dikalov S, Giddens DP, Griendling KK, Harrison DG, Jo H. Oscillatory shear stress stimulates endothelial production of O2- from p47phox-dependent NAD(P)H oxidases, leading to monocyte adhesion. J Biol Chem. 2003 Nov 21;278(47):47291–47298.

101. Pao M, Wiggs EA, Anastacio MM, Hyun J, DeCarlo ES, Miller JT, Anderson VL, Malech HL, Gallin JI, Holland SM. Cognitive function in patients with chronic granulomatous disease: a preliminary report. Psychosomatics. 2004 May–Jun;45(3):230–234.

102. Sumimoto H, Miyano K, Takeya R. Molecular composition and regulation of the Nox family NAD(P)H oxidases. Biochem Biophys Res Commun. 2005 Dec 9;338(1):677–686.

103. Anrather J, Racchumi G, Iadecola C. NF-kappaB regulates phagocytic NADPH oxidase by inducing the expression of gp91phox. J Biol Chem. 2006 Mar 3;281(9): 5657–5667.

104. Dang PM, Fontayne A, Hakim J, El Benna J, Perianin A. Protein kinase C zeta phosphorylates a subset of selective sites of the NADPH oxidase component p47phox and participates in formyl peptide-mediated neutrophil respiratory burst. J Immunol. 2001 Jan 15;166(2):1206–1213.

105. Maridonneau-Parini I, Tauber AI. Activation of NADPH-oxidase by arachidonic acid involves phospholipase A2 in intact human neutrophils but not in the cell-free system. Biochem Biophys Res Commun. 1986 Aug 14;138(3):1099–1105.

106. Odell EW, Segal AW. Killing of pathogens associated with chronic granulomatous disease by the non-oxidative microbicidal mechanisms of human neutrophils. J Med Microbiol. 1991 Mar;34(3):129–135.

107. Lafon-Cazal M, Pietri S, Culcasi M, Bockaert J. NMDA-dependent superoxide production and neurotoxicity. Nature. 1993 Aug 5;364(6437):535–537.

108. Karpova A, Sanna PP, Behnisch T. Involvement of multiple phosphatidylinositol 3-kinase-dependent pathways in the persistence of late-phase long term potentiation expression. Neuroscience. 2006 Feb;137(3):833–841.

109. Wang Q, Liu L, Pei L, Ju W, Ahmadian G, Lu J, Wang Y, Liu F, Wang YT. Control of synaptic strength, a novel function of Akt. Neuron. 2003 Jun 19;38(6):915–928.

110. Mizuno M, Yamada K, Takei N, Tran MH, He J, Nakajima A, Nawa H, Nabeshima T. Phosphatidylinositol 3-kinase: a molecule mediating BDNF-dependent spatial memory formation. Mol Psychiatry. 2003 Feb;8(2):217–224.

111. Chen Q, Powell DW, Rane MJ, Singh S, Butt W, Klein JB, McLeish KR. Akt phosphorylates p47phox and mediates respiratory burst activity in human neutrophils. J Immunol. 2003 May 15;170(10):5302–5308.

112. Mizuki K, Kadomatsu K, Hata K, Ito T, Fan QW, Kage Y, Fukumaki Y, Sakaki Y, Takeshige K, Sumimoto H. Functional modules and expression of mouse p40(phox) and p67(phox), SH3-domain-containing proteins involved in the phagocyte NADPH oxidase complex. Eur J Biochem. 1998 Feb 1;251(3):573–582.

113. Serrano F, Kolluri NS, Wientjes FB, Card JP, Klann E. NADPH oxidase immunoreactivity in the mouse brain. Brain Res. 2003 Oct 24;988(1–2):193–198.

114. Tejada-Simon M, Serrano F., Villasana LE, Kanterewicz BI, Wu G-Y, Quinn MT, Klann E. Synaptic localization of a functional NADPH oxidase in the mouse hippocampus. Mol Cell Neurosci. 2005 May;29(1):97–106.

115. Banfi B, Malgrange B, Knisz J, Steger K, Dubois-Dauphin M, Krause KH. NOX3, a superoxide-generating NADPH oxidase of the inner ear. J Biol Chem. 2004 Oct 29;279(44):46065–46072.

116. Cheng G, Cao Z, Xu X, van Meir EG, Lambeth JD. Homologs of gp91phox: cloning and tissue expression of Nox3, Nox4, and Nox5. Gene. 2001 May 16;269(1–2):131–140.

117. Vallet P, Charnay Y, Steger K, Ogier-Denis E, Kovari E, Herrmann F, Michel JP, Szanto I. Neuronal expression of the NADPH oxidase NOX4, and its regulation in mouse experimental brain ischemia. Neuroscience. 2005;132(2):233–238.

118. Rao PV, Maddala R, John F, Zigler JS, Jr. Expression of nonphagocytic NADPH oxidase system in the ocular lens. Mol Vis. 2004 Feb 19;10:112–121.

119. Diatchuk V, Lotan O, Koshkin V, Wikstroem P, Pick E. Inhibition of NADPH oxidase activation by 4-(2-aminoethyl)-benzenesulfonyl fluoride and related compounds. J Biol Chem. 1997 May 16;272(20):13292–13301.

120. Jackson SH, Gallin JI, Holland SM. The p47phox mouse knock-out model of chronic granulomatous disease. J Exp Med. 1995 Sep 1;182(3):751–758.
121. Pollock JD, Williams DA, Gifford MA, Li LL, Du X, Fisherman J, Orkin SH, Doerschuk CM, Dinauer MC. Mouse model of X-linked chronic granulomatous disease, an inherited defect in phagocyte superoxide production. Nat Genet. 1995 Feb;9(2):202–209.
122. Pearse DB, Dodd JM. Ischemia-reperfusion lung injury is prevented by apocynin, a novel inhibitor of leukocyte NADPH oxidase. Chest. 1999 Jul;116(1 Suppl):55S–56S.
123. Kishida KT, Hoeffer CA, Hu D, Pao M, Holland SM, Klann E. Synaptic Plasticity deficits and mild memory impairments in mouse models of chronic granulomatous disease. Mol Cell Biol. 2006 Aug;26(15):5908–5920.
124. Atlante A, Calissano P, Bobba A, Giannattasio S, Marra E, Passarella S. Glutamate neurotoxicity, oxidative stress and mitochondria. FEBS Lett. 2001 May 18;497(1):1–5.
125. Levin ED, Brucato FH, Crapo JD. Molecular overexpression of extracellular superoxide dismutase increases the dependency of learning and memory performance on motivational state. Behav Genet. 2000 Mar;30(2):95–100.
126. Banfi B, Tirone F, Durussel I, Knisz J, Moskwa P, Molnar GZ, Krause KH, Cox JA. Mechanism of Ca2+ activation of the NADPH oxidase 5 (NOX5). J Biol Chem. 2004 Apr 30;279(18):18583–18591.
127. Thiels E, Norman ED, Barrionuevo G, Klann E. Transient and persistent increases in protein phosphatase activity during long-term depression in the adult hippocampus in vivo. Neuroscience. 1998 Oct;86(4):1023–1029.
128. Kostrzewa RM, Kostrzewa JP, Brus R. Neuroprotective and neurotoxic roles of levodopa (L-DOPA) in neurodegenerative disorders relating to Parkinson's disease. Amino Acids. 2002;23(1–3):57–63.

Nitric Oxide Biochemistry: Pathophysiology of Nitric Oxide-Mediated Protein Modifications

Alba Rossi-George and Andrew Gow

Abstract Nitric oxide (NO) is a pluripotent signaling molecule, which has been proposed to be critically important in both physiological and pathological processes of the brain. The wide-ranging functionality of NO is in opposition to its relatively simple chemical structure. In this chapter we attempt to summarize the functional involvement of NO within the neurological system and then discuss how such complex signaling may be achieved via the differential post-translational modification of protein targets. It is our contention that the redox properties of NO allow this molecule to modify proteins in a variety of ways with physiological or pathological consequences. In this way the effects of NO production are dependent on the quantity produced, the redox environment in which it is synthesized, and the presence of reactive targets. A summary of known post-translational modifications is given as well as the functional consequences of their formation.

Keywords Nitric oxide · Nitrosothiol · Nitration · Nitroalkene · Thiol · Nitrosylation

1 Introduction

Nitric oxide (NO) plays a variety of physiological roles in the nervous system, including respiratory and blood flow control, immune defense, intracellular signaling, and neurotransmission. It is synthesized by all brain cells including neurons, endothelial cells, and glial cells and under physiological conditions, approximately 20 times more NO is produced in the brain than in the entire vasculature. The enzymes responsible for the production of NO are collectively known as nitric oxide synthases (NOS) with three distinct isoforms: endothelial (eNOS), inducible (iNOS), and neuronal (nNOS) [1]. Transient increases in

A. Rossi-George (✉)

Department of Pharmacology and Toxicology, Rutgers University, Piscataway, NJ 08664, USA

S.C. Veasey (ed.), *Oxidative Neural Injury,*
Contemporary Clinical Neuroscience, DOI 10.1007/978-1-60327-342-8_2,
© Humana Press, a part of Springer Science+Business Media, LLC 2009

intracellular calcium (Ca^{2+}) result in the release of nanomolar fluxes of NO by the Ca^{2+}/calmodulin-dependent nNOS and eNOS isoforms, which are essential for neurotransmission and the control of cerebral blood flow. However, iNOS (which is constitutively active) induces the release of fluxes of a higher order of magnitude of NO by glial and vascular cells [2]. iNOS function is principally controlled by its expression and degradation. The role of NO is paradoxical as it is not only critical in modulating immune function but also appears to play a role in mediating tissue damage. Indeed, a variety of neuropathological events including stroke and several neurodegenerative diseases are marked by elevated levels of NO [3]. Therefore, whether NO is protective or detrimental depends on the conditions under which it is released, including its flux and the cell environment. In order to provide a basis for understanding this complexity of NO-based mechanisms, this chapter will consider some of the basic biology surrounding NOS and NO within the brain and then provide an overview of the chemistry that underlies this biology.

2 NOS, NO, and the Brain

The nNOS isoform can be considered ubiquitous within the brain as it is found in a variety of neural structures including the cerebral cortex [4], the olfactory bulb [5], the nucleus accumbens, the striatum, the amygdala [6], the hippocampus (especially the CA1 and the dentate gyrus), the hypothalamus [7], the thalamus [8], and the cerebellum [4]. It is important to note that despite this generalized expression the distribution of nNOS is far from uniform but rather has areas of concentration. eNOS expression is also widespread as it is found in cerebral endothelial cells, but its expression in neurons appears to be limited to the hippocampus, where it is found in the granule cells of the dentate gyrus as well as those of the CA1, CA2, and CA3 regions [9]. Under physiological conditions, iNOS levels in the brain are low but are increased in response to glial activation. The mechanisms controlling iNOS induction within glial cells appear to be similar to those in other inflammatory cells, being responsive to toxins, such as lipopolysaccharide, and cytokines, such as interferon-γ. iNOS function is turned off either by its own degradation or by apoptosis of the induced cell. It is important to remember that the function of NOS is in fact more complicated even than just the production of NO. It is dependent on a number of cofactors including, NADPH, FMN, and tetrahydrobiopterin. In addition, substrate supply, namely, both oxygen and arginine, can affect both the products and rate of enzyme activity.

The importance of NO in signaling cannot be underestimated as NO can modify receptors and can activate intracellular messengers, ultimately affecting neurotransmitter release. Classically, NO interacts with soluble guanylyl cyclase (sGC, a cytosolic heme-containing enzyme that catalyzes the conversion of GTP into cGMP) ultimately affecting protein kinase G (PKG) and cyclic nucleotide-gated channels, both of which are important for neuronal transmission. For example, stimulation of *N*-methyl-D-aspartate (NMDA) receptors by glutamate results in the influx of Ca^{2+} with the subsequent activation of nNOS (which is physically associated with the NMDA receptor at the NR2B subunit) and the release of NO. The release of NO, in turn, results in the activation of nearby glutamatergic neurons that stimulate the release of acetylcholine from the nucleus accumbens [10]. NO also modulates the release of norepinephrine and glutamate in the hippocampus. NO donors increase norepinephrine and glutamate release, while NO scavengers (e.g., hemoglobin) inhibit the release of norepinephrine and glutamate. The effects on the GABAergic system also depend on NO concentration. Basal concentrations of NO reduce the release of GABA, whereas higher levels result in an increase of GABA [11]. Dopamine and serotonin release from the medial preoptic area in rats are similarly affected by NO levels [12].

Long-term potentiation (LTP), which is a cellular model of synaptic plasticity and purportedly a model of learning and memory, also appear to be modulated by NO. NO-mediated modulation of synaptic plasticity is an sGC-dependent mechanism: LTP is inhibited by the addition of an sGC inhibitor to hippocampal and amygdala slices, and potentiated when sGC is added [13]. It has been proposed that when produced post-synaptically, NO affects pre-synaptic transmission as a retrograde diffusible messenger in hippocampal and cortical LTP. One of the roles that NO plays in the mesencephalon is its involvement in the sleep–wake cycle [14]. Administration of L-arginine (the precursor of NO) into the pedunculopontine tegmentum during the light phase results is an increase in slow-wave sleep in rats [15], a result that was also corroborated in cats administered an NO donor [16].

The presence of nNOS in the PVN and the supraoptic nucleus of the hypothalamus attests to its involvement in the stress response. Both neural groups secrete corticotropic-releasing hormone (CRH) and arginine vasopressin (AVP). In response to stress, CRH and AVP are released in the median eminence where they stimulate the release of adrenocorticotropin-releasing hormone into the general circulation and, ultimately, stimulate the secretion of glucocorticoids from the adrenal gland. NO seems to exert diverging effects on different components of the stress response: the upregulation of CRH and AVP expression in response to a neurogenic stressor is decreased when rats are administered the NOS inhibitor L-NAME, and increased in response to intracerebroventricular administration of NO, but, basal levels of both CRH and AVP are unaffected by NO administration in rats [17]. Interestingly, the administration of L-NAME downstream of the PVN increases the release of ACTH from the pituitary gland suggesting that NO plays an inhibitory role possibly starting at the level of the median eminence [17].

NO plays an important role in neuroprotection, as it has been shown that inhibition of NO in cultured cerebellar granule cells results in enhanced apoptosis via activation of caspase-3, which can be reversed by the addition of NO donors [18]. In models of neurotoxicity where prolonged stimulation of NMDA receptors results in cell death by Ca^{2+} overload, NO has been shown to reduce the Ca^{2+} burden by S-nitrosylating the NR1 and NR2 subunits of the NMDA receptor [10]. A second mechanism by which NO might be neuroprotective is via the induction of heme oxygenase-1, which is an early marker of oxidative stress and has been shown to play a role in antioxidant protection pathways [19].

In contrast to its potential neuroprotective function, NO production in excess, potentially in response to inflammation, has been implicated in neuro-degenerative diseases. For example, patients afflicted by either Alzheimer's (AD) or Parkinson's disease show an increase in nitrotyrosine, a marker of nitrative stress [20–22]. NO release also activates cyclooxygenase, which is found in brain cells during the inflammatory state. The upregulation of cyclooxygenase, and other peripheral inflammatory signals, are potential biomarkers for the progression of AD [23, 24]. Postmortem examination of AD patients also reveals nitration of mitochondrial proteins implying a role for NO in energetic dys-function [25, 26]. The finding that both eNOS and iNOS expression is promoted by α-amyloid expression has been confirmed in autopsied AD patients, where increased NOS levels are observed [27]. It has also been suggested that nNOS expression is significantly altered in neurodegenerative disease [28]. Another neurodegenerative disorder where NO appears to play a central role is multiple sclerosis (MS), which is characterized by demyelinating lesions in the central and the peripheral nervous system. The cerebrospinal fluid of MS patients has elevated levels of nitrite [29]. Elevated levels of NO have also been confirmed in experimental autoimmune encephalomyelitis (EAE), an animal model of MS [30], although inhibition of NOS appears to play a role in inducing the disease [31]. This connection between NO production and neurodegeneration may stem from NO's function as a cell death mediator in neurodevelopment. During normal neurodevelopment, many more neurons are produced than those that survive. Programmed cell death or apoptosis is an important mechanism for the normal development of the central nervous system and in this regard NO has been shown to play an instrumental role. Peak levels of NOS in rats and guinea pigs are observed immediately before the period of maximal synaptogenesis, whereas both nNOS and iNOS expression is several fold higher during early postnatal development [32]. And, although nNOS is expressed in less than 4% of neurons, these cells are capable of killing neighboring cells.

Although NO is governed by the same rules that apply to other neurotrans-mitters, it appears to be more versatile and can be considered a non-conventional transmitter system. A few important differences between NO and other neuro-transmitters include the lower amount of NO needed, the fact that it is not stored in vesicles but directly produced on demand and the lack of a specific receptor, which confers it the ability to act on tissue volume and not necessarily in a defined cellular region [33]. Because of its properties, NO is not limited to any particular cell type or brain region, and, therefore, can potentially affect a variety of structures and functions. Among some of the important functions that have been so far identified are synaptic plasticity [34], neurodevelopment [35], sleep

[36], appetite [37], reproductive behavior [38], sensory [8] and motor involvement [39], thermal regulation [40], pain [41], as well as hormonal secretion [42].

3 Biological Chemistry of NO

A principal question that remains within NO biology is how can a simple diatomic molecule be utilized to obtain such wide-ranging signaling functions. The answer may lie in the unique biological chemistry of NO. Much discussion has been made of NO's gaseous nature, the fact it is a free radical, and that it is highly soluble especially in hydrophobic media. But perhaps the most interesting physical characteristic of NO is its ability to readily undergo redox transitions. This results primarily from the ability of nitrogen to adopt a variety of redox states. Within the field of oxidative stress research one is accustomed to consider partially reduced states of oxygen, such as superoxide, hydrogen peroxide, and peroxy radicals, and the relative facility with which such molecules can react within the biological system. The same can be said for nitrogen, which in its fully oxidized form, nitrate, has a redox state of $+5$ and its fully reduced form, ammonia, has a redox state of -3. However, like oxygen, nitrogen exists in reactive partially reduced states, such as nitroxyl anion (NO^-, oxidation state $+1$), nitric oxide (oxidation state $+2$), and nitrosonium cation (NO^+) or nitrite (oxidation state $+3$) [43]. Each of these partially reduced, or reactive nitrogen species (RNS), possesses its own particular reactivity [44].

NO and its isoforms are capable of reacting with a number of biological molecules, however, for the purpose of this review we will restrict our discussion to those that have been implicated in cellular signaling within the nervous system. NO reaction with proteins can be divided into three broad and basic pathways, namely reaction with (1) metals, (2) oxides, and (3) thiols. In the following section we will consider the reactivity involved in each of these pathways.

3.1 Metals

NO was first shown to interact with metal centers by Keilin and Hartree in the early 20th century when studying the reactions of various gaseous molecules with hemoglobin and cytochromes [45]. The binding of NO to the heme prosthetic group of soluble guanylate cyclase (sGC) along with the identification of the endothelium-derived relaxing factor raised the possibility of physiological relevance [46, 47]. Although the exact nature of this interaction is still undetermined, the generally accepted principle is that binding of NO to the iron within the heme group of sGC produces a conformational change and enzyme activation [48]. Peculiar to sGC, the binding of NO to the heme is approximately 400-fold faster than the binding of carbon monoxide, implying there are structural constraints, which focus the heme toward NO reactivity [49]. Clearly there are many other heme-containing proteins that bind NO avidly including hemoglobin [50]. Indeed, the possibility has been raised by recent research that redox interaction of hemoglobin with NO and its related oxides is an essential part of the hypoxic vasodilator response of the systemic vasculature [51–55]. The relevance of such reactions within the brain has been highlighted by the discovery of neuroglobin, which has been proposed as an antioxidant protein [56]. Cytochrome P450 has also been shown to be a target for binding of NO resulting in functional inhibition [57, 58]; the implications of this in xenobiotic degradation have not been clearly established [59].

The metalloproteins of the mitochondrion form another principal target for NO reactivity. Cytochrome oxidase has been shown to be reversibly inhibited by NO and it has been suggested that this is an important mechanism to control cellular respiration, especially under hypoxic conditions [60–63]. There are a number of iron–sulfur proteins within the mitochondria, such as aconitase and mitochondrial electron transport chain complex I. These proteins are targets for NO reactivity [64–66]. Zinc–thiolate clusters have also been shown to be the targets for NO reactivity. Metallothionein reacts with NO to release zinc converting a redox-based signal, NO, to a redox inactive one, zinc, protecting the cell against potentially harmful reactions [67–69]. Aberrant reactions of NO and other oxidants with metalloproteins have been proposed in the mechanisms of neurodegenerative diseases such as ALS [70, 71]. Critically important in these reactions is the potential involvement of copper [70], particularly in the + 1 redox state [72]. In this regard it is interesting to note that ceruloplasmin appears to react with NO to produce nitrosothiols in a copper-dependent manner [73, 74].

3.2 Oxides

As a result of the relative ease with which nitrogen can be altered in its oxidation state, NO is capable of participating in both reduction and oxidation reactions. Perhaps the best known of these reactions is its oxidation by molecular oxygen [75]. This trimolecular reaction, involving two NO molecules and one of oxygen, is slow at physiological concentrations of NO. However, under pathophysiological conditions of raised NO concentration this reaction becomes of greater relevance [76]. It is worth remembering that such reactions may become more relevant within microenvironments, such as the membrane or that surrounding NOS itself [77]. Although the autoxidation of NO is known to be bimolecular in NO and unimolecular in oxygen; there are a number of potential pathways. However, all of the potential intermediates are nitrosating agents, such as dinitrogen trioxide and dinitrogen tetroxide. Within biological systems there is a wide variety of targets for nitrosation, including reduced thiol. However, such nitrosation is by an entirely different chemical species to either that produced by transnitrosation or free radical NO. Therefore it may occur not only at different targets, such as amines, but also at different members of one class of targets, i.e. the same group of thiols may not be nitrosated.

Nitric oxide is also capable of interacting with other oxides, such as super-oxide and peroxy radicals [78]. Although there is clearly a significant role for these higher oxides of nitrogen in pathological processes, it is unclear what role they play in physiological systems. Much has been made of the potentially toxic consequences of the formation of peroxynitrite, the reactive product of NO and superoxide, and certainly it is capable of a number of potentially adverse reactions such as DNA damage and the generation of 8-hydroxyguanine, lipid peroxidation, and tyrosine nitration [78]. Peroxynitrite is also capable of oxidizing thiol residues and of nitrosating a number of targets including amine and thiol residues [79]. Tyrosine nitration is often used as a hallmark of this type of chemistry [80]. Invariably the presence of tyrosine nitration is taken as a sign of pathological NO chemistry [81], however, the potential for physiological signaling does exist either through the formation of nitrosothiols or tyrosine nitration [82–85].

Recently a novel set of signaling molecules, derived from higher oxides of nitrogen, has been identified, the nitroalkenes [86]. These molecules are formed by the combined action of nitrosative and oxidative stress upon unsaturated fatty acids [87]. Among the first such molecules identified was nitro-linoleic acid [88], which was identified as having an NO-like signaling in the vasculature. Nitroalkenes have been identified within human red cells and plasma [89] and have been suggested to operate as anti-inflammatory signals by activation of targets such as peripheral peroxisome activator-γ [90], inhibiting platelet aggregation [91], and neutrophil activation [92]. There are multiple mechanisms for such molecules to act as signaling molecules including NO release [93] and post-translational modification [94] via Michael addition to nucleophilic targets like cysteine [95]. The potential for these molecules to act within the brain has not been investigated.

3.3 Thiols

There are multiple mechanisms whereby NO can modify thiol residues including nitrosation via a higher oxide intermediate, direct reaction with a thiyl radical, direct reaction followed by electron abstraction, and via metal catalysis. It is also possible for NO to be involved in oxidative reactions with thiols either directly or through the decomposition of intermediates such as nitrosothiol. The chemistry of these reactions has been reviewed elsewhere [43, 96] and indeed in a recent kinetic model thiolates were predicted to form one of the principal targets for NO reactivity [97]. Although a complete discussion of the formation of nitrosothiols is beyond the scope of this submission, it is important to consider that different mechanisms may result in the synthesis of a different subset of nitrosothiols, i.e., the conditions under which NO is produced will alter which thiols are modified. These considerations have considerable importance with relevance to how higher oxides of nitrogen, such as peroxynitrite, might produce cellular signaling.

As well as considering the conditions surrounding NO production as factors in determining nitrosothiol formation, it is important to consider the thiol residues themselves. The pKa of thiol side chains vary greatly depending on the surrounding environment. For instance in cysteine itself the pKa is 8.75, while in glutathione, where there are two interacting amino groups, the pKa is 8.35. But not only the pH dependency can vary, as cysteines are often buried within extremely hydrophobic areas of the protein. These differences in environment can alter the ease with which a thiol may be nitrosylated; the stability of any nitrosothiols formed; and the mechanism whereby such nitrosothiols are formed (and hence which proximal chemical species will be the nitrosylating agent). Nowhere, is this variability of thiol environment more clearly demonstrated than in hemoglobin, in which the movement between relaxed and tense structure forms the basis of nitrosothiol formation and decay, allowing for delivery of NO equivalents in areas of hypoxia [53, 54, 98]. Consideration of nitrosothiols, which have been shown to be formed, such as those on the NMDA-receptor and caspase-3, has led to the development of two potential "consensus" sequences for formation. The first is the acid/base motif, in which the presence of proton donating and accepting groups promotes the ability to nitrosate a thiolate [99]. The second is the positioning of a cysteine in a hydrophobic region, which may promote the direct interaction between NO and thiol due to the hydrophobicity of dissolved NO [100, 101]. Two groups have attempted to use a proteomic approach in order to evaluate total nitrosothiol–protein formation in vivo [102, 103]. Utilizing modified biotin-switch [104] technology coupled with mass spectrometry, these groups have identified a wide range of proteins that can be nitrosylated under stimulated conditions. Greco et al. attempted to verify the two proposed consensus motifs for nitrosothiol formation and found that all of the modified cysteine residues conformed to one or another of the motifs [102]. The development of these new technologies for assessing nitrosylation within biological systems will rapidly accelerate our understanding of the importance of this form of post-translational modification within cellular signaling.

It has been proposed that nitrosothiols form a prevalent redox-based signaling post-translational modification [105]. Central to this proposal is the ability of nitrosothiol formation to functionally regulate a wide range of proteins. Within the literature proteins ranging from regulatory kinases [106, 107] to channel proteins [10], to transcription factors [108], to metabolic enzymes [109] have been suggested to be regulated by nitrosothiol formation at key cysteine residues. As one would predict the effect of nitrosylation, in terms of function, varies from protein to protein. For instance, nitrosylation of the NMDA receptor has been shown to inhibit calcium flow [10]; however, within the ryanodine receptor it increases ion current through the channel [110]. This observation of opposite regulatory effects on proteins of similar classes is reflected in the kinases, where p21ras is activated by the reaction of NO with its key cysteine residue [106, 107] while JNK1 is inhibited [111]. The pluripotency of S-nitrosylation as a mediator of post-translational signaling is underscored by the different mechanisms through which functional regulation is achieved. For instance, S-nitrosylation of the NMDA-receptor leads to a physical blockade of the calcium channel [10]; while nitrosothiol formation on Parkin appears to promote its association with E3 ubiquitin ligase leading to secondary enzyme activation [112]. In contrast, S-nitrosylation of GAPDH inhibits glycolytic activity while increasing its association with Siah-1 resulting in nuclear localization and initiation of a secondary signaling cascade [109].

These observations give some idea of the complexity of signaling that can be initiated by S-nitrosylation and the potential naivety of assigning nitrosothiol formation as either "good" or "bad". This would be similar to trying to determine whether an increase in total phosphorylation would either promote or inhibit cell death. Despite this complexity it is possible to develop models of nitrosothiol-based signaling as a result of NOS activation. Largely as a result of the work of Lipton and coworkers, the NMDA-receptor system forms one such model. The NMDA-receptor possesses key regulatory cysteine residues, whose identification formed the original basis of the acid/base motif for nitrosothiol formation [10]. The receptor is also physically linked to nNOS via a PDZ domain. Upon glutamate binding the calcium channel of the receptor opens allowing calcium entry to the cell, and among other signaling events activation of the calcium-dependent nNOS. The consequent production of NO leads to inhibition of the channel function of the receptor via nitrosylation at key reactive cysteine residues [113], thus providing a feedback inhibition mechanism. Recently it has been shown that COX-2 is also physically linked to nNOS, as well as iNOS in non-neuronal tissue [114], and that stimulation of NOS function leads to nitrosothiol-mediated activation, inducing the prostaglandin signaling cascade [115]. However, nNOS also activates Dex-ras, which is not directly linked to it but rather is associated via CAPON, providing another downstream target [116]. Therefore, both COX-2 and Dex-ras are potential feedforward activators of glutamate signaling via nitrosylation; but importantly they are targeted differently as a result of their proximity to nNOS. As work continues more and more potential protein targets are being identified for nitrosylation and hence the potential consequences expand often with diametrically opposed functions. For instance nitrosylation of GAPDH [109] appears to promote apoptosis while modification of caspase-3 is inhibitory [101]. Thus it has become important to not only evaluate the potential of a particular protein to be modified but also to measure its relevant degree of nitrosylation within cellular systems.

4 Summary

The work summarized here demonstrates the huge biological diversity that is achieved by NO, a simple diatomic molecule. It is our contention that this diversity is achieved through the redox chemistry of NO and its ability to post-translationally modify proteins through a variety of mechanisms, including metal nitrosylation, amino acid side chain oxidation and/or nitration, thiol alkylation, and/or nitrosylation. Thus it becomes a question not so much of when is NO made, but of how much, and under what conditions, and in the presence of what target molecules. In other words, physiological NO signaling requires the right amount of NO to be made at the right time and in the right place, which may explain the importance of NOS localization [117]. Disruption in quantity, time, or location is potentially an initiator of pathological processes and hence our often confused view of NO as both a "good" and a "bad" molecule.

References

1. Griffith OW, Stuehr DJ. Nitric oxide synthases: properties and catalytic mechanism. Annu Rev Physiol. 1995;57:707–736.
2. Stuehr DJ, Santolini J, Wang ZQ, Wei CC, Adak S. Update on mechanism and catalytic regulation in the NO synthases. J Biol Chem. 2004 Aug 27;279(35):36167–36170.
3. Bredt DS. Endogenous nitric oxide synthesis: biological functions and pathophysiology. Free Radic Res. 1999 Dec;31(6):577–596.
4. Campese VM, Sindhu RK, Ye S, Bai Y, Vaziri ND, Jabbari B. Regional expression of NO synthase, NAD(P)H oxidase and superoxide dismutase in the rat brain. Brain Res. 2007 Feb 23;1134(1):27–32.
5. Chen J, Tu Y, Moon C, Matarazzo V, Palmer AM, Ronnett GV. The localization of neuronal nitric oxide synthase may influence its role in neuronal precursor proliferation and synaptic maintenance. Dev Biol. 2004 May 1;269(1):165–182.
6. Menendez L, Insua D, Rois JL, Santamarina G, Suarez ML, Pesini P. The immunohistochemical localization of neuronal nitric oxide synthase in the basal forebrain of the dog. J Chem Neuroanat. 2006 Apr;31(3):200–209.
7. Ventura RR, Aguiar JF, Antunes-Rodrigues J, Varanda WA. Nitric oxide modulates the firing rate of the rat supraoptic magnocellular neurons. Neuroscience. 2008 Aug 13;155(2):359–365.
8. Yang S, Cox CL. Excitatory and anti-oscillatory actions of nitric oxide in thalamus. J Physiol. 2008 Aug 1;586(Pt 15):3617–3628.
9. Liu P, Smith PF, Appleton I, Darlington CL, Bilkey DK. Hippocampal nitric oxide synthase and arginase and age-associated behavioral deficits. Hippocampus. 2005;15(5):642–655.
10. Choi YB, Tenneti L, Le DA, Ortiz J, Bai G, Chen HS, Lipton SA. Molecular basis of NMDA receptor-coupled ion channel modulation by S-nitrosylation. Nat Neurosci. 2000 Jan;3(1):15–21.
11. Saransaari P, Oja SS. GABA release under normal and ischemic conditions. Neurochem Res. 2008 May;33(5):962–969.
12. Trabace L, Cassano T, Tucci P, Steardo L, Kendrick KM, Cuomo V. The effects of nitric oxide on striatal serotoninergic transmission involve multiple targets: an in vivo microdialysis study in the awake rat. Brain Res. 2004 May 22;1008(2):293–298.
13. Hopper RA, Garthwaite J. Tonic and phasic nitric oxide signals in hippocampal long-term potentiation. J Neurosci. 2006 Nov 8;26(45):11513–11521.
14. Kodama T, Koyama Y. Nitric oxide from the laterodorsal tegmental neurons: its possible retrograde modulation on norepinephrine release from the axon terminal of the locus coeruleus neurons. Neuroscience. 2006;138(1):245–256.
15. Hars B. Endogenous nitric oxide in the rat pons promotes sleep. Brain Res. 1999 Jan 16;816(1):209–219.
16. Datta S, Patterson EH, Siwek DF. Endogenous and exogenous nitric oxide in the pedunculopontine tegmentum induces sleep. Synapse. 1997 Sep;27(1):69–78.

17. Gadek-Michalska A, Bugajski J. Nitric oxide in the adrenergic-and CRH-induced activation of hypothalamic-pituitary-adrenal axis. J Physiol Pharmacol. 2008 Jun;59(2):365–378.
18. Contestabile A, Ciani E. Role of nitric oxide in the regulation of neuronal proliferation, survival and differentiation. Neurochem Int. 2004 Nov;45(6):903–914.
19. Mancuso C. Heme oxygenase and its products in the nervous system. Antioxid Redox Signal. 2004 Oct;6(5):878–887.
20. Good PF, Werner P, Hsu A, Olanow CW, Perl DP. Evidence of neuronal oxidative damage in Alzheimer's disease. Am J Pathol. 1996 July;149(1):21–28.
21. Duda JE, Giasson BI, Chen Q, Gur TL, Hurtig HI, Stern MB, Gollomp SM, Ischiropoulos H, Lee VM, Trojanowski JQ. Widespread nitration of pathological inclusions in neurodegenerative synucleinopathies. Am J Pathol. 2000 Nov;157(5):1439–1445.
22. Guix FX, Uribesalgo I, Coma M, Munoz FJ. The physiology and pathophysiology of nitric oxide in the brain. Prog Neurobiol. 2005 Jun;76(2):126–152.
23. Bermejo P, Martin-Aragon S, Benedi J, Susin C, Felici E, Gil P, Ribera JM, Villar AM. Differences of peripheral inflammatory markers between mild cognitive impairment and Alzheimer's disease. Immunol Lett. 2008 May 15;117(2):198–202.
24. Guerreiro RJ, Santana I, Bras JM, Santiago B, Paiva A, Oliveira C. Peripheral inflammatory cytokines as biomarkers in Alzheimer's disease and mild cognitive impairment. Neurodegener Dis. 2007;4(6):406–412.
25. Devi L, Prabhu BM, Galati DF, Avadhani NG, Anandatheerthavarada HK. Accumulation of amyloid precursor protein in the mitochondrial import channels of human Alzheimer's disease brain is associated with mitochondrial dysfunction. J Neurosci. 2006 Aug 30;26(35):9057–9068.
26. Elfering SL, Haynes VL, Traaseth NJ, Ettl A, Giulivi C. Aspects, mechanism, and biological relevance of mitochondrial protein nitration sustained by mitochondrial nitric oxide synthase. Am J Physiol Heart Circ Physiol. 2004 Feb;286(1):H22–H29.
27. Luth HJ, Holzer M, Gartner U, Staufenbiel M, Arendt T. Expression of endothelial and inducible NOS-isoforms is increased in Alzheimer's disease, in APP23 transgenic mice and after experimental brain lesion in rat: evidence for an induction by amyloid pathology. Brain Res. 2001 Sep 14;913(1):57–67.
28. Fernandez-Vizarra P, Fernandez AP, Castro-Blanco S, Encinas JM, Serrano J, Bentura ML, Munoz P, Martinez-Murillo R, Rodrigo J. Expression of nitric oxide system in clinically evaluated cases of Alzheimer's disease. Neurobiol Dis. 2004 Mar;15(2):287–305.
29. Rejdak K, Petzold A, Stelmasiak Z, Giovannoni G. Cerebrospinal fluid brain specific proteins in relation to nitric oxide metabolites during relapse of multiple sclerosis. Mult Scler. 2008 Jan;14(1):59–66.
30. O'Brien NC, Charlton B, Cowden WB, Willenborg DO. Nitric oxide plays a critical role in the recovery of Lewis rats from experimental autoimmune encephalomyelitis and the maintenance of resistance to reinduction. J Immunol. 1999 Dec 15;163(12):6841–6847.
31. O'Brien NC, Charlton B, Cowden WB, Willenborg DO. Inhibition of nitric oxide synthase initiates relapsing remitting experimental autoimmune encephalomyelitis in rats, yet nitric oxide appears to be essential for clinical expression of disease. J Immunol. 2001 Nov 15;167(10):5904–5912.
32. Lizasoain I, Weiner CP, Knowles RG, Moncada S. The ontogeny of cerebral and cerebellar nitric oxide synthase in the guinea pig and rat. Pediatr Res. 1996 May;39(5):779–783.
33. Boehning D, Snyder SH. Novel neural modulators. Annu Rev Neurosci. 2003;26:105–131.
34. Feil R, Kleppisch T. NO/cGMP-dependent modulation of synaptic transmission. Handb Exp Pharmacol. 2008(184):529–560.
35. Blomgren K, Leist M, Groc L. Pathological apoptosis in the developing brain. Apoptosis. 2007 May;12(5):993–1010.
36. Gautier-Sauvigne S, Colas D, Parmantier P, Clement P, Gharib A, Sarda N, Cespuglio R. Nitric oxide and sleep. Sleep Med Rev. 2005 Apr;9(2):101–113.

37. Yang SJ, Denbow DM. Interaction of leptin and nitric oxide on food intake in broilers and Leghorns. Physiol Behav. 2007 Nov 23;92(4):651–657.
38. Gregg AR. Mouse models and the role of nitric oxide in reproduction. Curr Pharm Des. 2003;9(5):391–398.
39. Fejgin K, Palsson E, Wass C, Svensson L, Klamer D. Nitric oxide signaling in the medial prefrontal cortex is involved in the biochemical and behavioral effects of phencyclidine. Neuropsychopharmacology. 2008 Jul;33(8):1874–1883.
40. Ding Z, Gomez T, Werkheiser JL, Cowan A, Rawls SM. Icilin induces a hyperthermia in rats that is dependent on nitric oxide production and NMDA receptor activation. Eur J Pharmacol. 2008 Jan 14;578(2–3):201–208.
41. Naik AK, Tandan SK, Kumar D, Dudhgaonkar SP. Nitric oxide and its modulators in chronic constriction injury-induced neuropathic pain in rats. Eur J Pharmacol. 2006 Jan 13;530(1–2):59–69.
42. Reis WL, Giusti-Paiva A, Ventura RR, Margatho LO, Gomes DA, Elias LL, Antunes-Rodrigues J. Central nitric oxide blocks vasopressin, oxytocin and atrial natriuretic peptide release and antidiuretic and natriuretic responses induced by central angiotensin II in conscious rats. Exp Physiol. 2007 Sep;92(5):903–911.
43. Gow AJ, Ischiropoulos H. Nitric oxide chemistry and cellular signaling. J Cell Physiol. 2001 June;187(3):277–282.
44. Wink DA, Hanbauer I, Grisham MB, Laval F, Nims RW, Laval J, Cook J, Pacelli R, Liebmann J, Krishna M, Ford PC, Mitchell JB. Chemical biology of nitric oxide: regulation and protective and toxic mechanisms. [Review]. Curr Top Cell Regul 1996. 1996;34:159–187.
45. Keilin D, Hartree EF. Reaction of nitric oxide with haemoglobin and methaemoglobin. Nature. 1937 March 27–32676;139:548–548.
46. Furchgott RF, Zawadzki JV. The obligatory role of endothelial cells in the relaxation of arterial smooth muscle by acetylcholine. Nature. 1980 Nov 27;288(5789):373–376.
47. Ignarro LJ, Buga GM, Wood KS, Byrns RE, Chaudhuri G. Endothelium-derived relaxing factor produced and released from artery and vein is nitric oxide. Proc Natl Acad Sci U S A. 1987 Dec;84(24):9265–9269.
48. Murad F. Regulation of cytosolic guanylyl cyclase by nitric oxide: the NO- cyclic GMP signal transduction system. Adv Pharmacol. 1994;26:19–33.
49. Sharma VS, Magde D. Activation of soluble guanylate cyclase by carbon monoxide and nitric oxide: a mechanistic model. Methods. 1999 Dec;19(4):494–505.
50. Doyle MP, Hoekstra JW. Oxidation of nitrogen oxides by bound dioxygen in hemoproteins. J Inorg Biochem. 1981 July;14(4):351–358.
51. Pluta RM, Dejam A, Grimes G, Gladwin MT, Oldfield EH. Nitrite infusions to prevent delayed cerebral vasospasm in a primate model of subarachnoid hemorrhage. JAMA. 2005;293(12):1477–1484.
52. Cosby K, Partovi KS, Crawford JH, Patel RP, Reiter CD, Martyr S, Yang BK, Waclawiw MA, Zalos G, Xu X, Huang KT, Shields H, Kim-Shapiro DB, Schechter AN, Cannon RO, III, Gladwin MT. Nitrite reduction to nitric oxide by deoxyhemoglobin vasodilates the human circulation. Nat Med. 2003 Dec;9(12):1498–1505.
53. McMahon TJ, Moon RE, Luschinger BP, Carraway MS, Stone AE, Stolp BW, Gow AJ, Pawloski JR, Watke P, Singel DJ, Piantadosi CA, Stamler JS. Nitric oxide in the human respiratory cycle. Nat Med. 2002 July;8(7):711–717.
54. Gow AJ, Stamler JS. Reactions between nitric oxide and haemoglobin under physiological conditions. Nature. 1998 Jan 08;391(6663):169–173.
55. Gow AJ, Luchsinger BP, Pawloski JP, Singel DJ, Stamler JS. The oxyhemoglobin reaction of nitric oxide. Proc Natl Acad Sci U S A. 1999;in press.
56. Herold S, Fago A, Weber RE, Dewilde S, Moens L. Reactivity studies of the Fe(III) and Fe(II)NO forms of human neuroglobin reveal a potential role against oxidative stress. J Biol Chem. 2004 May 28;279(22):22841–22847.

57. Tsubaki M, Hiwatashi A, Ichikawa Y, Fujimoto Y, Ikekawa N, Hori H. Electron paramagnetic resonance study of ferrous cytochrome P-450scc-nitric oxide complexes: effects of 20(R),22(R)-dihydroxycholesterol and reduced adrenodoxin. Biochemistry. 1988 June 28;27(13):4856–4862.

58. Takemura S, Minamiyama Y, Imaoka S, Funae Y, Hirohashi K, Inoue M, Kinoshita H. Hepatic cytochrome P450 is directly inactivated by nitric oxide, not by inflammatory cytokines, in the early phase of endotoxemia. J Hepatol. 1999 June;30(6):1035–1044.

59. Muller CM, Scierka A, Stiller RL, Kim YM, Cook DR, Lancaster JR, Jr., Buffington CW, Watkins WD. Nitric oxide mediates hepatic cytochrome P450 dysfunction induced by endotoxin. Anesthesiology. 1996 June;84(6):1435–1442.

60. Giuffre A, Barone MC, Mastronicola D, D'Itri E, Sarti P, Brunori M. Reaction of nitric oxide with the turnover intermediates of cytochrome c oxidase: reaction pathway and functional effects. Biochemistry. 2000 Dec 19;39(50):15446–15453.

61. Lancaster JR, Jr. Simulation of the diffusion and reaction of endogenously produced nitric oxide. Proc Natl Acad Sci U S A. 1994 Aug 16;91(17):8137–8141.

62. Shen W, Hintze TH, Wolin MS. Nitric oxide. An important signaling mechanism between vascular endothelium and parenchymal cells in the regulation of oxygen consumption. Circulation. 1995 Dec 15;92(12):3505–3512.

63. Clementi E, Brown GC, Foxwell N, Moncada S. On the mechanism by which vascular endothelial cells regulate their oxygen consumption. Proc Natl Acad Sci U S A. 1999 Feb 16;96(4):1559–1562.

64. Clementi E, Brown GC, Feelisch M, Moncada S. Persistent inhibition of cell respiration by nitric oxide: crucial role of S-nitrosylation of mitochondrial complex I and protective action of glutathione. Proc Natl Acad Sci U S A. 1998 June 23;95(13):7631–7636.

65. Gardner PR, Costantino G, Szabo C, Salzman AL. Nitric oxide sensitivity of the aconitases. J Biol Chem. 1997 Oct 03;272(40):25071–25076.

66. Hausladen A, Fridovich I. Superoxide and peroxynitrite inactivate aconitases, but nitric oxide does not. J Biol Chem. 1994 Nov 25;269(47):29405–29408.

67. St Croix CM, Wasserloos KJ, Dineley KE, Reynolds IJ, Levitan ES, Pitt BR. Nitric oxide-induced changes in intracellular zinc homeostasis are mediated by metallothionein/ thionein. Am J Physiol Lung Cell Mol Physiol. 2002 Feb;282(2):L185–L192.

68. Schwarz MA, Lazo JS, Yalowich JC, Allen WP, Whitmore M, Bergonia HA, Tzeng E, Billiar TR, Robbins PD, Lancaster JR, Jr. Metallothionein protects against the cytotoxic and DNA-damaging effects of nitric oxide. Proc Natl Acad Sci U S A. 1995 May 09;92(10):4452–4456.

69. Pearce LL, Gandley RE, Han W, Wasserloos K, Stitt M, Kanai AJ, McLaughlin MK, Pitt BR, Levitan ES. Role of metallothionein in nitric oxide signaling as revealed by a green fluorescent fusion protein. Proc Natl Acad Sci U S A. 2000 Jan 04;97(1):477–482.

70. Estevez AG, Crow JP, Sampson JB, Reiter C, Zhuang Y, Richardson GJ, Tarpey MM, Barbeito L, Beckman JS. Induction of nitric oxide-dependent apoptosis in motor neurons by zinc- deficient superoxide dismutase. Science. 1999 Dec 24;286(5449): 2498–2500.

71. Cassina P, Cassina A, Pehar M, Castellanos R, Gandelman M, de Leon A, Robinson KM, Mason RP, Beckman JS, Barbeito L, Radi R. Mitochondrial dysfunction in SOD1G93A-bearing astrocytes promotes motor neuron degeneration: prevention by mitochondrial-targeted antioxidants. J Neurosci. 2008 Apr 16;28(16):4115–4122.

72. Johnson MA, Macdonald TL, Mannick JB, Conaway MR, Gaston B. Accelerated S-nitrosothiol breakdown by amyotrophic lateral sclerosis mutant copper, zinc-super-oxide dismutase. J Biol Chem. 2001 Aug 22.

73. Inoue K, Akaike T, Miyamoto Y, Okamoto T, Sawa T, Otagiri M, Suzuki S, Yoshimura T, Maeda H. Nitrosothiol formation catalyzed by ceruloplasmin. Implication for cyto-protective mechanism in vivo. J Biol Chem. 1999 Sep 17;274(38):27069–27075.

74. Akaike T. Mechanisms of biological S-nitrosation and its measurement. Free Radic Res. 2000;33(5):461–469.
75. Kharitonov VG, Sundquist AR, Sharma VS. Kinetics of nitric oxide autoxidation in aqueous solution. J Biol Chem. 1994 Feb 25;269(8):5881–5883.
76. Wink DA, Grisham MB, Mitchell JB, Ford PC. Direct and indirect effects of nitric oxide in chemical reactions relevant to biology. Methods Enzymol. 1996;268:12–31.
77. Liu X, Miller MJS, Joshi MS, Thomas DD, Lancaster JR, Jr. Accelerated reaction of nitric oxide with O2 within the hydrophobic interior of biological membranes. Proc Natl Acad Sci U S A. 1998 Mar 03;95(5):2175–2179.
78. Beckman JS, Koppenol WH. Nitric oxide, superoxide, and peroxynitrite: the good, the bad, and ugly. Am J Physiol. 1996 Nov;271(5 Pt 1):C1424–C1437.
79. Beckman JS. Parsing the effects of nitric oxide, S-nitrosothiols, and peroxynitrite on inducible nitric oxide synthase-dependent cardiac myocyte apoptosis. Circ Res. 1999 Oct 29;85(9):870–871.
80. Ischiropoulos H. Biological selectivity and functional aspects of protein tyrosine nitration. Biochem Biophys Res Commun. 2003 June 06;305(3):776–783.
81. Schopfer FJ, Baker PR, Freeman BA. NO-dependent protein nitration: a cell signaling event or an oxidative inflammatory response? Trends Biochem Sci. 2003 Dec;28(12):646–654.
82. Kuo WN, Kocis JM. Nitration/S-nitrosation of proteins by peroxynitrite-treatment and subsequent modification by glutathione S-transferase and glutathione peroxidase. Mol Cell Biochem. 2002 Apr;233(1–2):57–63.
83. Yang Y, Loscalzo J. S-nitrosoprotein formation and localization in endothelial cells. Proc Natl Acad Sci U S A. 2005 Jan 04;102(1):117–122.
84. Greenacre SA, Ischiropoulos H. Tyrosine nitration: localisation, quantification, consequences for protein function and signal transduction. Free Radic Res. 2001 June;34(6):541–581.
85. Vadseth C, Souza JM, Thomson L, Seagraves A, Nagaswami C, Scheiner T, Torbet J, Vilaire G, Bennett JS, Murciano JC, Muzykantov V, Penn MS, Hazen SL, Weisel JW, Ischiropoulos H. Pro-thrombotic state induced by post-translational modification of fibrinogen by reactive nitrogen species. J Biol Chem. 2004 Mar 05;279(10):8820–8826.
86. Cui T, Schopfer FJ, Zhang J, Chen K, Ichikawa T, Baker PR, Batthyany C, Chacko BK, Feng X, Patel RP, Agarwal A, Freeman BA, Chen YE. Nitrated fatty acids: Endogenous anti-inflammatory signaling mediators. J Biol Chem. 2006 Aut 03.
87. O'Donnell VB, Eiserich JP, Chumley PH, Jablonsky MJ, Krishna NR, Kirk M, Barnes S, Darley-Usmar VM, Freeman BA. Nitration of unsaturated fatty acids by nitric oxide-derived reactive nitrogen species peroxynitrite, nitrous acid, nitrogen dioxide, and nitronium ion. Chem Res Toxicol. 1999 Jan;12(1):83–92.
88. Lim DG, Sweeney S, Bloodsworth A, White CR, Chumley PH, Krishna NR, Schopfer F, O'Donnell VB, Eiserich JP, Freeman BA. Nitrolinoleate, a nitric oxide-derived mediator of cell function: synthesis, characterization, and vasomotor activity. Proc Natl Acad Sci U S A. 2002 Dec 10;99(25):15941–15946.
89. Baker PR, Schopfer FJ, Sweeney S, Freeman BA. Red cell membrane and plasma linoleic acid nitration products: synthesis, clinical identification, and quantitation. Proc Natl Acad Sci U S A. 2004 Aug 10;101(32):11577–11582.
90. Schopfer FJ, Lin Y, Baker PR, Cui T, Garcia-Barrio M, Zhang J, Chen K, Chen YE, Freeman BA. Nitrolinoleic acid: an endogenous peroxisome proliferator-activated receptor gamma ligand. Proc Natl Acad Sci U S A. 2005 Feb 15;102(7):2340–2345.
91. Coles B, Bloodsworth A, Eiserich JP, Coffey MJ, McLoughlin RM, Giddings JC, Lewis MJ, Haslam RJ, Freeman BA, O'Donnell VB. Nitrolinoleate inhibits platelet activation by attenuating calcium mobilization and inducing phosphorylation of vasodilator-stimulated phosphoprotein through elevation of cAMP. J Biol Chem. 2002 Feb 22;277(8):5832–5840.

92. Coles B, Bloodsworth A, Clark SR, Lewis MJ, Cross AR, Freeman BA, O'Donnell VB. Nitrolinoleate inhibits superoxide generation, degranulation, and integrin expression by human neutrophils: novel antiinflammatory properties of nitric oxide-derived reactive species in vascular cells. Circ Res. 2002 Sep 06;91(5):375–381.

93. Schopfer FJ, Baker PRS, Giles G, Chumley P, Batthyany C, Crawford J, Patel RP, Hogg N, Branchaud BP, Lancaster JR, Freeman BA. Fatty acid transduction of nitric oxide signaling – Nitrolinoleic acid is a hydrophobically stabilized nitric oxide donor. J Biol Chem. 2005;280(19):19289–19297.

94. Batthyany C, Schopfer FJ, Baker PR, Duran R, Baker LM, Huang Y, Cervenansky C, Branchaud BP, Freeman BA. Reversible post-translational modification of proteins by nitrated fatty acids in vivo. J Biol Chem. 2006 July 21;281(29):20450–20463.

95. Baker LMS, Baker PRS, Golin-Bisello F, Schopfer FJ, Fink M, Woodcock SR, Branchaud BP, Radi R, Freeman BA. Nitro-fatty acid reaction with glutathione and cysteine: kinetic analysis of thiol alkylation by a Michael addition reaction. J Biol Chem. 2007 Oct 19, 2007;282(42):31085–31093.

96. Lancaster JR, Jr. Protein cysteine thiol nitrosation: maker or marker of reactive nitrogen species-induced nonerythroid cellular signaling? Nitric Oxide. 2008 Sep; 19(2):68–72.

97. Lancaster JR, Jr. Nitroxidative, nitrosative, and nitrative stress: kinetic predictions of reactive nitrogen species chemistry under biological conditions. Chem Res Toxicol. 2006 Sep;19(9):1160–1174.

98. Jia L, Bonaventura C, Bonaventura J, Stamler JS. S-nitrosohaemoglobin: a dynamic activity of blood involved in vascular control. Nature. 1996 Mar 21;380(6571):221–226.

99. Stamler JS, Toone EJ, Lipton SA, Sucher NJ. (S)NO signals: translocation, regulation, and a consensus motif. Neuron. 1997 May ;18(5):691–696.

100. Gow AJ, Buerk DG, Ischiropoulos H. A novel reaction mechanism for the formation of S-nitrosothiol in vivo. J Biol Chem. 1997 Jan 31;272(5):2841–2845.

101. Mannick JB, Hausladen A, Liu L, Hess DT, Zeng M, Miao QX, Kane LS, Gow AJ, Stamler JS. Fas-induced caspase denitrosylation. Science. 1999 Apr 23;284(5414):651–654.

102. Greco TM, Hodara R, Parastatidis I, Heijnen HFG, Dennehy MK, Liebler DC, Ischiropoulos H. Identification of S-nitrosylation motifs by site-specific mapping of the S-nitrosocysteine proteome in human vascular smooth muscle cells. PNAS. 2006 May 9, 2006;103(19):7420–7425.

103. Hao G, Derakhshan B, Shi L, Campagne F, Gross SS. SNOSID, a proteomic method for identification of cysteine S-nitrosylation sites in complex protein mixtures. Proc Natl Acad Sci. 2006 Jan 24;103(4):1012–1017.

104. Jaffrey SR, Snyder SH. The biotin switch method for the detection of S-nitrosylated proteins. Sci STKE. 2001 June 12;2001(86):L1.

105. Stamler JS, Lamas S, Fang FC. Nitrosylation. The prototypic redox-based signaling mechanism. Cell. 2001 Sep 21;106(6):675–683.

106. Williams JG, Pappu K, Campbell SL. Structural and biochemical studies of p21Ras S-nitrosylation and nitric oxide-mediated guanine nucleotide exchange. Proc Natl Acad Sci U S A. 2003 May 27;100(11):6376–6381.

107. Lander HM, Hajjar DP, Hempstead BL, Mirza UA, Chait BT, Campbell S, Quilliam LA. A molecular redox switch on p21(ras). Structural basis for the nitric oxide-p21(ras) interaction. J Biol Chem. 1997 Feb 14;272(7):4323–4326.

108. Marshall HE, Stamler JS. Inhibition of NF-kappa B by S-nitrosylation. Biochemistry. 2001 Feb 13;40(6):1688–1693.

109. Hara MR, Agrawal N, Kim SF, Cascio MB, Fujimuro M, Ozeki Y, Takahashi M, Cheah JH, Tankou SK, Hester LD, Ferris CD, Hayward SD, Snyder SH, Sawa A. S-nitrosylated GAPDH initiates apoptotic cell death by nuclear translocation following Siah1 binding. Nat Cell Biol. 2005 July ;7(7):665–674.

110. Eu JP, Xu L, Stamler JS, Meissner G. Regulation of ryanodine receptors by reactive nitrogen species. Biochem Pharmacol. 1999 May 15;57(10):1079–1084.
111. Park H-S, Huh S-H, Kim M-S, Lee SH, Choi E-J. Nitric oxide negatively regulates c-Jun N-terminal kinase/stress-activated protein kinase by means of S-nitrosylation. PNAS. 2000 Dec 19, 2000;97(26):14382–14387.
112. Yao D, Gu Z, Nakamura T, Shi ZQ, Ma Y, Gaston B, Palmer LA, Rockenstein EM, Zhang Z, Masliah E, Uehara T, Lipton SA. Nitrosative stress linked to sporadic Parkinson's disease: S-nitrosylation of parkin regulates its E3 ubiquitin ligase activity. Proc Natl Acad Sci U S A. 2004 Jul 20;101(29):10810–10814.
113. Choi YB, Chen HSV, Lipton SA. Three pairs of cysteine residues mediate both redox and Zn2+ modulation of the NMDA receptor. J Neurosci. 2001 Jan 15;21(2):392–400.
114. Kim SF, Huri DA, Snyder SH. Inducible nitric oxide synthase binds, S-Nitrosylates, and activates cyclooxygenase-2. Science. 2005 Dec 23;310(5756):1966–1970.
115. Tian J, Kim SF, Hester L, Snyder SH. S-nitrosylation/activation of COX-2 mediates NMDA neurotoxicity. Proc Natl Acad Sci U S A. 2008 Jul 29;105(30):10537–10540.
116. Fang M, Jaffrey SR, Sawa A, Ye K, Luo X, Snyder SH. Dexras1: a G protein specifically coupled to neuronal nitric oxide synthase via CAPON. Neuron. 2000 Oct;28(1):183–193.
117. Barouch LA, Harrison RW, Skaf MW, Rosas GO, Cappola TP, Kobeissi ZA, Hobai IA, Lemmon CA, Burnett AL, O'Rourke B, Rodriguez ER, Huang PL, Lima JA, Berkowitz DE, Hare JM. Nitric oxide regulates the heart by spatial confinement of nitric oxide synthase isoforms. Nature. 2002 Mar 21;416(6878):337–339.

Redox Imbalance in the Endoplasmic Reticulum

Gábor Bánhegyi, Éva Margittai, Miklós Csala, and József Mandl

Abstract The redox homeostasis of the endoplasmic reticulum critically depends on the concerted action of membrane transporters and local oxidoreductases that maintain the oxidized state of the thiol-disulfide and the reducing state of the pyridine nucleotide redox systems in the lumen. This situation, which is characteristically different from that of the other subcellular compartments, is a prerequisite for the normal functions of the organelle. The powerful thiol-oxidizing machinery allows oxidative protein folding but continuously challenges the local antioxidant defense. Alterations of both the cellular and the luminal redox environment either in oxidizing or reducing direction affect protein processing and may induce endoplasmic reticulum stress and the unfolded protein response. The activated signaling pathways attempt to restore the balance between protein loading and processing and induce apoptosis if the attempt fails. Recent observations strongly support the involvement of redox-based endoplasmic reticulum stress in a plethora of human diseases, including brain ischemia, neurodegenerative diseases, and traumatic injury. The aim of the present chapter is to overview those neurological diseases, in which the redox imbalance of the endoplasmic reticulum is an integral part of the pathomechanism either as a causative agent or as deleterious consequence.

Keywords Endoplasmic reticulum stress · Oxidative stress · Unfolded protein response · Brain ischemia · Neurodegenerative diseases · Traumatic injury · Apopstosis · Neuron

G. Bánhegyi (✉)
Department of Medical Chemistry, Molecular Biology and Pathobiochemistry, Semmelweis University, Budapest, Hungary
e-mail: banhegyi@puskin.sote.hu

S.C. Veasey (ed.), *Oxidative Neural Injury*,
Contemporary Clinical Neuroscience, DOI 10.1007/978-1-60327-342-8_3,
© Humana Press, a part of Springer Science+Business Media, LLC 2009

1 Introduction

Investigating the mechanisms that mediate neuronal cell death is particularly important since it is a major feature of such diseases as brain ischemia, neuronal degenerative diseases, and traumatic injury. It became evident that the signaling pathways involved in the control of neuronal cell death can be initiated and moderated by various organelles. The role of the endoplasmic reticulum (ER) dysfunction is suggested by recent molecular evidence. Although the stress response of the ER is regarded as a defense mechanism with a primary aim to maintain the function of the organelle, long-lasting or intensive ER stress ultimately leads to the induction of cell death. Such a lethal stress is often caused by a severe redox imbalance in the ER lumen. The consequent impairment of oxidative protein folding results in the luminal accumulation of unfolded/misfolded proteins, leading to the initiation of the unfolded protein response (UPR). It has been demonstrated that several luminal redox systems participate in the process. Since both hypoxia and neurodegenerative diseases result in the dysfunction of the ER due to a redox imbalance, studies on the malfunction of this cellular organelle may facilitate the development of novel strategies to prevent neuronal cell death by oxidative injuries. The aim of this chapter is to summarize the luminal redox systems and the redox sensing mechanisms of the ER, as well as their involvement in neurological diseases.

2 ER Stress and Unfolded Protein Response

The ER is not only an important metabolic compartment of the eukaryotic cell but also the principal site of the harmonization of the extra- and intracellular environment. Stimuli deriving from the neighborhood or from the interior of the cell are sensed by this organelle and tuned by ER-to-nucleus signaling pathways. It has been recently suggested that the ER can function as a sensor for electron donors and acceptors, i.e. nutrients and oxygen [1–3]. The imbalance of electron donors and acceptors, in other words, the discrepancy between demand and capacity, initiates an adaptation mechanism called ER stress. Since one of the most important functions of the ER is the synthesis and posttranslational modification of secretory and membrane proteins, the lumen of the organelle is equipped with a powerful protein-folding machine composed of chaperones, foldases and also with sensors that detect the presence of misfolded or unfolded proteins. Physiological and pathological effects or experimental agents that disturb the normal folding process provoke the UPR, an intracellular signaling pathway that coordinates ER protein-folding demand with protein-folding capacity and is essential to adapt to homeostatic alterations that cause protein misfolding. These include changes in intraluminal calcium, altered glycosylation, nutrient deprivation, pathogen infection, expression of folding-defective proteins, and changes in the redox status. Since there is a plethora of excellent reviews on ER stress and UPR [4–8], only the skeletonized events are summarized here.

There are three main proximal sensors of the UPR: the PKR-like ER protein kinase/pancreatic eIF2α (eukaryotic translation initiation factor 2, α subunit) kinase (PERK/PEK); the activating transcription factor 6 (ATF6); and the inositol-requiring enzyme 1 (IRE1). These sensors are integral proteins of the ER membrane; according to the dominant model of UPR, the association of their luminal domain with one of the most abundant ER resident chaperones BiP (immunoglobulin heavy chain-binding protein or Grp78, glucose-regulated protein of molecular weight 78 kDa) keeps them in inactive state. Perturbed ER homeostasis leads to the accumulation of unfolded/misfolded proteins in the ER lumen, which can preferentially bind BiP sequestering the chaperon in the lumen. The consequent dissociation of BiP from the transmembrane sensors permits their signaling. Signaling is principally based on dimerization–phosphorylation (IRE1 and PERK) or translocation–proteolysis (ATF6). The luminal domain of IRE1 and PERK regulates the protein kinase activity of these proteins by ER stress-regulated di- and oligomerization. The luminal domain of the transmembrane basic leucine zipper (bZIP) transcription factor ATF6 functions as a retention motif under normal conditions, but in ER stress, the protein is translocated to the Golgi complex, where it undergoes limited proteolysis. The cytosolic bZIP transcription factor domain is released from the membrane to allow its translocation to the nucleus.

These three mechanistically distinct arms of the UPR induce expression of chaperones, attenuate protein translation, promote the proliferation of ER membrane to enlarge the luminal compartment, and activate ER-associated degradation (ERAD) by not only regulating the expression of numerous genes related to protein folding but also affecting the metabolism of proteins, amino acids, and lipids. Alternatively, ER stress can also induce macroautophagy, a process whereby the cell recycles and remodels its macromolecules and organelles. Autophagy either counterbalances ER stress-induced ER expansion, enhances cell survival or commits the cell to nonapoptotic death [9]. The integrated UPR provides a tool to remodel the secretory apparatus and aligns cellular physiology to the demands imposed by ER stress. If the efforts of UPR are insufficient, prolonged ER stress can trigger mitochondria-dependent and mitochondria-independent forms of apoptosis [10]

Besides the classical sensors and signaling pathways of UPR, several other – sometimes cell-specific – mechanisms have been described. In the nervous system two sensors should be mentioned. OASIS (old astrocyte specifically induced substance) is an astrocyte-specific ER-resident transmembrane protein [11], while BBF2H7 (BBF2 human homolog on chromosome 7) [12] is a structurally homologous protein in neurons. Both proteins are bZIP transcription factors. Similar to ATF6, they are cleaved at the membrane in response to ER stress, and their cleaved cytoplasmic portions containing the bZIP domain, translocate into the nucleus and activate the transcription of target genes with ER stress-responsive and cyclic AMP-responsive elements. It should be noticed that BBF2H7 protein is not expressed under normal conditions but markedly induced during ER stress, suggesting that BBF2H7 might contribute only to the late phase of UPR signaling.

3 ER Stress and Redox Stress

The ER is particularly rich in oxygenases and oxidases (e.g., cytochrome P450s, flavin-containing monooxygenases, prolyl and lysyl hydroxylases), which often produce ROS as a by-product. Although ROS are formed in all cellular compartments, the ER seems to be a major place of production. Oxidative protein folding can be responsible for about one-fourth of ROS produced in a professional secretory cell [13]. ROS (presumably hydrogen peroxide) generated by Ero1p (endoplasmic reticulum oxidoreductin) is a putative end product of the electron transfer from protein thiols to molecular oxygen during disulfide bond formation. An ER-resident hydroxyl radical generation by an iron-dependent Fenton reaction has also been reported [14]. The oxidizing environment may cause a nonspecific oxidation of luminal proteins. Indeed, increased age-dependent carbonylation of BiP, protein disulfide isomerase (PDI), and calreticulin was observed in mouse liver [15]. The activity of BiP and PDI is inversely dependent on age and oxidative damage [16]. Increased carbonylation of these key proteins during ageing suggests an age-associated impairment in oxidative protein folding and posttranslational modifications in the liver.

Beside ROS, presumably RNS are also formed in the ER. This assumption is supported by the fact that S-nitrosylated PDI was found in brains manifesting sporadic Parkinson's or Alzheimer's disease. S-Nitrosylation of PDI inhibits its enzymatic activity, leads to the accumulation of polyubiquitinated proteins, and activates the UPR [17]. By this mechanism S-nitrosylation blocks the protective effect of PDI during UPR in neurodegenerative diseases. Although S-nitrosylation would imply the colocalization of an NO source (i.e., nitric oxide synthase), such enzyme has not yet been demonstrated in the ER [18].

The isomerization of misfolded proteins by PDI consumes GSH to reduce nonnative disulfide bonds; these reduced thiol groups will interact again with Ero1p to be reoxidized. This futile cycle continuously consumes GSH and produces ROS even under normal conditions; any factor disturbing the folding aggravates the situation. In fact, the GSH/GSSG redox system is present in more oxidized form in the ER lumen compared to the cytosolic, which means that the antioxidant mechanisms of the organelle cannot keep pace with the prooxidant effects.

GSH can also be utilized for the elimination of hydrogen peroxide by glutathione peroxidase, phospholipid hydroperoxide glutathione peroxidase [19], and peroxiredoxins [20]. From the latter family of enzymes the ER is equipped with peroxiredoxin IV, belonging to the class of 2-Cys peroxiredoxins. A redox-active cysteine in the active site of these proteins is oxidized to a sulfenic acid by the peroxide substrate. The recycling of the sulfenic acid back to a thiol occurs at the expense of glutathione.

Beyond the local redox imbalance, luminal accumulation of unfolded/misfolded proteins may cause generalized oxidative stress independent of disulfide bond formation. A typical accompaniment of the ER stress is the increased Ca^{2+} leak into the cytosol, which, in turn, stimulates ROS production in mitochondria.

Reductive stress is another known, albeit less intensively studied and less frequently mentioned, type of redox imbalance in the ER. It is defined as an excessive electron dumping or an overwhelming reducing power. Overreducing conditions prevent the formation of disulfide bonds; therefore, reducing agents (β-mercaptoethanol, dithiothreitol) belong to the group of prototypic ER stressors and UPR inducers [21, 22]. Recent observations suggest that over-nutrition by supplying an excess of reducing equivalents also leads to ER stress [1, 2, 23, 24]. Although the exact pathomechanism of ER stress is unknown in these conditions, the role of ER luminal redox alterations can be supposed.

Taking together, redox and ER stress can be developed on the ground of each other [25]. The activation of UPR upon exposure to redox and ER stress is an adaptive mechanism to preserve cell function and survival. However, persistent oxidative stress and protein misfolding ultimately initiate apoptotic cascades and are known to play eminent role in the pathomechanism of several human diseases including neurological diseases.

4 Redox Environment in the ER Lumen

The ratios of the oxidized and reduced components of the main redox couples in the ER lumen are fundamentally different from, yet not independent of, those in the cytosol. The luminal redox environment is determined by local oxidoreductions, by transmembrane fluxes of redox-active compounds, and by transmembrane electron fluxes. The major redox buffer of the ER lumen – similar to other compartments of the cell – is composed of glutathione and glutathione disulfide. Although the total concentration of the two components in the ER lumen and in the cytosol is similarly in the millimolar range, the ratio of [GSSG] to [GSH] is higher in the ER as compared to the cytosol, falling in the ranges of approximately 1:1–3 versus 1:30–100, respectively [26, 27]. The calculated redox potentials are −180 mV and −230 mV, respectively, i.e., the ER lumen can be regarded as a more thiol-oxidizing environment. Moreover, high percentage of luminal glutathione was shown to form mixed disulfides with proteins, which may play a role as a GSH reserve and also as a component of the thiol–disulfide redox buffering system [28].

It was supposed that this peculiar intraluminal ER milieu ensures disulfide bond formation in secretory proteins by maintaining an oxidizing power. However, it has been shown that protein disulfide formation is independent of glutathione, both in vivo and in vitro [29, 30]. The process is catalyzed by a protein-mediated electron relay system, in which the concerted action of PDI [31] and Ero1p [32–35] delivers the electrons from thiol groups to the ultimate acceptor, molecular oxygen. These studies demonstrated that, contrary to previous assumptions, glutathione is not the source of oxidizing equivalents for oxidative protein folding within the ER. It rather functions as a reductant for both PDI and secretory proteins, allowing the isomerization of disulfide bonds, until the correct conformation is achieved. Another important conclusion of these recent findings is that the relatively oxidized state of the glutathione redox buffer within the ER is not the actual cause, but rather the consequence of the oxidative protein folding. Nevertheless, the possible interferences between the GSSG/GSH system and both oxidation and reduction of protein thiols underscore the importance of a tight regulation of the ER redox milieu, since changes in *either* direction can drastically disturb ER functions.

The presence of other redox-active compounds has also been demonstrated in the ER lumen. However, their local concentration and redox state are presently unknown. Besides glutathione, the other antioxidant components of the Halliwell–Asada cycle are also present in the ER. The cycle may represent an alternative, possibly minor pathway for the transfer of electrons from (protein) thiols to oxygen. Ascorbate does not only act as an antioxidant in the lumen but also as a cofactor for various luminal enzymes (e.g., prolyl- and lysyl hydroxylases) and its oxidized form, dehydroascorbate, can accept electrons from PDI [36–38]. Ascorbate, in turn, can regenerate the oxidized form of tocopherol, a hydrophobic antioxidant in the ER membrane, which can mediate a transmembrane electron transfer [39]. Ascorbate is formed in the ER of ascorbate-synthesizing animals, while in species unable to produce this vitamin it is transported through the membrane as dehydroascorbic acid [40]. This circumstance might cause – presently unexplored – differences in the ascorbate-dependent processes of the ER.

The vitamin K-dependent glutamyl-γ-carboxylation of certain proteins in the ER involves the formation of reduced vitamin K, an essential cofactor for γ-carboxylase catalyzing this posttranslational protein modification. By accepting electrons from PDI, vitamin K can be regenerated during the protein γ-carboxylation-linked redox cycle [41]. FAD is known to act both as a tightly associated prosthetic group and as a relatively free cofactor of Ero1p, thereby making this enzyme highly responsive to small changes in the physiological levels of free FAD [30, 42]. Although FAD transport has been demonstrated in yeast and rat liver microsomes, the concentration and the redox state of luminal FAD are unknown.

Several oxidoreductases with intraluminal active site require pyridine nucleotides as coenzymes. Although their in vivo redox state and concentrations have not been elucidated, their presence has been demonstrated in ER-derived microsomal vesicles [43]. This luminal pyridine nucleotide pool is overwhelmingly composed of reduced NAD(P)H, which presumably mirrors the in vivo situation. The redox state is ensured by local dehydrogenases such as hexose-6-phosphate dehydrogenase (H6PDH) [44] and isocitrate dehydrogenase [45], because the permeability of the ER membrane to pyridine nucleotides is insignificant [46]. The reducing equivalents derive from the cytosol by means of a glucose-6-phosphate transporter [47]; the existence of an isocitrate transport has not been evidenced yet. The reduced state of the luminal pyridine nucleotides seems to be important for the functioning of luminal ER reductases such as NCB5OR [48] and short-chain dehydrogenases/reductases [49] including 11β-hydroxysteroid dehydrogenase type 1 (11βHSD1) [50]. 11βHSD1, an enzyme responsible for the prereceptorial activation of glucocorticoids, catalyzes the reversible interconversion of cortisone and cortisol in the lumen of the ER [51]. Under in vivo conditions, this reaction is predominantly shifted toward cortisone reduction, substantiating a high luminal [NADPH]/[NADP$^+$] ratio.

Colocalized redox couples are often linked to one another by oxidoreductases to form complex redox systems. However, if the linking enzymes are missing, the redox couples can coexist independently and have different redox potentials. For example, the luminal pyridine nucleotide pool and the GSH/GSSG system seem to be uncoupled. While both NADPH and glutathione are present predominantly in the reduced state in the cytosol, the ER lumen can be characterized by an oxidized glutathione and a reduced pyridine nucleotide system. The enzymatic coupling of the two redox pairs is catalyzed by glutathione reductase in the cytosol. Since glutathione reductase is hardly detectable in the lumen [46], pyridine nucleotides remain reduced in spite of the oxidizing power of the GSSG/GSH system. The peaceful coexistence of the two redox systems ensures the concurrent functioning of oxidative protein folding and local reductases.

Redox imbalance of these luminal redox systems and the shortage of their components have been reported to cause ER stress and ER-dependent apoptosis both in cellular systems and in vivo. Experimental agents affecting the redox state of glutathione are prototypic inducers of ER stress. The deficiency of the FAD precursor riboflavin impairs oxidative protein folding in various cell types [52, 53]. Inhibition or genetic absence of the ER glucose-6-phosphate transporter provokes apoptosis in glia cells [54, 55] and neutrophil granulocytes [56], presumably through the altered redox state of the luminal pyridine nucleotides. A very recent paper [57] supports this assumption, showing that the deletion of H6PDH [58] activates the UPR in skeletal muscle and leads to myopathy. These effects can be related to the antioxidant role of NADPH. Although in vivo models of ER stress are less numerous, some observations also support the view that ER redox imbalance is an important motive of ER stress. Ascorbate deficiency (scurvy) causes ER stress and apoptosis in the liver of guinea pigs [59]. The oxidation and depletion of luminal glutathione and the appearance of ER stress markers were also demonstrated in the background of acetaminophen hepatotoxicity [60].

5 Influence of Cytosolic Redox Changes on the ER and Vice Versa – Transport of Redox-Active Compounds

The continuous ER membrane isolates the inner and outer compartments, yet the luminal environment is not completely independent of the cytosol. The redox state of the ER lumen is greatly affected by the influx and efflux of potential oxidizing and reducing agents. The permeation of several redox-active compounds across the ER membrane has been demonstrated; nevertheless, the transporter proteins participating in this traffic remain to be identified. In general, the available information regarding the ER transporters [61] is relatively poor compared to other organelles, which is due to the technical difficulty of ER transport measurements and of the purification and reconstruction approach [62].

The inward transport of nascent polypeptides abundant in cysteinyl thiols – a process mediated by the translocon protein channel – is responsible for the majority of reducing equivalent import into the lumen. In addition, the temporarily open translocon was shown to allow the transport of various small molecules; hence, its further contribution to the permeation of redox-active compounds cannot be excluded [63, 64]. The hypothesis [27] that the relatively high [GSSG]/[GSH] ratio in the lumen is a consequence of the preferential import of GSSG from the cytosol was contradicted by the results of direct microsomal transport measurements. GSH has a slow protein-mediated transport, while the membrane is impermeable to GSSG [65]. Therefore, it seems more plausible that GSH entering the ER lumen is oxidized locally and gets entrapped in the form of GSSG, corresponding to the thiol-oxidizing environment in the compartment.

Reducing equivalents can be transported into the lumen also in the form of glucose-6-phosphate. The corresponding transporter (G6PT) is one of the few ER transporters characterized at molecular level [47]. Its concerted action with H6PDH seems to be the main source of the reduced NADPH in the ER lumen [44, 66, 67].

Although the generation and maintenance of the oxidative luminal environment can be attributed mostly to local oxidoreductases, the transport of oxidants can also contribute. The observation of dehydroascorbate and FAD transport across the ER membrane [40, 42, 68] is relevant from this aspect. It should be noticed that ascorbate and pyridine nucleotides were reported to be unable to enter the ER lumen at a significant rate [40, 46, 69].

6 BiP-Mediated Redox Sensing

Certain oxidative modifications can be considered as integral elements of the protein maturation process in the ER, which is, hence, often referred to as oxidative folding. The majority of the secretory and membrane proteins require the formation of disulfide bonds in proper positions for their appropriate and stable folding. Redox imbalance caused by either experimental agents or pathophysiological conditions leads to the accumulation of unfolded/misfolded proteins in the ER lumen. The ER stress component is involved in the putative pathomechanism of ageing, Alzheimer's, diabetes, etc. (for review see [70] and [71]) and similar danger can be posed by either oxidative or reductive stress. The accumulation of defective proteins can lead to the activation of UPR [5]. The present theory of the UPR premises that the various misfolded proteins are recognized and the diverse signaling pathways are initiated by a single master regulator, BiP.

This hypothesis is based on the observation that BiP is associated with the stress transducer proteins under stress-free conditions and is released upon accumulation of unfolded proteins. Moreover, BiP overexpression attenuates the initiation of all the three signaling branches of UPR. In conclusion, BiP functions as a sensor of unfolded proteins, which mechanistically links luminal redox imbalance with UPR activation. This attractive model, though valid in case of experimental, acute, and severe ER stress, cannot account for the selective activation of ER stress subpathways in pathophysiological conditions. Therefore, the in vivo mechanism seems much more complicated than previously thought. The passive competition model for BiP between unfolded proteins and transmembrane signal transducers has been challenged by recent observations. A relatively stable binding between ATF6 and BiP was observed, and a region within the luminal domain of ATF6 was identified as a specific ER stress-responsive sequence required for ER stress-triggered BiP release [72]. Furthermore, deletion of the BiP-binding site of IRE1 failed to alter the inducibility of ER stress, showing that BiP is not the principal determinant of IRE1 activity, but an adjustor for sensitivity to various stresses [73]. On the basis of these findings it can be supposed that additional regulators are involved in the initiation of UPR or that other tissue-specific adaptor proteins can moderate the signaling events and the biological responses once UPR is initiated by different physiological stimuli.

7 Direct Redox Sensing

The transmembrane proteins involved in ER stress signaling can also detect the changes in the luminal redox state directly, i.e., independently of the transiently luminal secretory proteins. For example, association/dissociation of BiP is not the sole regulatory mechanism in case of ATF6. It has been recently shown that owing to the presence of intra- and intermolecular disulfide bridges, ATF6 mono-, di-, and oligomers are formed in the unstressed ER. Various experimental ER stress inducers cause the reduction of these disulfide bonds, which increases the amount of reduced ATF6 monomers that are active in the UPR signaling. ER stress evoked by a more physiological mechanism, such as glucose starvation, also activates ATF6. Besides an enhanced ATF6 synthesis likely due to transcriptional induction, reduction of disulfide bridges and transport of reduced monomers to Golgi occurred in response to glucose starvation. The results show that at least two events are necessary for ATF6 activation, namely the dissociation of BiP and the reduction of disulfide bridges [74, 75]. Although the mechanism of ATF6 reduction is still enigmatic, it was supposed that ER luminal oxidoreductases, activated upon glucose starvation, may participate in the process. The enzymes responsible and the source of luminal reducing power remain to be clarified.

Ca^{2+} release from the luminal store is an important event of ER stress and ER-dependent apoptosis. Several observations indicate the role of luminal redox imbalance in the remarkable alteration of cellular calcium homeostasis in ER stress. Both ER/SR calcium channels and calcium pumps were proven to be greatly affected by intraluminal redox changes. Dynamic redox-sensitive thiols in the ryanodine receptor calcium channel are subject to reversible oxidoreduction, which, in turn, modulates the open probability of the channel. On this ground, the ryanodine receptor calcium channel has been postulated as a transmembrane redox sensor in the SR [76, 77]. Recent findings demonstrate a similar dependence of the InsP3 receptor (InsP3R) activity on the ER luminal redox status. It was found that ERp44, an ER luminal protein belonging to the thioredoxin family, interacts with the third luminal loop of InsP3R type 1 (InsP3R1) and directly inhibits the receptor. This interaction is dependent on the redox state (as well as on luminal pH and Ca^{2+} concentration). The presence of reduced cysteinyl thiols in the third loop is required for the interaction. Thus, ERp44 seems to sense the environment in the ER lumen and modulate the calcium homeostasis through InsP3R1 activity [78].

Calcium reuptake into the ER is also regulated by the redox state and calcium. The luminal protein ERp57 was shown to regulate SERCA2b activity [79]. ERp57 overexpression reduces the frequency of SERCA 2b-dependent Ca^{2+} oscillations; the effect is dependent on the presence of cysteinyl residues located in intraluminal loop 4. Store depletion results in ERp57 dissociation and a relief of SERCA2b inhibition. The results suggest that ERp57 modulates the redox state of luminal thiols in SERCA 2b in a Ca^{2+}-dependent manner, providing dynamic control of ER Ca^{2+} homeostasis.

These interactions between luminal redox and Ca^{2+} signaling may also be significant in the cellular response to stress, serving to protect the cell from apoptosis. Indeed, expression of both ERp57 and ERp44 is increased by cellular stress. ERp44 overexpression was shown to inhibit apoptosis [78]. In conclusion, these studies underline the interdependence of luminal redox state, oxidative protein folding, and calcium signaling. InsP3R-induced calcium release may be an important link between luminal redox imbalance and apoptosis.

8 ER Redox Changes in Neurological Diseases

8.1 Trauma

The signs of activation of the UPR indicative of ER dysfunction were demonstrated in animal models of brain or spinal cord trauma. Processed X-box protein 1 (xbp1) mRNA levels – generated by the endonuclease activity of IRE1 in UPR – dramatically rose 1 h after the trauma in the cortex and remained high till 24 h. A delayed elevation of processed xbp1 mRNA levels was observed in the hippocampus and striatum [80], as well as in case of spinal cord trauma. In the latter, the significant increase (peaking at 6 h of recovery) was observed not only at the site of the primary insult but also in the adjacent segments [81]. In addition to the processing of xbp1 mRNA, another event of UPR, the activation of caspase-12 was also detected in brain trauma, which further supports the presence of ER dysfunction [82]. The exact mechanisms leading to ER stress after the trauma are unexplored; however, local hypoxia, the excessive release of excitatory neurotransmitters, and the accumulation of reactive oxygen and nitrogen species alike can disturb the normal ER functions [83]. In summary, a trauma of the central nervous system induces ER dysfunction, which spreads from the primary spot to the surrounding areas. These circumstances may have clinical implications: ER stress can prove to be an important element in the pathomechanism of the secondary nonmechanical damage and hence should be considered as a target for therapeutic intervention of traumatic brain injury.

8.2 Neurodegenerative Diseases

The group of conditions referred to as neurodegenerative diseases is quite heterogeneous with regard to both their symptoms and the underlying pathomechanisms. However, they have certain common features like the deficiency of the ubiquitin–proteasome system (UPS) [84], the disturbance of mitochondrial functions [85], the increased oxidative damage [86], and the accumulation of misfolded, aggregated proteins [87]. These conditions secondarily lead to ER stress [88], which has been suggested to be involved in the pathomechanism of human neurodegenerative diseases such as Parkinson's, Alzheimer's, and prion disease. The contribution and the significance of ER stress, however, are matters of debate. All the three diseases are associated with the accumulation of misfolded proteins, which should trigger the UPS but its activity is depressed by oxidative effects or by the deposited protein aggregates. The further accumulation of misfolded proteins, in turn, leads to more serious ER stress and the aggravation of the disease.

8.2.1 Parkinson's Disease (PD)

The signs of ER stress and UPR are present in PD [89]; however, it is not clear whether the observed ER stress is secondary and largely neuroprotective or rather it contributes directly to the disease progression. The S-nitrosylation of PDI and parkin was reported [17, 90], suggesting the presence of reactive species in the luminal compartment of the ER. In accordance, the disease process can be mimicked by certain redox-active neurotoxins both in cell culture and in vivo. These model compounds, such as 6-hydroxydopamine and N-methyl-4-phenyl-1, 2,3,6-tetrahydropyridine or its active derivative, N-methyl-4-phenylpyridinium, induce oxidative stress, which can affect the ER. Studies on cultured neuronal cells, including dopaminergic neurons, showed that these compounds trigger ER stress with the upregulation of ER chaperones and the transcription factor CHOP/Gadd153, in addition to the phosphorylation of the ER stress kinases, IRE1, and PERK [91, 92]. Collectively, these observations suggest a redox-based ER stress in PD.

8.2.2 Alzheimer's Disease (AD)

Fibrillar Aβ accumulating in familial AD is toxic to neuronal cells and leads to the production of great amounts of ROS [93]. Recent studies concerning the role of the UPR revealed increased levels of BiP/Grp78 and the protein kinase PERK in the brain of the patients [94]. This shows the occurrence of ER stress in AD although the contribution of redox mechanisms remains to be clarified.

8.2.3 Amyotrophic Lateral Sclerosis (ALS)

Increased ROS production causing oxidative damage to crucial proteins and to other cell components may play a role in ALS. These changes can be attributed to a disturbed glutamate metabolism with prolonged stimulation of excitatory amino-acid receptors leading to increased intracellular calcium that can damage the mitochondria and the ER [95, 96]. It has been recently reported that cleavage of caspase-12 occurs in the spinal cord of transgenic ALS mice indicative of ER stress. The observed cleavage of caspase-12 in the ALS mice could be due to the activity of the calcium-dependent enzyme calpain that was also activated in the spinal cord of these mice [97]. It remains to be studied whether the ER-mediated caspase activation plays a role in the disease progression of ALS. Other markers of ER stress, such as the induction of BiP/Grp78 were also detected in the ALS mice, although not at the same extent [97]. Further evidence for ER stress in ALS comes from studies showing a large increase in BiP/Grp78 in spinal motorneurons of transgenic ALS mice prior to the onset of motor symptoms [98]. It was also reported that mutant, but not wild-type, SOD1 can aggregate and associate in the ER [99]. The possible role of the altered antioxidant protection of the ER and the consequent redox changes in the mechanism of ER stress need further investigations.

8.3 Hypoxia, Ischemia – Reperfusion

Oxidative protein folding in the ER requires oxygen as the ultimate electron acceptor; therefore, it is not surprising that the ER functions are sensitive to hypoxia. In addition to this direct effect, reduced oxygen availability can also influence ER functions through the adaptive and corrective cellular responses, controlled principally by the transcription factor, hypoxia-inducible factor-1 (HIF-1) [100]. HIF-1 target genes play key roles in multiple pathways, including angiogenesis, vascular reactivity and remodeling, glucose and energy metabolism, cell proliferation and survival, erythropoiesis, iron homeostasis, and others [for a review see [101]]. From our point of view, it is important that some ER proteins are known to be regulated by hypoxia, including Grp78, Grp94 [102], ORP150 (150 kDa oxygen-regulated protein) [103], PDI [104], and Ero1Lα [105]. Consequently, hypoxia can lead to ER stress and UPR activation by two means: impairment of the oxidative folding in the ER due to the shortage of electron acceptor and induction of certain ER proteins.

In agreement with the theoretical considerations, ER stress has been shown to be present in acute brain disorders complicated with ischemia. Focal cerebral ischemia in mice resulted in the activation of the UPR sensors, eIF2α and PERK, due to the detachment of BiP/Grp78 [106]. Global ischemia in mice also induced ER stress and the activation of the ER transcription factors, CHOP/Gadd153 and ATF4. It has been reported that hippocampal neurons from CHOP/Gadd153-deficient mice are more resistant to cell death induced by hypoxia reoxygenation compared with controls [107]. Fewer neurons degenerated in the CHOP–/– mice after ischemia, suggesting an important role for ER stress in ischemia/stroke. Autophagy was also observed in an animal model of cerebral ischemia [108]. It was proposed that the autophagy in neurons was stimulated by oxidative and ER stresses in cerebral hypoxia [109].

The hypoxia-induced ER chaperones, ORP150 and Grp94, have been shown to afford protection against ischemia-induced cell death in brain [110, 111]. In a mouse model of focal brain ischemia, the ER stress-inducible ER stress sensor BBF2H7 protein is prominently induced in neurons in the peri-infarction region [12]. The overexpression of the protein in neuroblastoma cell culture prevents ER stress-induced cell death.

Redox mechanisms are possibly underlying the ischemia-induced impairment of ER functions [112]. First, it was observed that the ischemia-induced PERK-dependent ER stress response is markedly less pronounced in animals overexpressing SOD1. Therefore, superoxide radicals play a role in provoking ER stress probably by PERK activation [113]. Furthermore, ischemia-induced activation of PERK and phosphorylation of eIF2α was blocked in animals with targeted deletion of the endothelial or neuronal nitric oxide synthase gene [114]. The role of NO in ischemia-induced ER stress is further strengthened by the observation that ER calcium stores are depleted after ischemia and that recovery of ER calcium homeostasis was observed in the presence of NO synthase inhibitor in neuroprotective level [115]. Moreover, exposure of primary neuronal cell cultures to NO suppressed ER calcium pump activity, depleted ER calcium stores, and suppressed protein synthesis [116]. Collectively, these findings suggest that the neuronal toxicity of NO – excessively produced after ischemia – involves the dysfunction of the intracellular Ca^{2+} transport systems including ER and may contribute to ER stress.

9 Concluding Remarks

Recent investigations provide an insight into the redox circumstances in the ER lumen. This compartment is equipped with powerful oxidizing mechanisms, but its antioxidant defense is relatively weak. Consequently, the lumen can be regarded as an oxidizing environment. While these conditions favor and permit oxidative protein folding, one of the most important functions of the ER, they also make the organelle vulnerable to redox insults. Either oxidative or reductive stress may affect protein synthesis and induce ER stress and UPR, albeit by different mechanisms. ER redox changes are integral parts of the pathomechanisms of neurological diseases, either as causative agents or as complications of the disorders due to other primary factors.

Acknowledgments Thanks are due to the Szentágothai János Knowledge Center (RET) and the János Bolyai Research Scholarship of the Hungarian Academy of Sciences.

References

1. Ozcan U, Cao Q, Yilmaz E, Lee AH, Iwakoshi NN, Ozdelen E, Tuncman G, Gorgun C, Glimcher LH, Hotamisligil GS. Endoplasmic reticulum stress links obesity, insulin action, and type 2 diabetes. Science. 2004 Oct 15;306(5695):457–461.
2. Hotamisligil GS. Role of endoplasmic reticulum stress and c-Jun NH2-terminal kinase pathways in inflammation and origin of obesity and diabetes. Diabetes. 2005 Dec;54 Suppl 2:S73–S78.
3. Banhegyi G, Baumeister P, Benedetti A, Dong D, Fu Y, Lee AS, Li J, Mao C, Margittai E, Ni M, Paschen W, Piccirella S, Senesi S, Sitia R, Wang M, Yang W. Endoplasmic reticulum stress. Ann N Y Acad Sci. 2007 Oct;1113:58–71.
4. Sitia R, Molteni SN. Stress, protein (mis)folding, and signaling: the redox connection. Sci STKE. 2004 Jun 29;2004(239):e27.
5. Schroder M, Kaufman RJ. ER stress and the unfolded protein response. Mutat Res. 2005 Jan 6;569(1–2):29–63.
6. Szegezdi E, Logue SE, Gorman AM, Samali A. Mediators of endoplasmic reticulum stress-induced apoptosis. EMBO Rep. 2006 Sep;7(9):880–885.
7. Ron D, Walter P. Signal integration in the endoplasmic reticulum unfolded protein response. Nat Rev Mol Cell Biol. 2007 Jul;8(7):519–529.
8. Schroder M. Endoplasmic reticulum stress responses. Cell Mol Life Sci. 2008 Mar;65(6):862–894.
9. Hoyer-Hansen M, Jaattela M. Connecting endoplasmic reticulum stress to autophagy by unfolded protein response and calcium. Cell Death Differ. 2007 Sep;14(9):1576–1582.
10. Hetz CA. ER stress signaling and the BCL-2 family of proteins: from adaptation to irreversible cellular damage. Antioxid Redox Signal. 2007 Dec;9(12):2345–2355.
11. Kondo S, Murakami T, Tatsumi K, Ogata M, Kanemoto S, Otori K, Iseki K, Wanaka A, Imaizumi K. OASIS, a CREB/ATF-family member, modulates UPR signalling in astrocytes. Nat Cell Biol. 2005 Feb;7(2):186–194.
12. Kondo S, Saito A, Hino S, Murakami T, Ogata M, Kanemoto S, Nara S, Yamashita A, Yoshinaga K, Hara H, Imaizumi K. BBF2H7, a novel transmembrane bZIP transcription factor, is a new type of endoplasmic reticulum stress transducer. Mol Cell Biol. 2007 Mar;27(5):1716–1729.
13. Tu BP, Weissman JS. Oxidative protein folding in eukaryotes: mechanisms and consequences. J Cell Biol. 2004 Feb 2;164(3):341–346.
14. Liu Q, Berchner-Pfannschmidt U, Moller U, Brecht M, Wotzlaw C, Acker H, Jungermann K, Kietzmann T. A Fenton reaction at the endoplasmic reticulum is involved in the redox control of hypoxia-inducible gene expression. Proc Natl Acad Sci U S A. 2004 Mar 23;101(12):4302–4307.
15. Rabek JP, Boylston WH, 3rd, Papaconstantinou J. Carbonylation of ER chaperone proteins in aged mouse liver. Biochem Biophys Res Commun. 2003 Jun 6;305(3):566–572.
16. Nuss JE, Choksi KB, DeFord JH, Papaconstantinou J. Decreased enzyme activities of chaperones PDI and BiP in aged mouse livers. Biochem Biophys Res Commun. 2008 Jan 11;365(2):355–361.
17. Uehara T, Nakamura T, Yao D, Shi ZQ, Gu Z, Ma Y, Masliah E, Nomura Y, Lipton SA. S-nitrosylated protein-disulphide isomerase links protein misfolding to neurodegeneration. Nature. 2006 May 25;441(7092):513–517.
18. Iwakiri Y, Satoh A, Chatterjee S, Toomre DK, Chalouni CM, Fulton D, Groszmann RJ, Shah VH, Sessa WC. Nitric oxide synthase generates nitric oxide locally to regulate compartmentalized protein S-nitrosylation and protein trafficking. Proc Natl Acad Sci U S A. 2006 Dec 26;103(52):19777–19782.
19. Marinho HS, Antunes F, Pinto RE. Role of glutathione peroxidase and phospholipid hydroperoxide glutathione peroxidase in the reduction of lysophospholipid hydroperoxides. Free Radic Biol Med. 1997;22(5):871–883.

20. Rhee SG, Chae HZ, Kim K. Peroxiredoxins: a historical overview and speculative preview of novel mechanisms and emerging concepts in cell signaling. Free Radic Biol Med. 2005 Jun 15;38(12):1543–1552.

21. Brostrom MA, Prostko CR, Gmitter D, Brostrom CO. Independent signaling of grp78 gene transcription and phosphorylation of eukaryotic initiator factor 2 alpha by the stressed endoplasmic reticulum. J Biol Chem. 1995 Feb 24;270(8):4127–4132.

22. Kim YK, Kim KS, Lee AS. Regulation of the glucose-regulated protein genes by beta-mercaptoethanol requires de novo protein synthesis and correlates with inhibition of protein glycosylation. J Cell Physiol. 1987 Dec;133(3):553–559.

23. Ozcan U, Yilmaz E, Ozcan L, Furuhashi M, Vaillancourt E, Smith RO, Gorgun CZ, Hotamisligil GS. Chemical chaperones reduce ER stress and restore glucose homeostasis in a mouse model of type 2 diabetes. Science. 2006 Aug 25;313(5790):1137–1140.

24. Gregor MF, Hotamisligil GS. Thematic review series: adipocyte biology. adipocyte stress: the endoplasmic reticulum and metabolic disease. J Lipid Res. 2007 Sep;48(9):1905–1914.

25. Malhotra JD, Kaufman RJ. Endoplasmic reticulum stress and oxidative stress: a vicious cycle or a double-edged sword? Antioxid Redox Signal. 2007 Dec;9(12):2277–2293.

26. Braakman I, Helenius J, Helenius A. Manipulating disulfide bond formation and protein folding in the endoplasmic reticulum. EMBO J. 1992 May;11(5):1717–1722.

27. Hwang C, Sinskey AJ, Lodish HF. Oxidized redox state of glutathione in the endoplasmic reticulum. Science. 1992 Sep 11;257(5076):1496–1502.

28. Bass R, Ruddock LW, Klappa P, Freedman RB. A major fraction of endoplasmic reticulum-located glutathione is present as mixed disulfides with protein. J Biol Chem. 2004 Feb 13;279(7):5257–5262.

29. Cuozzo JW, Kaiser CA. Competition between glutathione and protein thiols for disulphide-bond formation. Nat Cell Biol. 1999 Jul;1(3):130–135.

30. Tu BP, Ho-Schleyer SC, Travers KJ, Weissman JS. Biochemical basis of oxidative protein folding in the endoplasmic reticulum. Science. 2000 Nov 24;290(5496):1571–1574.

31. Kersteen EA, Raines RT. Catalysis of protein folding by protein disulfide isomerase and small-molecule mimics. Antioxid Redox Signal. 2003 Aug;5(4):413–424.

32. Frand AR, Kaiser CA. The ERO1 gene of yeast is required for oxidation of protein dithiols in the endoplasmic reticulum. Mol Cell. 1998 Jan;1(2):161–170.

33. Freudenthal R, Romano A. Participation of Rel/NF-kappaB transcription factors in long-term memory in the crab Chasmagnathus. Brain Res. 2000 Feb 14;855(2):274–281.

34. Pollard MG, Travers KJ, Weissman JS. Ero1p: a novel and ubiquitous protein with an essential role in oxidative protein folding in the endoplasmic reticulum. Mol Cell. 1998 Jan;1(2):171–182.

35. Mezghrani A, Fassio A, Benham A, Simmen T, Braakman I, Sitia R. Manipulation of oxidative protein folding and PDI redox state in mammalian cells. EMBO J. 2001 Nov 15;20(22):6288–6296.

36. Wells WW, Xu DP, Yang YF, Rocque PA. Mammalian thioltransferase (glutaredoxin) and protein disulfide isomerase have dehydroascorbate reductase activity. J Biol Chem. 1990 Sep 15;265(26):15361–15364.

37. Nardai G, Braun L, Csala M, Mile V, Csermely P, Benedetti A, Mandl J, Banhegyi G. Protein-disulfide isomerase- and protein thiol-dependent dehydroascorbate reduction and ascorbate accumulation in the lumen of the endoplasmic reticulum. J Biol Chem. 2001 Mar 23;276(12):8825–8828.

38. Banhegyi G, Csala M, Szarka A, Varsanyi M, Benedetti A, Mandl J. Role of ascorbate in oxidative protein folding. Biofactors. 2003;17(1–4):37–46.

39. Csala M, Szarka A, Margittai E, Mile V, Kardon T, Braun L, Mandl J, Banhegyi G. Role of vitamin E in ascorbate-dependent protein thiol oxidation in rat liver endoplasmic reticulum. Arch Biochem Biophys. 2001 Apr 1;388(1):55–59.

40. Banhegyi G, Marcolongo P, Puskas F, Fulceri R, Mandl J, Benedetti A. Dehydroascorbate and ascorbate transport in rat liver microsomal vesicles. J Biol Chem. 1998 Jan 30;273(5):2758–2762.
41. Wajih N, Hutson SM, Wallin R. Disulfide-dependent protein folding is linked to operation of the vitamin K cycle in the endoplasmic reticulum. A protein disulfide isomerase-VKORC1 redox enzyme complex appears to be responsible for vitamin K1 2,3-epoxide reduction. J Biol Chem. 2007 Jan 26;282(4):2626–2635.
42. Tu BP, Weissman JS. The FAD- and O(2)-dependent reaction cycle of Ero1-mediated oxidative protein folding in the endoplasmic reticulum. Mol Cell. 2002 Nov;10(5): 983–994.
43. Bublitz C, Lawler CA. The levels of nicotinamide nucleotides in liver microsomes and their possible significance to the function of hexose phosphate dehydrogenase. Biochem J. 1987 Jul 1;245(1):263–267.
44. Hewitt KN, Walker EA, Stewart PM. Minireview: hexose-6-phosphate dehydrogenase and redox control of 11{beta}-hydroxysteroid dehydrogenase type 1 activity. Endocrinology. 2005 Jun;146(6):2539–2543.
45. Margittai E, Banhegyi G. Isocitrate dehydrogenase: a NADPH-generating enzyme in the lumen of the endoplasmic reticulum. Arch Biochem Biophys. 2008 Mar 15;471(2):184–190.
46. Piccirella S, Czegle I, Lizak B, Margittai E, Senesi S, Papp E, Csala M, Fulceri R, Csermely P, Mandl J, Benedetti A, Banhegyi G. Uncoupled redox systems in the lumen of the endoplasmic reticulum. Pyridine nucleotides stay reduced in an oxidative environment. J Biol Chem. 2006 Feb 24;281(8):4671–4677.
47. van Schaftingen E, Gerin I. The glucose-6-phosphatase system. Biochem J. 2002 Mar 15;362(Pt 3):513–532.
48. Zhu H, Larade K, Jackson TA, Xie J, Ladoux A, Acker H, Berchner-Pfannschmidt U, Fandrey J, Cross AR, Lukat-Rodgers GS, Rodgers KR, Bunn HF. NCB5OR is a novel soluble NAD(P)H reductase localized in the endoplasmic reticulum. J Biol Chem. 2004 Jul 16;279(29):30316–30325.
49. Hoffmann F, Maser E. Carbonyl reductases and pluripotent hydroxysteroid dehydrogenases of the short-chain dehydrogenase/reductase superfamily. Drug Metab Rev. 2007;39(1):87–144.
50. Draper N, Stewart PM. 11beta-hydroxysteroid dehydrogenase and the pre-receptor regulation of corticosteroid hormone action. J Endocrinol. 2005 Aug;186(2):251–271.
51. Odermatt A, Atanasov AG, Balazs Z, Schweizer RA, Nashev LG, Schuster D, Langer T. Why is 11beta-hydroxysteroid dehydrogenase type 1 facing the endoplasmic reticulum lumen? Physiological relevance of the membrane topology of 11beta-HSD1. Mol Cell Endocrinol. 2006 Mar 27;248(1–2):15–23.
52. Camporeale G, Zempleni J. Oxidative folding of interleukin-2 is impaired in flavin-deficient jurkat cells, causing intracellular accumulation of interleukin-2 and increased expression of stress response genes. J Nutr. 2003 Mar;133(3):668–672.
53. Manthey KC, Chew YC, Zempleni J. Riboflavin deficiency impairs oxidative folding and secretion of apolipoprotein B-100 in HepG2 cells, triggering stress response systems. J Nutr. 2005 May;135(5):978–982.
54. Belkaid A, Copland IB, Massillon D, Annabi B. Silencing of the human microsomal glucose-6-phosphate translocase induces glioma cell death: potential new anticancer target for curcumin. FEBS Lett. 2006 Jun 26;580(15):3746–3752.
55. Belkaid A, Currie JC, Desgagnes J, Annabi B. The chemopreventive properties of chlorogenic acid reveal a potential new role for the microsomal glucose-6-phosphate translocase in brain tumor progression. Cancer Cell Int. 2006;6:7.
56. Leuzzi R, Banhegyi G, Kardon T, Marcolongo P, Capecchi PL, Burger HJ, Benedetti A, Fulceri R. Inhibition of microsomal glucose-6-phosphate transport in human neutrophils results in apoptosis: a potential explanation for neutrophil dysfunction in glycogen storage disease type 1b. Blood. 2003 Mar 15;101(6):2381–2387.

57. Lavery GG, Walker EA, Turan N, Rogoff D, Ryder JW, Shelton JM, Richardson JA, Falciani F, White PC, Stewart PM, Parker KL, McMillan DR. Deletion of hexose-6-phosphate dehydrogenase activates the unfolded protein response pathway and induces skeletal myopathy. J Biol Chem. 2008 Mar 28;283(13):8453–8461.

58. Lavery GG, Walker EA, Draper N, Jeyasuria P, Marcos J, Shackleton CH, Parker KL, White PC, Stewart PM. Hexose-6-phosphate dehydrogenase knock-out mice lack 11 beta-hydroxysteroid dehydrogenase type 1-mediated glucocorticoid generation. J Biol Chem. 2006 Mar 10;281(10):6546–6551.

59. Margittai E, Banhegyi G, Kiss A, Nagy G, Mandl J, Schaff Z, Csala M. Scurvy leads to endoplasmic reticulum stress and apoptosis in the liver of Guinea pigs. J Nutr. 2005 Nov;135(11):2530–2534.

60. Nagy G, Kardon T, Wunderlich L, Szarka A, Kiss A, Schaff Z, Banhegyi G, Mandl J. Acetaminophen induces ER dependent signaling in mouse liver. Arch Biochem Biophys. 2007 Mar 15;459(2):273–279.

61. Csala M, Marcolongo P, Lizak B, Senesi S, Margittai E, Fulceri R, Magyar JE, Benedetti A, Banhegyi G. Transport and transporters in the endoplasmic reticulum. Biochim Biophys Acta. 2007 Jun;1768(6):1325–1341.

62. Csala M, Banhegyi G, Benedetti A. Endoplasmic reticulum: a metabolic compartment. FEBS Lett. 2006 Apr 17;580(9):2160–2165.

63. Lizak B, Csala M, Benedetti A, Banhegyi G. The translocon and the non-specific transport of small molecules in the endoplasmic reticulum (Review). Mol Membr Biol. 2008 Feb;25(2):95–101.

64. Lizak B, Czegle I, Csala M, Benedetti A, Mandl J, Banhegyi G. Translocon pores in the endoplasmic reticulum are permeable to small anions. Am J Physiol Cell Physiol. 2006 Sep;291(3):C511–517.

65. Banhegyi G, Lusini L, Puskas F, Rossi R, Fulceri R, Braun L, Mile V, di Simplicio P, Mandl J, Benedetti A. Preferential transport of glutathione versus glutathione disulfide in rat liver microsomal vesicles. J Biol Chem. 1999 Apr 30;274(18):12213–12216.

66. Banhegyi G, Benedetti A, Fulceri R, Senesi S. Cooperativity between 11beta-hydroxysteroid dehydrogenase type 1 and hexose-6-phosphate dehydrogenase in the lumen of the endoplasmic reticulum. J Biol Chem. 2004 Jun 25;279(26):27017–27021.

67. Marcolongo P, Piccirella S, Senesi S, Wunderlich L, Gerin I, Mandl J, Fulceri R, Banhegyi G, Benedetti A. The glucose-6-phosphate transporter-hexose-6-phosphate dehydrogenase-11beta-hydroxysteroid dehydrogenase type 1 system of the adipose tissue. Endocrinology. 2007 May;148(5):2487–2495.

68. Varsanyi M, Szarka A, Papp E, Makai D, Nardai G, Fulceri R, Csermely P, Mandl J, Benedetti A, Banhegyi G. FAD transport and FAD-dependent protein thiol oxidation in rat liver microsomes. J Biol Chem. 2004 Jan 30;279(5):3370–3374.

69. Csala M, Mile V, Benedetti A, Mandl J, Banhegyi G. Ascorbate oxidation is a prerequisite for its transport into rat liver microsomal vesicles. Biochem J. 2000 Jul 15;349(Pt 2):413–415.

70. Yoshida H. ER stress and diseases. FEBS J. 2007 Feb;274(3):630–658.

71. Marciniak SJ, Ron D. Endoplasmic reticulum stress signaling in disease. Physiol Rev. 2006 Oct;86(4):1133–1149.

72. Shen J, Snapp EL, Lippincott-Schwartz J, Prywes R. Stable binding of ATF6 to BiP in the endoplasmic reticulum stress response. Mol Cell Biol. 2005 Feb;25(3):921–932.

73. Kimata Y, Oikawa D, Shimizu Y, Ishiwata-Kimata Y, Kohno K. A role for BiP as an adjustor for the endoplasmic reticulum stress-sensing protein Ire1. J Cell Biol. 2004 Nov 8;167(3):445–456.

74. Nadanaka S, Okada T, Yoshida H, Mori K. Role of disulfide bridges formed in the luminal domain of ATF6 in sensing endoplasmic reticulum stress. Mol Cell Biol. 2007 Feb;27(3):1027–1043.

75. Nadanaka S, Yoshida H, Mori K. Reduction of disulfide bridges in the lumenal domain of ATF6 in response to glucose starvation. Cell Struct Funct. 2006;31(2):127–134.

76. Zable AC, Favero TG, Abramson JJ. Glutathione modulates ryanodine receptor from skeletal muscle sarcoplasmic reticulum. Evidence for redox regulation of the Ca2+ release mechanism. J Biol Chem. 1997 Mar 14;272(11):7069–7077.

77. Feng W, Liu G, Allen PD, Pessah IN. Transmembrane redox sensor of ryanodine receptor complex. J Biol Chem. 2000 Nov 17;275(46):35902–35907.

78. Higo T, Hattori M, Nakamura T, Natsume T, Michikawa T, Mikoshiba K. Subtype-specific and ER lumenal environment-dependent regulation of inositol 1,4,5-trisphosphate receptor type 1 by ERp44. Cell. 2005 Jan 14;120(1):85–98.

79. Li Y, Camacho P. Ca2+-dependent redox modulation of SERCA 2b by ERp57. J Cell Biol. 2004 Jan 5;164(1):35–46.

80. Paschen W, Yatsiv I, Shoham S, Shohami E. Brain trauma induces X-box protein 1 processing indicative of activation of the endoplasmic reticulum unfolded protein response. J Neurochem. 2004 Feb;88(4):983–992.

81. Aufenberg C, Wenkel S, Mautes A, Paschen W. Spinal cord trauma activates processing of xbp1 mRNA indicative of endoplasmic reticulum dysfunction. J Neurotrauma. 2005 Sep;22(9):1018–1024.

82. Larner SF, Hayes RL, McKinsey DM, Pike BR, Wang KK. Increased expression and processing of caspase-12 after traumatic brain injury in rats. J Neurochem. 2004 Jan;88(1):78–90.

83. Werner C, Engelhard K. Pathophysiology of traumatic brain injury. Br J Anaesth. 2007 Jul;99(1):4–9.

84. Halliwell B. Proteasomal dysfunction: a common feature of neurodegenerative diseases? Implications for the environmental origins of neurodegeneration. Antioxid Redox Signal. 2006 Nov–Dec;8(11–12):2007–2019.

85. Lin MT, Beal MF. Mitochondrial dysfunction and oxidative stress in neurodegenerative diseases. Nature. 2006 Oct 19;443(7113):787–795.

86. Halliwell B. Oxidative stress and neurodegeneration: where are we now? J Neurochem. 2006 Jun;97(6):1634–1658.

87. Bence NF, Sampat RM, Kopito RR. Impairment of the ubiquitin-proteasome system by protein aggregation. Science. 2001 May 25;292(5521):1552–1555.

88. Lindholm D, Wootz H, Korhonen L. ER stress and neurodegenerative diseases. Cell Death Differ. 2006 Mar;13(3):385–392.

89. Hoozemans JJ, van Haastert ES, Eikelenboom P, de Vos RA, Rozemuller JM, Scheper W. Activation of the unfolded protein response in Parkinson's disease. Biochem Biophys Res Commun. 2007 Mar 16;354(3):707–711.

90. Nakamura T, Lipton SA. Emerging roles of S-nitrosylation in protein misfolding and neurodegenerative diseases. Antioxid Redox Signal. 2008 Jan;10(1):87–101.

91. Holtz WA, O'Malley KL. Parkinsonian mimetics induce aspects of unfolded protein response in death of dopaminergic neurons. J Biol Chem. 2003 May 23;278(21):19367–19377.

92. Ryu EJ, Harding HP, Angelastro JM, Vitolo OV, Ron D, Greene LA. Endoplasmic reticulum stress and the unfolded protein response in cellular models of Parkinson's disease. J Neurosci. 2002 Dec 15;22(24):10690–10698.

93. Kadowaki H, Nishitoh H, Urano F, Sadamitsu C, Matsuzawa A, Takeda K, Masutani H, Yodoi J, Urano Y, Nagano T, Ichijo H. Amyloid beta induces neuronal cell death through ROS-mediated ASK1 activation. Cell Death Differ. 2005 Jan;12(1):19–24.

94. Hoozemans JJ, Veerhuis R, Van Haastert ES, Rozemuller JM, Baas F, Eikelenboom P, Scheper W. The unfolded protein response is activated in Alzheimer's disease. Acta Neuropathol. 2005 Aug;110(2):165–172.

95. Verkhratsky A. Physiology and pathophysiology of the calcium store in the endoplasmic reticulum of neurons. Physiol Rev. 2005 Jan;85(1):201–279.

96. Bruijn LI, Miller TM, Cleveland DW. Unraveling the mechanisms involved in motor neuron degeneration in ALS. Annu Rev Neurosci. 2004;27:723–749.

97. Wootz H, Hansson I, Korhonen L, Napankangas U, Lindholm D. Caspase-12 cleavage and increased oxidative stress during motoneuron degeneration in transgenic mouse model of ALS. Biochem Biophys Res Commun. 2004 Sep 10;322(1):281–286.

98. Tobisawa S, Hozumi Y, Arawaka S, Koyama S, Wada M, Nagai M, Aoki M, Itoyama Y, Goto K, Kato T. Mutant SOD1 linked to familial amyotrophic lateral sclerosis, but not wild-type SOD1, induces ER stress in COS7 cells and transgenic mice. Biochem Biophys Res Commun. 2003 Apr 4;303(2):496–503.

99. Kikuchi H, Almer G, Yamashita S, Guegan C, Nagai M, Xu Z, Sosunov AA, McKhann GM, 2nd, Przedborski S. Spinal cord endoplasmic reticulum stress associated with a microsomal accumulation of mutant superoxide dismutase-1 in an ALS model. Proc Natl Acad Sci U S A. 2006 Apr 11;103(15):6025–6030.

100. Czyzyk-Krzeska MF. Molecular aspects of oxygen sensing in physiological adaptation to hypoxia. Respir Physiol. 1997 Nov;110(2–3):99–111.

101. Semenza GL. HIF-1: mediator of physiological and pathophysiological responses to hypoxia. J Appl Physiol. 2000 Apr;88(4):1474–1480.

102. Sciandra JJ, Subjeck JR, Hughes CS. Induction of glucose-regulated proteins during anaerobic exposure and of heat-shock proteins after reoxygenation. Proc Natl Acad Sci U S A. 1984 Aug;81(15):4843–4847.

103. Kuwabara K, Matsumoto M, Ikeda J, Hori O, Ogawa S, Maeda Y, Kitagawa K, Imuta N, Kinoshita T, Stern DM, Yanagi H, Kamada T. Purification and characterization of a novel stress protein, the 150-kDa oxygen-regulated protein (ORP150), from cultured rat astrocytes and its expression in ischemic mouse brain. J Biol Chem. 1996 Mar 1;271(9):5025–5032.

104. Tanaka S, Uehara T, Nomura Y. Up-regulation of protein-disulfide isomerase in response to hypoxia/brain ischemia and its protective effect against apoptotic cell death. J Biol Chem. 2000 Apr 7;275(14):10388–10393.

105. Gess B, Hofbauer KH, Wenger RH, Lohaus C, Meyer HE, Kurtz A. The cellular oxygen tension regulates expression of the endoplasmic oxidoreductase ERO1-Lalpha. Eur J Biochem. 2003 May;270(10):2228–2235.

106. Hayashi T, Saito A, Okuno S, Ferrand-Drake M, Dodd RL, Chan PH. Damage to the endoplasmic reticulum and activation of apoptotic machinery by oxidative stress in ischemic neurons. J Cereb Blood Flow Metab. 2005 Jan;25(1):41–53.

107. Tajiri S, Oyadomari S, Yano S, Morioka M, Gotoh T, Hamada JI, Ushio Y, Mori M. Ischemia-induced neuronal cell death is mediated by the endoplasmic reticulum stress pathway involving CHOP. Cell Death Differ. 2004 Apr;11(4):403–415.

108. Adhami F, Liao G, Morozov YM, Schloemer A, Schmithorst VJ, Lorenz JN, Dunn RS, Vorhees CV, Wills-Karp M, Degen JL, Davis RJ, Mizushima N, Rakic P, Dardzinski BJ, Holland SK, Sharp FR, Kuan CY. Cerebral ischemia-hypoxia induces intravascular coagulation and autophagy. Am J Pathol. 2006 Aug;169(2):566–583.

109. Adhami F, Schloemer A, Kuan CY. The roles of autophagy in cerebral ischemia. Autophagy. 2007 Jan–Feb;3(1):42–44.

110. Bando Y, Katayama T, Kasai K, Taniguchi M, Tamatani M, Tohyama M. GRP94 (94 kDa glucose-regulated protein) suppresses ischemic neuronal cell death against ischemia/reperfusion injury. Eur J Neurosci. 2003 Aug;18(4):829–840.

111. Tamatani M, Matsuyama T, Yamaguchi A, Mitsuda N, Tsukamoto Y, Taniguchi M, Che YH, Ozawa K, Hori O, Nishimura H, Yamashita A, Okabe M, Yanagi H, Stern DM, Ogawa S, Tohyama M. ORP150 protects against hypoxia/ischemia-induced neuronal death. Nat Med. 2001 Mar;7(3):317–323.

112. Banhegyi G, Benedetti A, Csala M, Mandl J. Stress on redox. FEBS Lett. 2007 Jul 31;581(19):3634–3640.

113. Hayashi T, Saito A, Okuno S, Ferrand-Drake M, Dodd RL, Nishi T, Maier CM, Kinouchi H, Chan PH. Oxidative damage to the endoplasmic reticulum is implicated in ischemic neuronal cell death. J Cereb Blood Flow Metab. 2003 Oct;23(10):1117–1128.

114. DeGracia DJ, Montie HL. Cerebral ischemia and the unfolded protein response. J Neurochem. 2004 Oct;91(1):1–8.
115. Kohno K, Higuchi T, Ohta S, Kumon Y, Sakaki S. Neuroprotective nitric oxide synthase inhibitor reduces intracellular calcium accumulation following transient global ischemia in the gerbil. Neurosci Lett. 1997 Mar 7;224(1):17–20.
116. Doutheil J, Althausen S, Treiman M, Paschen W. Effect of nitric oxide on endoplasmic reticulum calcium homeostasis, protein synthesis and energy metabolism. Cell Calcium. 2000 Feb;27(2):107–115.

Exocytosis, Mitochondrial Injury and Oxidative Stress in Neurodegenerative Diseases

Mark P. Zanin and Damien J. Keating

Abstract Common features seen in the early stages of many neurodegenerative diseases include increases in oxidative stress and mitochondrial dysfunction, ultimately leading to defects in cellular energy production. These changes particularly affect cells that are highly active, such as neurons. As such, reduced synaptic transmission is a common early feature associated with neurodegenerative diseases, such as Parkinson s disease, Alzheimer s disease, Huntington s disease and Amyotrophic Lateral Sclerosis. Many genes associated with neurodegenerative diseases are now known to regulate either mitochondrial function, redox state or the exocytosis of neurotransmitters. Mitochondria are a significant source of cellular ATP and reactive oxygen species and are prevalent in synapses at areas of exocytosis. Therefore, it follows that reductions in mitochondrial function and/or increases in oxidative stress will impact on neurotransmission.

Keywords Mitochondria · Oxidative stress · Exocytosis · Neurotransmission · Synaptic transmission · SNARE proteins · Long-term potentiation · β-amyloid · Alzheimer s disease · Parkinson s disease · Amyotrophic lateral sclerosis · Huntington s disease

1 Introduction

It is important to recognize that neurodegenerative disorders involve not only loss of neurons, but significant dysfunction in select populations of remaining neurons. This impaired function of remaining but injured neurons manifests in disruption of neurotransmission. This chapter describes the impact of oxidative stress and deficits in cellular energy and mitochondrial function on neurotransmission, particularly in the context of neurodegenerative disease. A functional

D.J. Keating (✉)
Department of Human Physiology and Centre for Neuroscience, Flinders University, Adelaide, Australia
e-mail: damien.keating@flinders.edu.au

S.C. Veasey (ed.), *Oxidative Neural Injury,*
Contemporary Clinical Neuroscience, DOI 10.1007/978-1-60327-342-8_4,

window into neural injury enables elucidation of early injuries. The roles these deficits are thought to play in the early aetiology of several neurodegenerative diseases are highlighted. The theory underlying pertinent techniques used to study exocytosis and neurotransmission and the application of these techniques in the study of neurodegenerative disease are also described.

2 Synaptic Transmission, Oxidative Stress and Mitochondrial Dysfunction

2.1 Gene Responses to Normal Ageing in the Brain

While the factors underlying cognitive decline in both normal ageing and neuro-degenerative diseases are not clearly understood, reductions in the amount and efficiency of synaptic transmission are obvious focal points. Synaptic transmission decreases during normal ageing [1, 2] and in the early stages of disease pathology in animal models of some neurodegenerative diseases [3–8]. A post-mortem analysis of gene expression in the frontal cortex of human subjects aged 26–106 revealed that the expression of genes involved in synaptic plasticity, vesicular transport and mitochondrial function decrease after 40 years of age [9]. However, genes involved in the stress response, antioxidation and DNA repair are induced after 40 years of age [9]. DNA damage is also markedly increased in the promoters of genes that show reduced expression in the aged cortex and these same genes are less capable of undergoing DNA repair [9].

2.2 Oxidative Injury in Normal Ageing with Impact on Synaptic Transmission

Oxidative damage is a likely contributing factor to the cognitive decline that accompanies ageing. Mitochondrial dysfunction, amounts of reactive oxygen species (ROS) and prolonged periods of oxidative stress increase with age (reviewed in [10]). Sources of ROS, in the form of $O_2^{\bullet-}$, and hence ultimately oxidative stress, include mitochondria, NADPH oxidase, xanthine oxidase and P450 oxidase. Throughout this text, the term ROS will be used in reference to hydrogen peroxide (H_2O_2). The usual reference is to the superoxide anion ($O_2^{\bullet}$), which is produced during the reduction of O_2 to water in the mitochondrial electron transport chain. As $O_2^{\bullet-}$ is spontaneously or enzymatically converted to H_2O_2 at rapid rates, the term ROS usually refers ultimately to H_2O_2, as it is far more long-lived than $O_2^{\bullet-}$ within cells.

   Increasing amounts of ROS have been attributed to impairing long-term potentiation (LTP), an increase in the strength of a neuronal synapse, in hippo-campal area CA1 in an age-related manner as determined by measuring excitatory postsynaptic potentials (EPSPs) from hippocampal slices [1, 11]. Hippocampal slices from older rats produce more H_2O_2 than younger rats (reviewed in [12]) and age is associated with decreases in the hippocampal concentration of vitamins C and E and increased activity of superoxide dismutase [13]. The concentrations of ROS applied to these slices have differential effects on LTP. For example, low concentrations (1 μM) of H_2O_2 potentiate LTP in hippocampus of young rats by up to two-fold by affecting internal calcium (Ca^{2+}) stores [14]. However, high concentrations (20 μM or above) of H_2O_2 impair synaptic transmission and LTP via calcineurin-dependent mechanisms [14]. These findings indicate that optimal amounts of ROS are required for the highest levels of LTP to occur and too few or too many ROS are detrimental to synaptic transmission. The requirement for an optimal amount of ROS is also evident in other aspects of normal cellular function, such as the regulation of neuronal excitability via redox-sensitive ion channels [15–18], synaptic plasticity [12], gene transcription [19], multiple signal transduction pathways [19] and the activity of enzymes controlling protein phosphorylation [20].

2.3 Redox Regulation of Exocytosis

The synaptic fusion machinery can be directly and acutely regulated by redox state. This was illustrated in frog and mouse motor nerve endings, where evoked and spontaneous quantal release was reduced by physiological levels of ROS [21]. Antioxidants also increased the quantal level of release, consistent with the tonic inhibition of exocytosis by endogenous levels of ROS [21]. The use of a Ca^{2+} ionophore subsequently revealed that this effect of ROS was due to action directly on the fusion machinery [21]. Subsequent investigations established that some SNARE proteins are sensitive to oxidative stress, SNAP-25 being the most sensitive of those studied [21]. Specific cysteine residues in SNAP-25 are required for SNARE disassembly and exocytosis but not for membrane targeting [22]. As cysteine residues are commonly affected by redox state, crucial alterations in SNAP-25 structure may underlie the lack of SNARE complex assembly and reduced exocytosis seen in motor nerve endings during conditions of oxidative stress. Interestingly, the expression of some exocytotic proteins, including SNARE proteins, is altered in disease-relevant brain areas in neurodegenerative diseases such as AD and HD [23, 24]. Therefore, the redox modulation of SNARE complex formation is a mechanism by which mitochondrial dysfunction occurring early in some neurodegenerative diseases might reduce synaptic activity.

2.4 Energy-Dependence of Exocytosis

Certain steps in the exocytotic pathway are also regulated by ATP in the cell. The two major ATP-regulated steps in exocytosis have been identified as a reversible ATP-dependent priming of docked granules, which is followed by a second, Ca^{2+}-dependent step involving vesicle fusion and the release of vesicle contents [25, 26]. The ATP-dependent step comprises both vesicle recruitment to the plasma membrane and vesicle priming [27, 28]. The role of ATP here is to enable the phosphorylation of phosphatidylinositol groups via phosphatidylinositol kinases localized to both vesicle and plasma membranes. These kinases include phosphatidylinositol-4-phosphate 5-kinase [29], phosphatidylinositol 4-kinase [30] and phosphatidylinositol 3-kinase C2α [31]. While there is some association between neurodegenerative diseases and the effects of ATP on exocytosis, most of this evidence is indirect. An example of this is the overexpression of the PD-associated protein α-synuclein inhibiting a vesicle-priming step prior to Ca^{2+}-dependent vesicle fusion but after secretory vesicle trafficking to docking sites [32]. While it is unknown how this inhibition occurs, the step at which it takes place is also where ATP-dependent priming occurs.

Two recent independent studies on *Drosophila* mutants with synapses containing few or no mitochondria support the hypothesis that proper synaptic mitochondrial ATP production is vital for normal synaptic transmission and that mitochondria play a specific role in regulating synaptic strength [33, 34]. Flies with loss-of-function mutations in *dynamin-related protein* (*drp*), which is implicated in the fission of the outer mitochondrial membrane, display almost no synaptic mitochondria. These animals sometimes develop into adult flies and the mitochondria in cell bodies remain functional, allowing the role of synaptic mitochondria in neurotransmission to be assessed. In the neuromuscular junction of these mutants, resting Ca^{2+} levels are slightly increased and basal synaptic properties are hardly altered. However, during intense stimulation, normal neurotransmission cannot be maintained [34]. This is not due to alterations in the kinetics of vesicle fusion or fission, but rather stems from a reduced ability to mobilize the reserve pool of vesicles, an effect that is partially rescued by exogenous ATP [34].

Drosophila mutants of Miro (*dmiro*), a mitochondrial GTPase with an essential role in mitochondrial trafficking, exhibit defects in locomotion and die prematurely [33]. Neuronal and muscular mitochondria in *dmiro* mutants are not transported into axons and dendrites but accumulate in neuronal somata. Similarly to *drp* mutants, *dmiro* mutants only display changes in neurotransmission during prolonged stimulation [33]. Given the considerable reduction in synaptic mitochondria, it might be presumed that amounts of ROS would be considerably reduced in these mutants. As previously stated, a significant reduction in ROS can be detrimental to multiple facets of neuronal function. Unfortunately, neither of the studies mentioned [33, 34] measured synaptic ROS levels in these mutant flies, so assessment of the separate effect of mitochondrial ATP and ROS on neurotransmission in this setting cannot be made. However, these studies highlight the requirement for synaptic mitochondrial ATP during periods of intense activity for the recruitment and transport of vesicles from the reserve vesicle pool (Fig. 1).

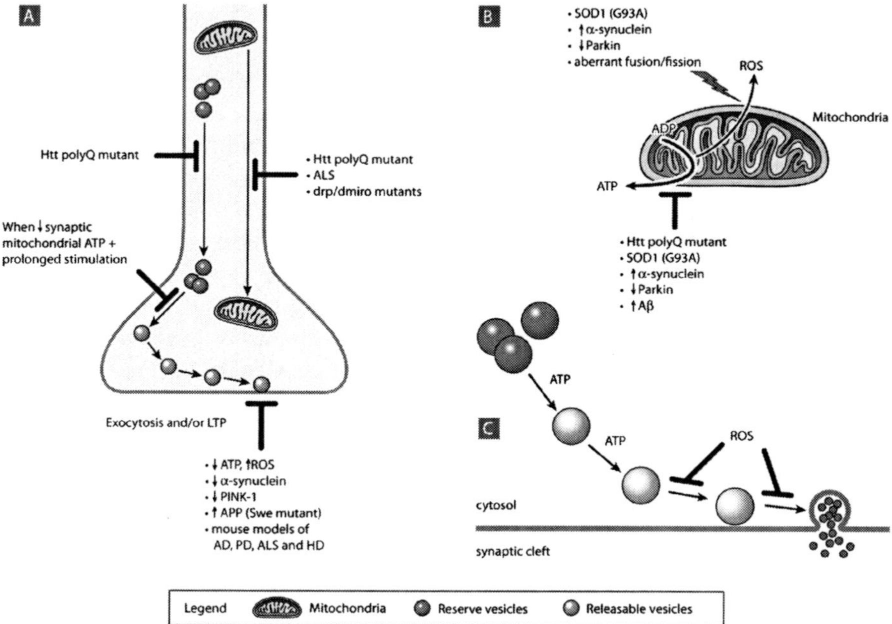

Fig. 1 Mitochondrial and vesicular trafficking, mitochondrial function, exocytosis and LTP are altered in various neurodegenerative diseases. (**A**) Axonal mitochondria and vesicle trafficking are regulated by mutant Htt in ALS patients and in *Drosophila* mutants of *dmiro* and *drp* in which a lack of mitochondrial ATP results in reduced recruitment and transport of vesicles from the reserve vesicle pool during prolonged stimulation. Exocytosis and/or LTP are reduced by the factors indicated in the *white box* (**B**) Disease-associated proteins which either reduce mitochondrial ATP production or increase mitochondrial ROS levels. (**C**) Simple schematic illustrating the points at which ATP and ROS regulate exocytosis. ATP is needed for trafficking of vesicles towards the plasma membrane and then for 'priming of vesicles for release. ROS inhibit the formation of SNARE protein complexes and therefore inhibit the final steps in vesicle fusion. Increased ROS may either reduce the amount of SNARE proteins available and may also reduce its function

3 Neurodegenerative Disorders

There is extensive evidence supporting a role for mitochondrial dysfunction and oxidative stress in the pathogenesis of multiple neurodegenerative disorders and in animal models of these diseases. Mitochondria are significantly more likely to be localized at synaptic sites [35] and the speed of mitochondrial motility is greatest in younger, developing neurons [36]. Mitochondria are increasingly localized to dendritic spines during early developmental stages or following repeated stimulation [37]. This mitochondrial localization is necessary for the normal development and morphological plasticity of dendritic spines [37]. The following section describes studies that provide evidence for roles of mitochondrial dysfunction, oxidative stress and cell signalling in the early aetiologies of some neurodegenerative diseases.

3.1 Alzheimer s Disease

In human Alzheimer s disease (AD) patients, signs of oxidative damage are observed earlier in the disease course than the characteristic plaques [38]. Similar observations were also made in transgenic amyloid precursor protein (APP) mice, in which the upregulation of genes involved in mitochondrial metabolism and apoptosis precede Aβ deposition [39, 40]. Oxidative stress may be a causative agent of AD as it induces the intracellular accumulation of Aβ. Secretion of neuronal Aβ into the extracellular space is positively correlated with synaptic activity [41–43] and intraneuronal Aβ accumulation is linked with early electrophysiological, synaptic and pathological abnormalities (reviewed in [44]). Intracellular Aβ accumulation is reduced by antibodies against Aβ in transgenic mice [45] and this reduction currently provides the best correlate for improving cognitive function [46]. It is possible that Aβ accumulation caused by oxidative stress is due to reduced exocytosis from neurons. Various mouse models of AD manifest synaptic dysfunction, including LTP deficits, well before classical pathologies such as neurodegeneration and plaque and tangle formation are evident [3, 6, 7]. Similar results are also seen in nonmammalian models of AD, such as in *Drosophila*, where the deletion or overexpression of the *Drosophila* APP orthologue, *App-like* gene, results in reduced axonal transport of synaptic vesicles [47].

3.2 Parkinson s Disease

Mitochondrial dysfunction and vesicular abnormalities have also been implicated in Parkinson s disease (PD). Several genes associated with PD have roles in mitochondrial function, such as Parkin and PINK-1. The inhibition of complex I of the mitochondrial electron transport chain has been shown to cause widespread neuronal damage [48]. This damage includes PD-like pathology, such as nigrostriatal dopaminergic degeneration and α-synuclein-positive cytoplasmic inclusions [48]. The mechanisms underlying these changes are thought to involve oxidative stress [49].

The role of α-synuclein, a presynaptic protein, is thought to be central in PD pathogenesis. However it remains unclear exactly what role α-synuclein plays in the development of PD. In transgenic mice overexpressing α-synuclein, mitochondrial function is impaired, oxidative stress is increased and nigral degeneration, induced by inhibitors of mitochondrial complex I, is enhanced [50]. α-synuclein at normal expression levels appears to be neuroprotective in non-dopaminergic human cortical neurons but induces apoptosis when overexpressed in the dopaminergic neurons affected in PD [51]. Ablation or reduction of α-synuclein expression reduces transmitter release, LTP and the size of vesicle pools [32, 52, 53]. Mice overexpressing a dominant-negative mutation of α-synuclein exhibit decreased dopamine release in response to prolonged stimulation [54, 55]. However, while the mutations in α-synuclein that cause genetic forms of PD are clinically of interest, no clear evidence has been published to clearly illustrate that cellular levels of free α-synuclein are increased or decreased in PD brains and thus it is difficult to assess exactly what role α-synuclein plays in the pathogenesis of sporadic forms of PD.

Parkin, a PD-related ubiquitin ligase, is thought to be protective of mitochondria. In Parkin-null mice, oxidative stress and mitochondrial dysfunction are evident [56]. Parkin also seems to be susceptible to oxidative stress, as S-nitrosylation of Parkin reduces its ubiquitin-ligase activity as well as its protective function [57]. Similar effects are also seen in vivo in a mouse model of PD and in the brains of patients with PD [57]. PD mouse models exhibit various alterations in dopaminergic neuronal function prior to the degeneration of dopaminergic brain areas [55, 58, 59].

The role of PINK-1 has also been studied using PINK-1 null mice. The numbers of dopaminergic neurons and dopamine content in the striatum are unaltered in PINK-1 null mice [60]. However, evoked dopamine release from striatal slices is reduced, as is LTP in striatal medium spiny neurons, the main target of dopaminergic neurons [60]. These results, and those mentioned previously, indicate that altered dopaminergic exocytosis may be a pathogenic precursor to nigrostriatal degeneration in PD.

3.3 Amyotrophic Lateral Sclerosis

Amyotrophic Lateral Sclerosis (ALS), commonly known as motor neuron disease, is another neurodegenerative disease with a mitochondrial and oxidative stress component. Mutations in the Cu/Zn-superoxide dismutase gene account for approximately 20% of familial ALS (FALS) cases. Spinal cords and brains obtained from mice that overexpress a SOD1 mutation (G93A) found in some FALS patients display defects in mitochondrial respiration, electron transfer chain activity and ATP synthesis. Signs of oxidative damage to mitochondrial proteins and lipids, indicative of oxidative stress at the time of disease onset, are also evident [61]. Changes in mitochondrial function are seen before this stage though, as mitochondria in the spinal cord and brain display reduced Ca^{2+} loading ability in these mice [62]. Mitochondria in the proximal axons of anterior horn neurons also appear to be aberrantly localized to the axon hillock in sporadic ALS patients and mouse models of ALS [63, 64]. This was observed in transgenic mice expressing the G93A mutant human SOD1 during presymptomatic stages, indicating that alterations in mitochondrial localization may occur prior to the onset of ALS [64]. Early changes in synaptic transmission are also seen in mouse models of ALS prior to the onset of neurodegeneration [65, 66].

3.4 Huntington s Disease

Huntington s Disease (HD) is also thought to have a mitochondrial component. In neurons of *huntingtin* knock-in mice, ATP production and mitochondrial respiration are decreased [67], as are the activities of complexes I and III of the electron transport chain in human HD brain samples [68]. These effects may be caused by direct interaction of the *huntingtin* gene with mitochondria. Mutant *huntingtin* is localized to neuronal mitochondrial membranes and incubation of mutant *huntingtin* with normal mitochondria results in depolarization of the mitochondrial membrane at lower Ca^{2+} loads. This defect has also been observed in mitochondria obtained from HD patients and mice that transgenically overexpress mutant *huntingtin* [69].

Huntingtin is involved in axonal trafficking in mammals and mutant huntingtin impairs the in vitro and in vivo trafficking of vesicles and mitochondria in mammalian neurons [70]. Importantly, these changes are seen prior to the development of observed neurological abnormalities [70], implying a possible causative role for this altered mitochondrial localization in axons in HD. Huntingtin also interacts with the transcription factor p53 [71], which is known to regulate genes that are responsive to oxidative stress as well as genes which regulate mitochondrial function. In both *Drosophila* and mouse models of HD, p53 has been found to link nuclear and mitochondrial pathologies characteristic of HD [72]. Altered synaptic plasticity in the R6/2 mouse model of HD occurs prior to the more overt disease phenotypes [5], indicating that synaptic pathology also occurs early in this disease.

4 Methods Used to Study Exocytosis

The following sections attempt to provide a detailed explanation of some of the more common techniques used in the study of exocytosis and synaptic transmission. These methods are relevant to many of the studies mentioned in this chapter. Along with an explanation of the methodology itself, we have also attempted to profile the pros and cons associated with the use of individual techniques and the types of preparations they can be applied to.

4.1 Carbon Fibre Amperometry and Cyclic Voltammetry

Carbon fibre amperometry is a technique used to detect exocytosis in cells that secrete oxidizable molecules such as adrenaline, noradrenaline, dopamine and serotonin. It involves placing a carbon fibre electrode on a single cell in culture or in a brain slice preparation. The cell is then stimulated to trigger exocytosis, releasing vesicle contents that are then oxidized upon contact with the charged

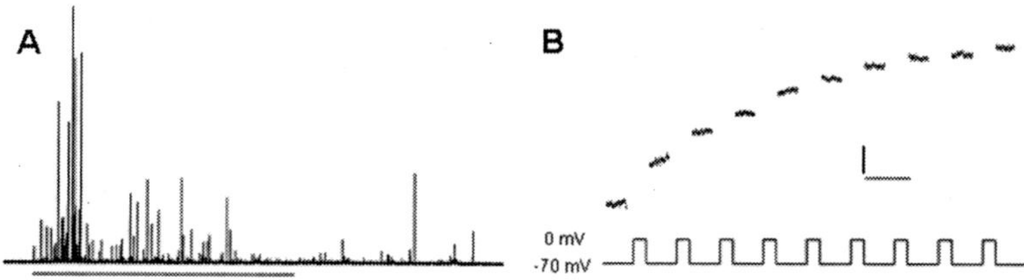

Fig. 2 Example recordings of carbon fibre amperometry and whole cell membrane capacitance measurements from mouse chromaffin cells. (**A**) Carbon fibre amperometry recording of exocytosis. The line beneath the trace indicates where the cell has been stimulated. Current spikes indicate the release of catecholamines from an individual vesicle. (**B**) Membrane capacitance measurement of exocytosis in which exocytosis is elicited by a repeated set of voltage pulses from –70 to 0 mV. The membrane capacitance is measured in between these pulses and increases accordingly. Scale bar in (**B**) represents 50 pA and 10 s in (**A**) and 200 fF and 1 s in (**B**)

carbon fibre electrode. This oxidation results in a flow of current which can be recorded as an amperometric spike (Fig. 2A).

Amperometry is useful as it is noninvasive and does not alter the intracellular environment. It is possible to visualize single exocytotic events providing the stimulus to the cell is not too great and does not cause overlap of individual amperometric spikes. This allows the shape of individual spikes to be studied, providing insight into the kinetic parameters of individual fusion events. Such parameters include the amount of transmitter released from a single vesicle, the speed of fusion pore opening and closing and the kinetics associated with the formation of the transient fusion pore.

Studies of exocytosis using amperometry have shown that, in addition to the classical model of full collapse fusion, where the vesicle membrane fully collapses into the plasma membrane, transient fusion events occur that also result in secretion [73–75]. Amperometry has also provided insight into the kinetics of vesicle fusion and the mechanisms behind it in a diverse array of cell types, including chromaffin cells, dopaminergic neurons, PC12 cells and pancreatic β-cells [76–80]. Whilst pancreatic β-cells do not secrete endogenous compounds which can be oxidized, they can be loaded with a 'false oxidizable transmitter such as dopamine. However, in our experience, this loading of false transmitter works poorly and is highly unreliable.

Amperometry is limited to cells that secrete oxidizable molecules and by the geometry of the electrode and the cell, as exocytotic events will only be detected when they occur within the proximity of the electrode. This necessitates the use of large sample sizes to reduce the random effects of signal diffusion from exocytotic events more distant from the electrode and also due to the Gaussian distribution of vesicle size within a cell. Amperometry has been utilized in the studies described in this section to assess the effects of PINK1 inactivation on catecholamine release from dissociated adrenal chromaffin cells and the release of dopamine from acute striatal slices obtained from *PINK1*$^{-/-}$ mice [60].

Amperometry has also been used to study secretion from dissociated chromaffin cells isolated from mice overexpressing α-synuclein [32].

It is not possible to determine the identity of the molecules secreted from cells using carbon fibre amperometry. This information is discernable using a similar technique called cyclic voltammetry, which involves constantly scanning the electrical potential of the electrode through a range of voltages [80]. The identity of the molecule is determined according to the potential at which it is oxidized or reduced [80]. Cyclic voltammerty has been used to study dopamine release from neurons in vivo, in mice where the human α-synuclein gene has been overexpressed, mutated or knocked out, and in vitro, such as in brain slices obtained from DJ-1 null mice [54, 55, 58].

4.2 Patch Clamp Capacitance and Postsynaptic Potential Measurement

Measurement of cell membrane capacitance via whole cell patch clamp provides another means to study exocytosis (Fig. 2B) [81]. Vesicles that fuse with the cell membrane cause an increase in membrane area, which can be detected as an increase in membrane capacitance (a capacitor being any material that separates two charged environments). Carrying out capacitance measurements in 'Cell-Attached mode allows single exocytotic events to be resolved on a millisecond timescale. Unlike amperometry and voltammetry, patch clamping is not limited to cells that secrete oxidizable molecules. However, patch clamping is invasive and can disturb the intracellular environment. Most importantly, it is not applicable to the direct study of exocytosis in most types of neurons as the synapse is too small to place a microelectrode onto and obtain a seal. Recordings can be made at the neuronal soma but this places the microelectrode too electrically distant from the site of synaptic release, precluding direct measurement of presynaptic exocytosis. Such recordings are limited to neurons with relatively large synapses, such as the Calyx of Held and ribbon synapses in the inner ear and retina. Studies using these cells have provided a greater understanding of the mechanisms underlying neuronal exocytosis.

Another application of patch clamping is in the study of postsynaptic currents in neurons. Such experiments involve patch clamping a neuron in culture or in a brain slice preparation then applying electrical or pharmacological stimuli to neurons that synapse onto the patched neuron. Measuring the postsynaptic response elicited by the stimulated neuron provides insights into the exocytotic activity elicited in the presynaptic neuron [60]. This method is advantageous as it allows recordings to be obtained in a system that is relatively more representative of the in vivo environment. However, recording from postsynaptic neurons following a stimulus applied presynaptically in order to gauge presynaptic levels of exocytosis assumes that the applied experimental conditions have only affected presynaptic vesicle release and not the postsynaptic response. This is a popular assumption but one in which extreme care should be taken when making.

Insights into synaptic dysfunction and memory deficits in murine models of neurodegenerative diseases have been obtained by measuring field excitatory postsynaptic potentials. These measurements are obtained using field electrodes to detect action potentials from a population of neurons in hippocampal slices in response to electrical stimulation. Such measurements give insights into synaptic plasticity, learning and memory by providing a comparison of stimulus intensity and the speed and strength of the resulting postsynaptic potential [3, 5–7, 52, 65, 66].

4.3 Fluorescent Dyes and Live Cell Imaging

Fluorescent styryl compound dyes, commonly known as FM dyes (such as FM1-43], reversibly stain membranes and can be used to give a fluorescent readout of exocytosis and endocytosis. These compounds emit a low fluorescence in aqueous solutions which increases upon membrane binding. Stimulating cells in the presence of FM dyes in the extracellular solution results in dye becoming trapped in recycling vesicles, which can be visualized as punctate fluorescence within cells. Upon exocytosis, the trapped dye is released and the rate and the amount of decrease are indicative of the kinetics of exocytosis and the size of the vesicle recycling pool, respectively. Studies using FM dyes have yielded information on the rate and regulation of vesicle exocytosis and recycling in numerous cell types, such as frog neuromuscular junctions, hippocampal neurons and chromaffin cells [82–84].

pHluorins are another set of fluorescent molecules used to study exocytosis. These molecules consist of a mutant form of green fluorescent protein that displays pH-dependent fluorescence [85]. One such molecule is synapto-pHlourin consisting of a pHlourin molecule fused to VAMP2, a synaptic vesicle protein involved in SNARE formation and vesicle exocytosis [85]. Synapto-pHluorin fluorescence is quenched in the relatively low pH of the vesicle lumen (approximately 5.6]. During exocytosis, exposure of the vesicle lumen to the extracellular environment increases the luminal pH to approximately 7.4, resulting in increased fluorescence of the synapto-pHluorin and a visual indication of exocytosis [85]. Synapto-pHluorins have been used to study exocytosis and vesicle recycling in real time, in both astrocytes and neurons [85–88].

Fluorescent dyes are also available that specifically bind to Ca^{2+} within cells. These are important tools in the study of exocytosis as Ca^{2+} entry into the cell is the major physiological trigger of vesicle exocytosis and recycling. These dyes, such as fura-2, are fluorophores linked to calcium chelators. When the calcium chelator binds to calcium in a cell, the absorption spectrum of the fluorophore is altered, leading to fluorescence and a visual readout of changes in intracellular Ca^{2+} levels. This technique is useful in experiments using groups of cells, as the excitation of all cells present can be assessed simultaneously. However, it is not sensitive enough to detect small influxes of calcium that could still be sufficient to stimulate exocytosis [89, 90]. It is also technically difficult, but not impossible, to measure changes in intracellular Ca^{2+} levels in structurally limited areas such as synaptic boutons. Thus, imaging synaptic Ca^{2+} changes while simultaneously measuring neuronal exocytosis is not frequently performed.

5 Conclusion

Identifying the steps which instigate human neurodegenerative diseases is extremely complex and continues to elude researchers within the field. Studies described in this chapter provide evidence for an early disease course involving mitochondrial dysfunction leading to aberrant ROS and ATP levels. These abnormalities then lead to reduced synaptic activity and altered neuronal communication, underlying cognitive deficits and sometimes motor dysfunction, and possibly neuronal death. Whilst there is evidence supporting this hypothesis, it would be imprudent to believe it is the only compelling hypothesis on this topic or that it might be responsible for all facets of these diseases. The use of animal models for these diseases will likely prove to be invaluable in identifying the precise sequence of events leading to neurodegeneration and the subsequent verification of these findings in human samples. Proper use of techniques which enable neuronal exocytosis and vesicle recycling and trafficking to be measured will also play a vital part in increasing our understanding of the pathogenesis of such diseases. An improved understanding of the pathogenesis of neurodegenerative diseases will hopefully lead to treatments that are increasingly effective and disease specific.

References

1. Auerbach J, Segal M. Peroxide modulation of slow onset potentiation in rat hippocampus. Journal of Neuroscience 1997;17(22):8695–8701.
2. Murray C, Lynch M. Dietary supplementation with vitamin E reverses the age-related deficit in long term potentiation in dentate gyrus. Journal of Biological Chemistry 1998;273(20):12161–12168.
3. Jacobsen J, Wu C, Redwine J, et al. Early-onset behavioral and synaptic deficits in a mouse model of Alzheimer s disease. Proceedings of the National Academy of Sciences of the United States of America 2006;103(13):5161–5166.
4. Kleschevnikov A, Belichenko P, Villar A, Epstein C, Malenka R, Mobley W. Hippocampal long-term potentiation suppressed by increased inhibition in the Ts65Dn mouse, a genetic model of Down syndrome. Journal of Neuroscience 2004;24(37):8153–8160.
5. Murphy K, Carter R, Lione L, et al. Abnormal synaptic plasticity and impaired spatial cognition in mice transgenic for exon 1 of the human Huntington s disease mutation. Journal of Neuroscience 2000;20(13):5115–5123.

6. Oddo S, Caccamo A, Shepherd J, et al. Triple-transgenic model of Alzheimer s disease with plaques and tangles: intracellular Abeta and synaptic dysfunction. Neuron 2003;39(3):409–421.

7. Saura C, Choi S, Beglopoulos V, et al. Loss of presenilin function causes impairments of memory and synaptic plasticity followed by age-dependent neurodegeneration. Neuron 2004;42(1):23–36.

8. Siarey R, Stoll J, Rapoport S, Galdzicki Z. Altered long-term potentiation in the young and old Ts65Dn mouse, a model for Down syndrome. Neuropharmacology 1997;36(11–12):1549–1554.

9. Lu T, Pan Y, Kao S, et al. Gene regulation and DNA damage in the ageing human brain. Nature 2004;429(6994):883–891.

10. Lin M, Beal M. Mitochondrial dysfunction and oxidative stress in neurodegenerative diseases. Nature 2006;443(7113):787–795.

11. Kamsler A, Segal M. Paradoxical actions of hydrogen peroxide on long-term potentiation in transgenic superoxide dismutase-1 mice. Journal of Neuroscience 2003;23(32):10359–10367.

12. Serrano F, Klann E. Reactive oxygen species and synaptic plasticity in the aging hippocampus. Ageing Research Reviews 2004;3(4):431–443.

13. McGahon B, Murray C, Horrobin D, Lynch M. Age-related changes in oxidative mechanisms and LTP are reversed by dietary manipulation. Neurobiology of Aging 1999;20(6):643–653.

14. Kamsler A, Segal M. Hydrogen peroxide modulation of synaptic plasticity. Journal of Neuroscience 2003;23(1):269–276.

15. Conforti L, Takimoto K, Petrovic M, Pongs O, Millhorn D. The pore region of the Kv1.2alpha subunit is an important component of recombinant Kv1.2 channel oxygen sensitivity. Biochemical and Biophysical Research Communications 2003;306(2):450–456.

16. Keating D, Rychkov G, Giacomin P, Roberts M. Oxygen-sensing pathway for SK channels in the ovine adrenal medulla. Clinical and Experimental Pharmacology and Physiology 2005;32(10):882–887.

17. Singh H, Ashley R. Redox regulation of CLIC1 by cysteine residues associated with the putative channel pore. Biophysical Journal 2006;90(5):1628–1638.

18. Zeng X, Xia X, Lingle C. Redox-sensitive extracellular gates formed by auxiliary beta subunits of calcium-activated potassium channels. Nature Structural Biology 2003;10(6):448–454.

19. Sen C, Packer L. Antioxidant and redox regulation of gene transcription. FASEB Journal 1996;10(7):709–720.

20. Klann E, Thiels E. Modulation of protein kinases and protein phosphatases by reactive oxygen species: implications for hippocampal synaptic plasticity. Progress in Neuro-Psychopharmacology and Biological Psychiatry 1999;23(3):359–376.

21. Giniatullin A, Darios F, Shakirzyanova A, Davletov B, Giniatullin R. SNAP25 is a presynaptic target for the depressant action of reactive oxygen species on transmitter release. Journal of Neurochemistry 2006;98(6):1789–1797.

22. Washbourne P, Cansino V, Mathews J, Graham M, Burgoyne R, Wilson M. Cysteine residues of SNAP-25 are required for SNARE disassembly and exocytosis, but not for membrane targeting. Biochemical Journal 2001;357(Pt 3):625–634.

23. Morton A, Faull R, Edwardson J. Abnormalities in the synaptic vesicle fusion machinery in Huntington s disease. Brain Research Bulletin 2001;56(2):111–117.

24. Sze C, Bi H, Kleinschmidt-DeMasters B, Filley C, Martin L. Selective regional loss of exocytotic presynaptic vesicle proteins in Alzheimer s disease brains. Journal of the Neurological Sciences 2000;175(2):81–90.

25. Hay J, Martin T. Resolution of regulated secretion into sequential MgATP-dependent and calcium-dependent stages mediated by distinct cytosolic proteins. Journal of Cell Biology 1992;119(1):139–151.

26. Holz R, Bittner M, Peppers S, Senter R, Eberhard D. MgATP-independent and MgATP-dependent exocytosis. Evidence that MgATP primes adrenal chromaffin cells to undergo exocytosis. Journal of Biological Chemistry 1989;264(10):5412–5419.

27. Dunn L, Holz R. Catecholamine secretion from digitonin-treated adrenal medullary chromaffin cells. Journal of Biological Chemistry 1983;258(8):4989–4993.

28. Sarafian T, Aunis D, Bader M. Loss of proteins from digitonin-permeabilized adrenal chromaffin cells essential for exocytosis. Journal of Biological Chemistry 1987; 262(34):16671–16676.

29. Hay J, Fisette P, Jenkins G, et al. ATP-dependent inositide phosphorylation required for Ca^{2+}-activated secretion. Nature 1995;374(6518):173–177.

30. Wiedemann C, Schafer T, Burger M. Chromaffin granule-associated phosphatidylinositol 4-kinase activity is required for stimulated secretion. EMBO Journal 1996; 15(9):2094–2101.

31. Meunier F, Osborne S, Hammond G, et al. Phosphatidylinositol 3-kinase C2alpha is essential for ATP-dependent priming of neurosecretory granule exocytosis. Molecular Biology of the Cell 2005;16(10):4841–4851.

32. Larsen K, Schmitz Y, Troyer M, et al. Alpha-synuclein overexpression in PC12 and chromaffin cells impairs catecholamine release by interfering with a late step in exocytosis. Journal of Neuroscience 2006;26(46):11915–11922.

33. Guo X, Macleod G, Wellington A, et al. The GTPase dMiro is required for axonal transport of mitochondria to *Drosophila* synapses. Neuron 2005;47(3):379–393.

34. Verstreken P, Ly C, Venken K, Koh T, Zhou Y, Bellen H. Synaptic mitochondria are critical for mobilization of reserve pool vesicles at *Drosophila* neuromuscular junctions. Neuron 2005;47(3):365–378.

35. Chang D, Honick A, Reynolds I. Mitochondrial trafficking to synapses in cultured primary cortical neurons. Journal of Neuroscience 2006;26(26):7035–7045.

36. Chang D, Reynolds I. Differences in mitochondrial movement and morphology in young and mature primary cortical neurons in culture. Neuroscience 2006;141(2):727–736.

37. Li Z, Okamoto K, Hayashi Y, Sheng M. The importance of dendritic mitochondria in the morphogenesis and plasticity of spines and synapses. Cell 2004;119(6):873–887.

38. Nunomura A, Perry G, Aliev G, et al. Oxidative damage is the earliest event in Alzheimer disease. Journal of Neuropathology and Experimental Neurology 2001;60(8):759–767.

39. Pratico D, Uryu K, Leight S, Trojanoswki J, Lee V. Increased lipid peroxidation precedes amyloid plaque formation in an animal model of Alzheimer amyloidosis. Journal of Neuroscience 2001;21(12):4183–4187.

40. Reddy P, McWeeney S, Park B, et al. Gene expression profiles of transcripts in amyloid precursor protein transgenic mice: up-regulation of mitochondrial metabolism and apoptotic genes is an early cellular change in Alzheimer s disease. Human Molecular Genetics 2004;13(12):1225–1240.

41. Cirrito J, Yamada K, Finn M, et al. Synaptic activity regulates interstitial fluid amyloid-beta levels *in vivo*. Neuron 2005;48(6):913–922.

42. Kamenetz F, Tomita T, Hsieh H, et al. APP processing and synaptic function. Neuron 2003;37(6):925–937.

43. Ohyagi Y, Yamada T, Nishioka K, et al. Selective increase in cellular A beta 42 is related to apoptosis but not necrosis. Neuroreport 2000;11(1):167–171.

44. Gouras G, Almeida C, Takahashi R. Intraneuronal Abeta accumulation and origin of plaques in Alzheimer s disease. Neurobiology of Aging 2005;26(9):1235–1244.

45. Oddo S, Billings L, Kesslak J, Cribbs D, LaFerla F. Abeta immunotherapy leads to clearance of early, but not late, hyperphosphorylated tau aggregates via the proteasome. Neuron 2004;43(3):321–332.

46. Billings L, Oddo S, Green K, McGaugh J, LaFerla F. Intraneuronal Abeta causes the onset of early Alzheimer s disease-related cognitive deficits in transgenic mice. Neuron 2005;45(5):675–688.

47. Gunawardena S, Goldstein L. Disruption of axonal transport and neuronal viability by amyloid precursor protein mutations in *Drosophila*. Neuron 2001;32(3):389–401.

48. Betarbet R, Sherer T, MacKenzie G, Garcia-Osuna M, Panov A, Greenamyre J. Chronic systemic pesticide exposure reproduces features of Parkinson s disease. Nature Neuroscience 2000;3(12):1301–1306.

49. Sherer T, Betarbet R, Testa CM, et al. Mechanism of toxicity in rotenone models of Parkinson s disease. Journal of Neuroscience 2003;23(34):10756–10764.

50. Song DD, Shults CW, Sisk A, Rockenstein E, Masliah E. Enhanced substantia nigra mitochondrial pathology in human alpha-synuclein transgenic mice after treatment with MPTP. Experimental Neurology 2004;186(2):158–172.

51. Xu J, Kao S, Lee F, Song W, Jin L, Yankner B. Dopamine-dependent neurotoxicity of alpha-synuclein: a mechanism for selective neurodegeneration in Parkinson disease. Nature Medicine 2002;8(6):600–606.

52. Cabin D, Shimazu K, Murphy D, et al. Synaptic vesicle depletion correlates with attenuated synaptic responses to prolonged repetitive stimulation in mice lacking alpha-synuclein. Journal of Neuroscience 2002;22(20):8797–8807.

53. Murphy D, Rueter S, Trojanowski J, Lee V. Synucleins are developmentally expressed, and alpha-synuclein regulates the size of the presynaptic vesicular pool in primary hippocampal neurons. Journal of Neuroscience 2000;20(9):3214–3220.

54. Yavich L, Oksman M, Tanila H, et al. Locomotor activity and evoked dopamine release are reduced in mice overexpressing A30P-mutated human alpha-synuclein. Neurobiology of Disease 2005;20(2):303–313.

55. Yavich L, Tanila H, Vepsalainen S, Jakala P. Role of alpha-synuclein in presynaptic dopamine recruitment. Journal of Neuroscience 2004;24(49): 11165–11170.

56. Palacino J, Sagi D, Goldberg M, et al. Mitochondrial dysfunction and oxidative damage in parkin-deficient mice. Journal of Biological Chemistry 2004;279(18):18614–18622.

57. Chung K, Thomas B, Li X, et al. S-nitrosylation of parkin regulates ubiquitination and compromises parkin s protective function. Science 2004;304(5675):1328–1331.

58. Chen L, Cagniard B, Mathews T, et al. Age-dependent motor deficits and dopaminergic dysfunction in DJ-1 null mice. Journal of Biological Chemistry 2005;280(22):21418–21426.

59. Goldberg M, Fleming S, Palacino J, et al. Parkin-deficient mice exhibit nigrostriatal deficits but not loss of dopaminergic neurons. Journal of Biological Chemistry 2003;278(44):43628–43635.

60. Kitada T, Pisani A, Porter D, et al. Impaired dopamine release and synaptic plasticity in the striatum of PINK-1 deficient mice. Proceedings of the National Academy of Sciences of the United States of America 2007;104(27):11441–11446.

61. Mattiazzi M, D Aurelio M, Gajewski C, et al. Mutated human SOD1 causes dysfunction of oxidative phosphorylation in mitochondria of transgenic mice. Journal of Biological Chemistry 2002;277(33):29626–29633.

62. Damiano M, Starkov A, Petri S, et al. Neural mitochondrial Ca^{2+} capacity impairment precedes the onset of motor symptoms in G93A Cu/Zn-superoxide dismutase mutant mice. Journal of Neurochemistry 2006;96(5):1349–1361.

63. Sasaki S, Iwata M. Impairment of fast axonal transport in the proximal axons of anterior horn neurons in amyotrophic lateral sclerosis. Neurology 1996;47(2):535–540.

64. Sasaki S, Warita H, Abe K, Iwata M. Impairment of axonal transport in the axon hillock and the initial segment of anterior horn neurons in transgenic mice with a G93A mutant SOD1 gene. Acta Neuropathologica 2005;110(1):48–56.

65. Kuo J, Schonewille M, Siddique T, et al. Hyperexcitability of cultured spinal motoneurons from presymptomatic ALS mice. Journal of Neurophysiology 2004;91(1):571–575.

66. Spalloni A, Geracitano R, Berretta N, et al. Molecular and synaptic changes in the hippocampus underlying superior spatial abilities in pre-symptomatic $G93A^{+/+}$ mice overexpressing the human Cu/Zn superoxide dismutase (Gly93 $\rightarrow$ ALA) mutation. Experimental Neurology 2006;197(2):505–514.

67. Milakovic T, Johnson G. Mitochondrial respiration and ATP production are significantly impaired in striatal cells expressing mutant huntingtin. Journal of Biological Chemistry 2005;280(35):30773–30782.
68. Gu M, Gash M, Mann V, Javoy-Agid F, Cooper J, Schapira A. Mitochondrial defect in Huntington s disease caudate nucleus. Annals of Neurology 1996;39(3):385–389.
69. Panov A, Gutekunst C, Leavitt B, et al. Early mitochondrial calcium defects in Huntington s disease are a direct effect of polyglutamines. Nature Neuroscience 2002;5(8):731–736.
70. Trushina E, Dyer R, Badger J, et al. Mutant huntingtin impairs axonal trafficking in mammalian neurons *in vivo* and *in vitro*. Molecular and Cellular Biology 2004; 24(18):8195–8209.
71. Steffan J, Kazantsev A, Spasic-Boskovic O, et al. The Huntington s disease protein interacts with p53 and CREB-binding protein and represses transcription. Proceedings of the National Academy of Sciences of the United States of America 2000; 97(12):6763–6768.
72. Bae BI, Xu H, Igarashi S, et al. p53 mediates cellular dysfunction and behavioral abnormalities in Huntington s disease. Neuron 2005;47(1):29–41.
73. Zhou Z, Misler S, Chow R. Rapid fluctuations in transmitter release from single vesicles in bovine adrenal chromaffin cells. Biophysical Journal 1996;70(3):1543–1552.
74. Alvarez de Toledo G, Fernández-Chacón R, Fernández J. Release of secretory products during transient vesicle fusion. Nature 1993;363(6429):554–558.
75. Bruns D, Jahn R. Real-time measurement of transmitter release from single synaptic vesicles. Nature 1995;377(6544):62–65.
76. Chen T, Luo G, Ewing A. Amperometric monitoring of stimulated catecholamine release from rat pheochromocytoma (PC12) cells at the zeptomole level. Analytical Chemistry 1994;66(19):3031–3035.
77. Huang L, Shen H, Atkinson M, Kennedy R. Detection of exocytosis at individual pancreatic beta cells by amperometry at a chemically modified microelectrode. Proceedings of the National Academy of Sciences of the United States of America 1995; 92(21):9608–9612.
78. Leszczyszyn D, Jankowski J, Viveros O, Diliberto E, Near J, R W. Nicotinic receptor-mediated catecholamine secretion from individual chromaffin cells. Chemical evidence for exocytosis. Journal of Biological Chemistry 1990;265(25):14736–14737.
79. Pothos E, Davila V, Sulzer D. Presynaptic recording of quanta from midbrain dopamine neurons and modulation of the quantal size. Journal of Neuroscience 1998; 18(11):4106–4118.
80. Wightman R, Jankowski J, Kennedy R, et al. Temporally resolved catecholamine spikes correspond to single vesicle release from individual chromaffin cells. Proceedings of the National Academy of Sciences of the United States of America 1991;88(23):10754–10758.
81. Neher E, Sakmann B. Single-channel currents recorded from membrane of denervated frog muscle fibres. Nature 1976;260(5554):799–802.
82. Betz W, Bewick G. Optical analysis of synaptic vesicle recycling at the frog neuromuscular junction. Science 1992;255(5041):200–203.
83. Ryan T, Reuter H, Wendland B, Schweizer F, Tsien R, Smith S. The kinetics of synaptic vesicle recycling measured at single presynaptic boutons. Neuron 1993;11(4):713–724.
84. Smith C, Betz W. Simultaneous independent measurement of endocytosis and exocytosis. Nature 1996;380(6574):531–534.
85. Miesenböck G, De Angelis D, Rothman J. Visualizing secretion and synaptic transmission with pH-sensitive green fluorescent proteins. Nature 1998;394(6689):192–195.
86. Sankaranarayanan S, De Angelis D, Rothman J, Ryan T. The use of pHluorins for optical measurements of presynaptic activity. Biophysical Journal 2000;79(4):2199–2208.
87. Sankaranarayanan S, Ryan T. Real-time measurements of vesicle-SNARE recycling in synapses of the central nervous system. Nature Cell Biology 2000;2(4):197–204.

88. Bowser D, Khakh B. Two forms of single-vesicle astrocyte exocytosis imaged with total internal reflection fluorescence microscopy. Proceedings of the National Academy of Sciences of the United States of America 2007;104(10):4212–4217.
89. Dai Z, Peng H. Fluorescence microscopy of calcium and synaptic vesicle dynamics during synapse formation in tissue culture. The Histochemical Journal 1998;30(3):189–196.
90. Wong R. Calcium imaging and multielectrode recordings of global patterns of activity in the developing nervous system. The Histochemical Journal 1998;30(3):217–229.

Neuronal Vulnerability to Oxidative Damage in Aging

Eitan Okun and Mark P. Mattson

Abstract The aging process in the brain is as robust as it is in other body organs and manifests as decrements in cognition, sensory and motor abilities, and autonomic control of various organ systems. This is due to the fact that, with advancing age, brain cells are exposed to increasing levels of oxidative stress, disturbed energy homeostasis, and accumulation of damage to protein, lipids, and nucleic acids. While these changes occur during normal aging, they are exacerbated in neurons that are susceptible to neurodegenerative disorders. The final outcome of the balance between a person's own genetic background and the environmental changes that affect him or her determines if and when a neurodegenerative disorder will occur. Oxidative molecular alterations that occur during normal aging and that are amplified in the neurons that are affected in neurodegenerative disorders include protein nitrosylation, oxidation of amino acids and DNA bases, lipid peroxidation, and increased amounts of neurotoxic amino acid derivates such as homocysteine. Oxidative damage may render neurons vulnerable to metabolic stress, excitotoxicity, and apoptosis in Alzheimer's disease (AD), Parkinson's disease (PD), Huntington's disease (HD), and amyotrophic lateral sclerosis (ALS). Oxidative stress contributes to the abnormal protein aggregations specific to each disorder – amyloid β-peptide (Aβ) in AD, α-synuclein in PD, huntingtin in HD, and Cu/Zn-superoxide dismutase (Cu/Zn-SOD) in ALS. These protein inclusions may arise, in part, from impaired proteasome function and autophagy. Although the utility of antioxidant ingestion as a preventative strategy for age-related neurological disorders has not yet been demonstrated, age- and disease-related oxidative damage to neurons can be decreased by dietary energy restriction and exercise and accelerated by overeating and diabetes.

M.P. Mattson (✉)
Laboratory of Neurosciences, National Institute on Aging Intramural Research Program, Baltimore, MD 21224, USA
e-mail: mattsonm@grc.nia.nih.gov

S.C. Veasey (ed.), *Oxidative Neural Injury*,
Contemporary Clinical Neuroscience, DOI 10.1007/978-1-60327-342-8_5,
© Humana Press, a part of Springer Science+Business Media, LLC 2009

Keywords ALS · Alzheimer's disease · Free radicals · Hormesis · Huntington's disease · Hydroxynonenal · Lipid peroxidation · Parkinson's disease · Peroxynitrite

1 Introduction

In a simplified sense, aging may be considered as progressive dyshomeostasis in many biological processes across time. The inability to maintain biological processes over time occurs as a direct consequence of repeated physiological challenges to cellular homeostasis where at least some of the challenges are countered by incomplete repair. The net result is cumulative injury across time that translates into both impaired function and impaired energy homeostasis. The dysfunction and dyshomeostasis then render cells more susceptible to future perturbances, perpetuating a vicious cycle. While this seems straightforward, it remains perplexing why within species some age faster or more slowly than others, why within an organism some tissues age faster or more slowly than others, and why within a specific tissue or organ, specific cells age more rapidly than others.

Within the nervous system, aging influences functions in most groups of neurons. Yet not all groups are affected at the same time and to the same degree. This selective vulnerability to aging across neural groups can be accentuated in neurodegenerative disorders. Yet each neurodegenerative disorder first impairs function and destroys cells in unique populations of neurons, adding a second layer to the selective vulnerability. Recent studies have begun to identify molecular tags to the selective vulnerability to aging and selective vulnerability to neuronal groups in neurodegenerative disorders. Oxidative stress is present in affected groups across aging and the oxidative modifications are accentuated in neurodegenerative disorders.

In this chapter, we describe the effects of aging on specific molecules and cellular processes and describe how genetics, environment, and cellular substrate combine to predict the effect of aging on neuronal function and on risk for neurodegenerative processes.

2 Aging and Oxidative Stress

2.1 Oxidatively Damaged Molecules Accumulate in Neurons During Aging

Accumulation of damage to DNA, proteins, and lipids that is caused by reactive oxygen species (ROS) and their byproducts is believed to contribute to the functional decline of the brain during aging. Lipid peroxidation products, such as 4-hydroxy-2-nonenal (4-HNE) and acrolein, react with DNA and proteins to produce further damage to neurons during aging [1]. It has been estimated that 10,000 oxidative interactions occur between endogenously generated free radicals and DNA in each cell of the body every day, and at least one of every three proteins in the cell of older animals is rendered dysfunctional through oxidative modification [2]. Although these estimates reveal that free radical-mediated protein and DNA modification play significant roles in the deterioration of the brain, they do not imply that free radical damage is the only cause of functional decline. However, the effects of oxidative stress on postmitotic cells such as neurons are cumulative [3], and neurons may therefore be particularly prone to age-related oxyradical-mediated dysfunction and degeneration.

2.2 Reactive Oxygen Species Involved in Brain Aging

Reactive oxygen species (ROS) include oxygen-containing molecules that are either free radicals with an unpaired electron (for example, superoxide anion radical, hydroxyl radical, and peroxynitrite) or reactive molecules such as hydrogen peroxide that can readily generate free radicals [4]. ROS form as a natural byproduct of the normal mitochondrial metabolism of oxygen and have important roles in cell signaling [5]. However, during aging and in times of environmental stress ROS levels can increase dramatically, which can result in significant damage to macromolecules (Fig. 1). Cells are normally able to defend themselves against ROS damage through the use of enzymes such as Cu/Zn-SOD, catalases and glutathione peroxidases, and free radical-scavenging molecules such as glutathione, bilirubin, coenzyme Q10, and tocopherols [6–8]. The oxygen-rich environments in which cells exist tend to produce a variety of chemical reactions in proteins. ROS are responsible for deamidation, racemization, and isomerization of protein residues; these oxidatively modified proteins are not repaired and must be removed by means of protein degradation [9, 10].

Nitric oxide (NO) can induce protein modifications in neurons including nitrosylation of metal and thiol centers and nitration of tyrosine residues; levels of these NO modifications increase in brain cells during aging [11]. These modifications, in turn, have their own specific effects on the function of the proteins that are modified; exacerbation of these effects may be involved in the pathogenesis of neurodegenerative disorders. At least 115 different proteins have so far been identified as targets for S-nitrosylation; these proteins include kinases, ion channels, transcription factors, structural proteins, proteases, and respiratory enzymes [12, 13]. Protein tyrosine nitration is a covalent protein modification resulting from the interaction of peroxynitrite or nitrogen dioxide with tyrosine residues resulting in the addition of a nitro group onto one of the two equivalent *ortho*carbons of the aromatic ring of tyrosine residues. Nitro-tyrosine residues have commonly been used as a marker of pathological processes and of oxidative stress [12, 13].

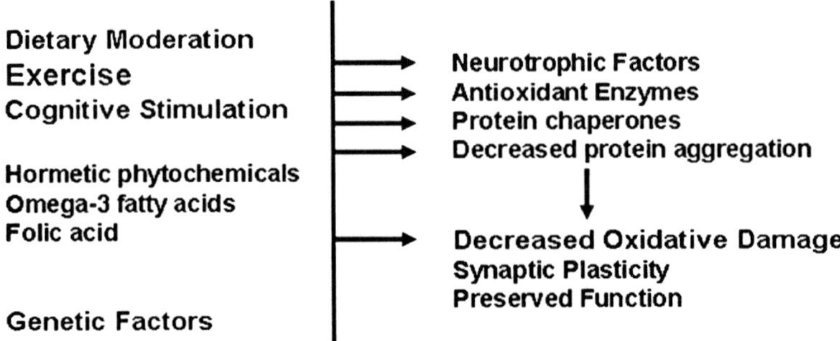

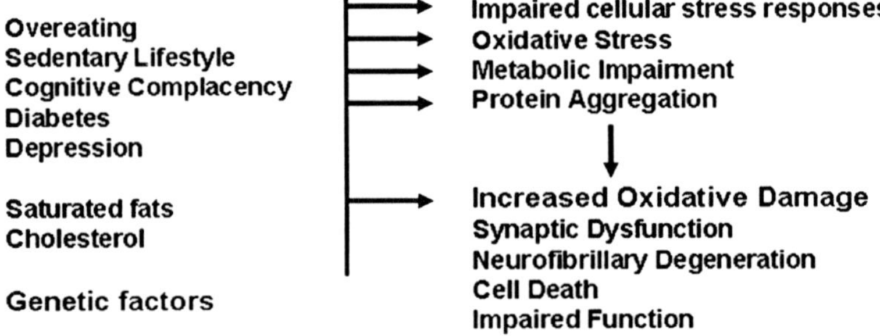

Fig. 1 Factors that may influence the amount of oxidative damage to neurons that occurs during aging, and thereby determine whether brain health is maintained or a neurodegenerative disorder occurs. Factors that promote brain health during aging include a low dietary energy (calorie) intake, regular exercise, and a cognitively stimulating lifestyle. Specific dietary components that may benefit neurons include omega-3 fatty acids, folic acid, and certain phytochemicals (e.g., curcumin and resveratrol). The latter factors may reduce levels of oxidative damage in neurons by increasing the expression of neurotrophic factors, antioxidant enzymes, and protein chaperones, thereby decreasing the accumulation of proteins such as β-amyloid and α-synuclein. Low levels of oxidative damage ensure that synapses function properly and that brain functions (cognitive, motor, and behavioral) are maintained. Factors that may endanger neurons during aging by exacerbating oxidative damage to neurons include overeating, a sedentary lifestyle, diabetes, and depression. Certain dietary factors (e.g., saturated fats and cholesterol) may promote oxidative stress during aging. Finally, genetic factors also influence one's risk for age-related neurodegeneration

Two ROS that are particularly prominent inducers of membrane lipid peroxidation are hydroxyl radical (generated as the result of interaction of hydrogen peroxide with Fe^{2+} and Cu^+) and peroxynitrite (formed in a reaction of superoxide anion radical with nitric oxide). Lipid peroxidation is increased in brain cells during normal aging and to a much greater extent in AD and PD [14, 15]. While direct damage to membranes by lipid peroxidation may adversely affect neurons, recent findings suggest that toxic aldehydes such as 4-hydroxynonenal (4-HNE), liberated from peroxidized membrane lipids, mediate many of the adverse effects of lipid peroxidation on neuronal function and viability [15]. Although highly reactive with macromolecules, the aldehydes are much more stable than the ROS, so they can spread from the site of origin and act at more distant sites. One of the most important products of lipid peroxidation is the highly reactive aldehyde 4-HNE, which has been shown to be generated during various physiological and pathophysiological conditions [15]. The presence of 4-HNE is increased in brain tissues and cerebrospinal fluid of AD patients and in spinal cord of ALS patients [16, 17]. Immunohistochemical studies show the presence of 4-HNE in neurofibrillary tangles and in senile plaques in AD, in the cytoplasm of the residual motor neurons in sporadic ALS, in Lewy bodies in neocortical and brain stem neurons in Parkinson's disease (PD) and in diffuse Lewy bodies disease [18–21]. Thus, increased levels of 4-HNE in neurodegenerative disorders and immunohistochemical distribution of 4-HNE in brain tissues indicate pathophysiological roles for neuronal oxidative stress and membrane lipid peroxidation in these diseases.

2.3 Aging Tips the Redox Balance

Quantities of oxidatively modified proteins, lipids, and nucleic acids increase in many brain regions with age [12]. In addition to oxidative modifications in the brain, the redox state shifts with aging toward a more oxidant state [22]. Glutathione (GSH) is the most abundant nonprotein thiol to buffer oxidative stress in brain tissues. Amounts of reduced:oxidized glutathione (GSH:GSSG) may be used as an index of oxidative stress. GSH/GSSG falls significantly in the cortex, striatum, and cerebellum, where declines in GSH:GSSG paralleled protein oxidation [23]. In contrast, glutathione levels are much lower in the brainstem and do not show an overall age-related decline [23].

Antioxidant enzymatic activity would be expected to increase in response to increased oxidative stress. In select brain regions, however, manganese superoxide dismutase activity is reduced in aged animals in the midbrain, the caudate putamen, and the hippocampus, while other regions show no change [24]. Methionine sulfoxide reductase activity is reduced in the aged forebrain in parallel with declining gene transcription [25]. In contrast, enzymes involved in increased oxidation of GSH (e.g., glutathione peroxidase and glutathione S-transferase) are increased with age, while the major enzyme involved in GSH synthesis (γ-glutamylcysteine synthetase) declines [26]. In addition to an impaired antioxidant response, the production of superoxide increases with age in the brain [27].

3 Oxidative Stress and Neuronal Cell Death in the Aging Brain

3.1 ROS and Neuronal Loss

Most brain regions in rodents and mammals exhibit little or no loss of neurons during normal aging [28]. However, in age-related neurodegenerative disorders neurons in brain regions affected by the disease process die [see reference [29] for review]. In AD, pyramidal neurons in the hippocampus, layer II entorhinal neurons, and large neurons in temporal and frontal lobe association cortices die. In PD, dopaminergic neurons in the substantia nigra and monaminergic neurons in the brainstem degenerate, while in HD medium spiny neurons in the striatum and cortical neurons die. ALS involves the death of lower motor neurons in the spinal cord and, in late stages of the disease, upper motor neurons in the motor cortex. Although a few populations of neurons may be replaced by new cells generated from stem cells (dentate gyrus granule neurons and olfactory bulb interneurons), the neurons affected in neurodegenerative disorders are not replaced [30].

Because oxidative stress may play a major role in the death of neurons in age-related neurodegenerative disorders, it is believed that the selectivity of neuronal vulnerability may result, at least in part, from differences amongst neurons in the amount of oxidative stress they encounter and/or their ability to protect themselves against oxidative stress [29] (Fig. 1). In all of the major age-related neurodegenerative disorders, there is evidence that oxidative stress results in the dysregulation of cellular Ca^{2+} homeostasis, metabolic (mitochondrial) impairment, the accumulation of damaged proteins, and neuronal apoptosis [31].

3.2 Oxidative Stress, Neuronal Calcium Dysregulation, and Excitotoxicity

Regulation of the Ca^{2+} concentration in the cytosol is tightly controlled in all cell types. In neurons, the Ca^{2+} concentration is typically in the range of 75–200 nM and may transiently reach 1–10 μM in response to membrane depolarization due to opening of voltage-dependant Ca^{2+} and NMDA receptor channels [32]. Ca^{2+} may also be released from the endoplasmic reticulum in response to extracellular signals or increased cytosolic Ca^{2+} concentrations. Ca^{2+} removal from the cytosol is mediated by plasma membrane and endoplasmic reticulum Ca^{2+} ATPases, and calcium-binding proteins that may buffer Ca^{2+}. One or more of the systems that control Ca^{2+} influx, efflux, and buffering may be perturbed during the normal aging of the brain [33, 34]. It is well established that prolonged elevations of intracellular Ca^{2+} levels can cause synapse and neurite degeneration, and eventually cell death by activating proteases and inducing ROS production [33, 34].

Oxidative stress may occur upstream or downstream of perturbed neuronal Ca^{2+} regulation in neurodegenerative disorders. Studies of the physiological consequences of mutations in the genes that cause familial forms of AD suggest that disturbances in Ca^{2+} homeostasis are pivotal to the disease-causing effects of the mutations [33]. Mutations in the β-amyloid precursor protein (APP) result in increased production and accumulation of amyloid β-peptide (Aβ) at synapses. Oligomers of Aβ may increase Ca^{2+} influx by forming pores in the plasma membrane and by inducing membrane lipid peroxidation, which impairs the function of the ion-motive ATPases and glucose and glutamate transporters [35]. Mutations in presenilin-1, which cause many cases of early onset inherited AD, may compromise a Ca^{2+}-regulating function of the presenilin-1 protein in the endoplasmic reticulum [31]. There is evidence that differential expression of Ca^{2+}-regulating proteins may contribute to the selective vulnerability of neurons in AD. For example, in the hippocampus CA1 pyramidal neurons which express high levels of NMDA receptors and relatively low levels of calcium-binding proteins are vulnerable, whereas dentate granule neurons which express high levels of calcium-binding proteins are resistant [see [29]].

Glutamate is the main excitatory neurotransmitter in the central nervous system. Excessive activation of glutamate receptors causes continuous Ca^{+2} influx and ROS production. This, in turn, leads to excitotoxicity that can harm dendrites and cause cell death. The overall changes that aging neurons undergo render them more susceptible to excitotoxicity. Disease-specific abnormalities, such as misfolded proteins or aggregated proteins, might enhance the adverse effects of glutamate on neurons. For example, Aβ, dopamine, mutant huntingtin, and mutant Cu/Zn-SOD were all shown to sensitize neurons to excitotoxic death [29, 33–35].

4 Mitochondrial Dysfunction

Mitochondrial function declines during aging in general and aging of the nervous system, in particular [36]. Analysis of mitochondria from different brain regions of young, middle aged, and old rats revealed that mitochondria from the cerebral cortex of old rats produce more ROS and exhibit mitochondrial swelling due to increased Ca^{+2} loads when compared to younger rats [37]. However, the sensitivity of brain mitochondria to Ca^{+2} was relatively unaffected by the aging process. Reduced buffering of voltage-gated Ca^{+2} influx in basal forebrain neurons from aged rats suggests that the capacity of mitochondria to appropriately respond to excitation might be impaired during aging [38]. In addition, neurons show impaired glucose uptake during the normal aging process, which may further compromise their ability to maintain ion homeostasis and other energy-consuming cellular processes. The perturbed glucose metabolism may contribute to hyperphosphorylation of the microtubule-associated protein tau due to reduced tau GlcNacylation, a cascade that may contribute to tau protein aggregation in AD and other tauopathies [39].

Mitochondrial abnormalities have been widely reported in studies of post-mortem brain tissue samples from AD and PD patients and in studies of animal and cell culture models of these disorders [40, 41]. For example, measurements of the activity of mitochondrial enzymes in various brain tissues from AD patients showed significant decreases in the activity of pyruvate dehydrogenase, isocitrate dehydrogenase, and α-keto glutarate dehydrogenase. Increased levels of oxidative stress are associated with mitochondrial abnormalities in mice expressing mutant forms of APP and/or presenilin-1 which cause AD in humans [42–44]. Mitochondria-associated oxidative stress is also strongly implicated in the pathogenesis of neuronal dysfunction and death in stroke, PD, HD, and ALS [45–47].

5 Oxidative Stress, Misfolded Proteins, and Damaged DNA

An obvious alteration that occurs in neurons during aging is the accumulation of misfolded, oxidatively damaged molecules within and outside of the cells. Normally, damaged proteins are degraded by lysosomes, proteasomes, and autophagy; data suggest that all three of these clearance mechanisms for damaged proteins are impaired in neurodegenerative conditions [see [48–50] for review]. The effect of aging on the proteasomal activity can be mimicked by using pharmacological proteasomal inhibitors, which induce some of the adverse consequences of aging on the neuronal function. Proteasome functionality in rats is decreased with advancing age in the cerebral cortex, the hippocampus, and the spinal cord, but not in the cerebellum or the brainstem [51]. This adverse effect of aging on proteasome function may be exacerbated in AD, wherein Aβ and tau are the two main proteins that are prone to aggregate. Impairment of one or more of the cellular protein degradation systems for Aβ (proteasome, insulin-degrading enzyme, or neprilysin) by age-related increases in oxidative stress or by protein damage might contribute to the formation of Aβ deposits in aging and AD [52]. The aggregation of Aβ might also be promoted by oxidative processes that involve hydrogen peroxide, Fe^{2+}, and Cu^+ [53, 54]. Tau normally plays a critical role in regulating cytoskeletal plasticity in neurons. During aging, and more so in AD, this function of tau is compromised as a result of oxidative stress-induced aggregation of tau to form neurofibrillary tangles [55, 56].

Aging is also associated with increasing amount of DNA damage in neurons. The amount of oxidative DNA damage to neurons is greatly increased in age-related neurodegenerative disorders such as AD [57, 58]. In addition to increased oxidative stress causing more DNA damage, the ability of neurons to repair damaged DNA may be compromised in aging and neurodegenerative disorders [59]. Age-related oxidative DNA damage to promoters of genes may cause a widespread dysregulation of gene expression in neurons resulting in their dysfunction [60]. Indeed, mutations in DNA repair proteins such as Werner and Cockayne proteins cause premature aging phenotypes in humans [61].

6 Implications for Promoting Healthy Brain Aging

One of the major goals of research in the field of aging is to identify ways of preventing or delaying age-related morbidity including neurodegenerative disorders. As described above, oxidative stress and impaired cellular stress adaptation are among the factors that render neurons vulnerable to dysfunction and degeneration during aging. One general approach for reducing oxidative damage to neurons during aging is to administer antioxidants, either by consuming fruits and vegetables rich in antioxidants or by supplementing the diet with specific antioxidants. While much epidemiological data have suggested health benefits of diets rich in fruits, vegetables, and spices, therapeutic trials of supplementation with antioxidants such as vitamins E, A, and C have been disappointing, typically with no beneficial effects or even adverse effects [62, 63]. It is possible that in diseases such as AD, clinical trials with antioxidants usually begin when the patients already exhibit early or late symptoms, whereas the ability of lifelong administration of antioxidants to slow down the progress of AD in susceptible populations was never tested. However, an alternative explanation comes from recent findings which suggest that, rather than acting as free radical scavengers, phytochemicals may exert their beneficial effects on the nervous system by activating adaptive stress response pathways in neurons [64]. Examples of such phytochemicals and the pathways they activate include sulforaphane (in broccoli), Nrf-2−antioxidant response element pathway; resveratrol (in red grapes and wine), sirtuin−FOXO pathway; and allicin (in garlic and onions).

Perhaps the most effective approaches for reducing oxidative damage to neurons and reducing the risk of age-related functional decline are dietary energy restriction and regular exercise [65, 66] (Fig. 1). Dietary energy restriction may reduce oxidative damage to neurons by stimulating adaptive stress response pathways involving neurotrophic factors and protein chaperones that enhance antioxidant defense mechanisms [65–69]. Similarly, exercise induces the expression of neurotrophic factors and suppresses oxidative stress [70]. In support of antioxidant effects of dietary energy restriction and exercise are data showing that diabetes (which is typically caused by overeating and lack of exercise) accelerates age-related neuronal dysfunction [71, 72]. Alas, there is no easy solution to the problems created by oxidative stress during aging. However, if you have the will to maintain a healthy brain as you age, then a Spartan lifestyle will do the trick.

References

1. Poon HF, Calabrese V, Scapagnini G, Butterfield DA. Free radicals and brain aging. Clin Geriatr Med. 2004 May;20(2):329–359.
2. Wagner JR, Motchnik PA, Stocker R, Sies H, Ames BN. The oxidation of blood plasma and low density lipoprotein components by chemically generated singlet oxygen. J Biol Chem. 1993 Sep 5;268(25):18502–18506.
3. Fishel ML, Vasko MR, Kelley MR. DNA repair in neurons: so if they don't divide what's to repair? Mutat Res. 2007 Jan 3;614(1–2):24–36.
4. Reynolds A, Laurie C, Mosley RL, Gendelman HE. Oxidative stress and the pathogenesis of neurodegenerative disorders. Int Rev Neurobiol. 2007;82:297–325.
5. Gutierrez J, Ballinger SW, Darley-Usmar VM, Landar A. Free radicals, mitochondria, and oxidized lipids: the emerging role in signal transduction in vascular cells. Circ Res. 2006 Oct 27;99(9):924–932.
6. Powers SK, Lennon SL. Analysis of cellular responses to free radicals: focus on exercise and skeletal muscle. Proc Nutr Soc. 1999 Nov;58(4):1025–1033.
7. Fridovich I. Fundamental aspects of reactive oxygen species, or what's the matter with oxygen? Ann N Y Acad Sci. 1999;893:13–18.
8. Hyun DH, Hernandez JO, Mattson MP, de Cabo R. The plasma membrane redox system in aging. Ageing Res Rev. 2006 May;5(2):209–220.
9. Hipkiss AR. Accumulation of altered proteins and ageing: causes and effects. Exp Gerontol. 2006 May;41(5):464–473.
10. Galletti P, De Bonis ML, Sorrentino A, Raimo M, D'Angelo S, Scala I, Andria G, D'Aniello A, Ingrosso D, Zappia V. Accumulation of altered aspartyl residues in erythrocyte proteins from patients with Down's syndrome. FEBS J. 2007 Oct; 274(20):5263–5277.
11. Stamler JS, Lamas S, Fang FC. Nitrosylation the prototypic redox-based signaling mechanism. Cell. 2001 Sep 21;106(6):675–683.
12. Floyd RA. Antioxidants, oxidative stress, and degenerative neurological disorders. Proc Soc Exp Biol Med. 1999 Dec;222(3):236–245.
13. Ischiropoulos H. Biological tyrosine nitration: a pathophysiological function of nitric oxide and reactive oxygen species. Arch Biochem Biophys. 1998 Aug 1;356(1):1–11.
14. Mattson MP. Modification of ion homeostasis by lipid peroxidation: roles in neuronal degeneration and adaptive plasticity. Trends Neurosci. 1998 Feb;21(2):53–57.
15. Zarkovic K. 4-hydroxynonenal and neurodegenerative diseases. Mol Aspects Med. 2003 Aug–Oct;24(4–5):293–303.
16. Cutler RG, Kelly J, Storie K, Pedersen WA, Tammara A, Hatanpaa K, Troncoso JC, Mattson MP. Involvement of oxidative stress-induced abnormalities in ceramide and cholesterol metabolism in brain aging and Alzheimer's disease. Proc Natl Acad Sci U S A. 2004 Feb 17;101(7):2070–2075.
17. Cutler RG, Pedersen WA, Camandola S, Rothstein JD, Mattson MP. Evidence that accumulation of ceramides and cholesterol esters mediates oxidative stress-induced death of motor neurons in amyotrophic lateral sclerosis. Ann Neurol. 2002 Oct;52(4):448–457.
18. Castellani RJ, Perry G, Siedlak SL, Nunomura A, Shimohama S, Zhang J, Montine T, Sayre LM, Smith MA. Hydroxynonenal adducts indicate a role for lipid peroxidation in neocortical and brainstem Lewy bodies in humans. Neurosci Lett. 2002 Feb 8;319(1):25–28.
19. Pedersen WA, Fu W, Keller JN, Markesbery WR, Appel S, Smith RG, Kasarskis E, Mattson MP. Protein modification by the lipid peroxidation product 4-hydroxynonenal in the spinal cords of amyotrophic lateral sclerosis patients. Ann Neurol. 1998 Nov;44(5):819–824.

20. Sayre LM, Zelasko DA, Harris PL, Perry G, Salomon RG, Smith MA. 4-Hydroxynonenal-derived advanced lipid peroxidation end products are increased in Alzheimer's disease. J Neurochem. 1997 May;68(5):2092–2097.

21. Dalfo E, Portero-Otin M, Ayala V, Martinez A, Pamplona R, Ferrer I. Evidence of oxidative stress in the neocortex in incidental Lewy body disease. J Neuropathol Exp Neurol. 2005 Sep;64(9):816–830.

22. Rebrin I, Sohal RS. Pro-oxidant shift in glutathione redox state during aging. Adv Drug Deliv Rev. 2008 Jul 4.

23. Rebrin I, Forster MJ, Sohal RS. Effects of age and caloric intake on glutathione redox state in different brain regions of C57BL/6 and DBA/2 mice. Brain Res. 2007 Jan 5;1127(1):10–18.

24. Cardozo-Pelaez F, Song S, Parthasarathy A, Hazzi C, Naidu K, Sanchez-Ramos J. Oxidative DNA damage in the aging mouse brain. Mov Disord. 1999 Nov;14(6):972–980.

25. Petropoulos I, Mary J, Perichon M, Friguet B. Rat peptide methionine sulphoxide reductase: cloning of the cDNA, and down-regulation of gene expression and enzyme activity during aging. Biochem J. 2001 May 1;355(Pt 3):819–825.

26. Zhu Y, Carvey PM, Ling Z. Age-related changes in glutathione and glutathione-related enzymes in rat brain. Brain Res. 2006 May 23;1090(1):35–44.

27. Sasaki T, Unno K, Tahara S, Shimada A, Chiba Y, Hoshino M, Kaneko T. Age-related increase of superoxide generation in the brains of mammals and birds. Aging Cell. 2008 May 9;7(4):459–469.

28. West MJ, Coleman PD, Flood DG, Troncoso JC. Differences in the pattern of hippocampal neuronal loss in normal ageing and Alzheimer's disease. Lancet. 1994 Sep 17;344(8925):769–772.

29. Mattson MP, Magnus T. Ageing and neuronal vulnerability. Nat Rev Neurosci. 2006 Apr;7(4):278–294.

30. Sohur US, Emsley JG, Mitchell BD, Macklis JD. Adult neurogenesis and cellular brain repair with neural progenitors, precursors and stem cells. Philos Trans R Soc Lond B Biol Sci. 2006 Sep 29;361(1473):1477–1497.

31. Mattson MP, Duan W, Pedersen WA, Culmsee C. Neurodegenerative disorders and ischemic brain diseases. Apoptosis. 2001 Feb-Apr;6(1–2):69–81.

32. Arundine M, Tymianski M. Molecular mechanisms of calcium-dependent neurodegeneration in excitotoxicity. Cell Calcium. 2003 Oct–Nov;34(4–5):325–337.

33. Bezprozvanny I, Mattson MP. Neuronal calcium mishandling and the pathogenesis of Alzheimer's disease. Trends Neurosci. 2008;31(9):454–463.

34. Mattson MP. Excitotoxic and excitoprotective mechanisms: abundant targets for the prevention and treatment of neurodegenerative disorders. Neuromolecular Med. 2003;3(2):65–94.

35. Mattson MP. Pathways towards and away from Alzheimer's disease. Nature. 2004 Aug 5;430(7000):631–639.

36. Melov S. Modeling mitochondrial function in aging neurons. Trends Neurosci. 2004 Oct;27(10):601–606.

37. Brown MR, Geddes JW, Sullivan PG. Brain region-specific, age-related, alterations in mitochondrial responses to elevated calcium. J Bioenerg Biomembr. 2004 Aug;36(4):401–406.

38. Murchison D, Griffith WH. Calcium buffering systems and calcium signaling in aged rat basal forebrain neurons. Aging Cell. 2007 Jun;6(3):297–305.

39. Liu F, Iqbal K, Grundke-Iqbal I, Hart GW, Gong CX. O-GlcNAcylation regulates phosphorylation of tau: a mechanism involved in Alzheimer's disease. Proc Natl Acad Sci U S A. 2004 Jul 20;101(29):10804–10809.

40. Reddy PH. Mitochondrial dysfunction in aging and Alzheimer's disease: strategies to protect neurons. Antioxid Redox Signal. 2007 Oct;9(10):1647–1658.

41. Schapira AH. Mitochondrial involvement in Parkinson's disease, Huntington's disease, hereditary spastic paraplegia and Friedreich's ataxia. Biochim Biophys Acta. 1999 Feb 9;1410(2):159–170.
42. Begley JG, Duan W, Chan S, Duff K, Mattson MP. Altered calcium homeostasis and mitochondrial dysfunction in cortical synaptic compartments of presenilin-1 mutant mice. J Neurochem. 1999 Mar;72(3):1030–1039.
43. Guo Q, Sebastian L, Sopher BL, Miller MW, Ware CB, Martin GM, Mattson MP. Increased vulnerability of hippocampal neurons from presenilin-1 mutant knock-in mice to amyloid beta-peptide toxicity: central roles of superoxide production and caspase activation. J Neurochem. 1999 Mar;72(3):1019–1029.
44. Hauptmann S, Scherping I, Drose S, Brandt U, Schulz KL, Jendrach M, Leuner K, Eckert A, Muller WE. Mitochondrial dysfunction: an early event in Alzheimer pathology accumulates with age in AD transgenic mice. Neurobiol Aging. 2008 Feb 21. (Epub ahead of print)
45. Keller JN, Kindy MS, Holtsberg FW, St Clair DK, Yen HC, Germeyer A, Steiner SM, Bruce-Keller AJ, Hutchins JB, Mattson MP. Mitochondrial manganese superoxide dismutase prevents neural apoptosis and reduces ischemic brain injury: suppression of peroxynitrite production, lipid peroxidation, and mitochondrial dysfunction. J Neurosci. 1998 Jan 15;18(2):687–697.
46. Mattson MP, Gleichmann M, Cheng A. Mitochondria in neuroplasticity and neurological disorders. Neuron. 2008;60(5):748–766.
47. Poon HF, Frasier M, Shreve N, Calabrese V, Wolozin B, Butterfield DA. Mitochondrial associated metabolic proteins are selectively oxidized in A30P alpha-synuclein transgenic mice – a model of familial Parkinson's disease. Neurobiol Dis. 2005 Apr;18(3):492–498.
48. Martinez-Vicente M, Cuervo AM. Autophagy and neurodegeneration: when the cleaning crew goes on strike. Lancet Neurol. 2007 Apr;6(4):352–361.
49. Nixon RA, Cataldo AM. Lysosomal system pathways: genes to neurodegeneration in Alzheimer's disease. J Alzheimers Dis. 2006;9(3 Suppl):277–289.
50. Szweda PA, Camouse M, Lundberg KC, Oberley TD, Szweda LI. Aging, lipofuscin formation, and free radical-mediated inhibition of cellular proteolytic systems. Ageing Res Rev. 2003 Oct;2(4):383–405.
51. Keller JN, Hanni KB, Markesbery WR. Possible involvement of proteasome inhibition in aging: implications for oxidative stress. Mech Ageing Dev. 2000 Jan 24;113(1):61–70.
52. Miners JS, Baig S, Palmer J, Palmer LE, Kehoe PG, Love S. Abeta-degrading enzymes in Alzheimer's disease. Brain Pathol. 2008 Apr;18(2):240–252.
53. Hensley K, Carney JM, Mattson MP, Aksenova M, Harris M, Wu JF, Floyd RA, Butterfield DA. A model for beta-amyloid aggregation and neurotoxicity based on free radical generation by the peptide: relevance to Alzheimer disease. Proc Natl Acad Sci U S A. 1994 Apr 12;91(8):3270–3274.
54. Mattson MP. Metal-catalyzed disruption of membrane protein and lipid signaling in the pathogenesis of neurodegenerative disorders. Ann N Y Acad Sci. 2004 Mar;1012:37–50.
55. Mattson MP, Fu W, Waeg G, Uchida K. 4-Hydroxynonenal, a product of lipid peroxidation, inhibits dephosphorylation of the microtubule-associated protein tau. Neuroreport. 1997 Jul 7;8(9–10):2275–2281.
56. Munch G, Kuhla B, Luth HJ, Arendt T, Robinson SR. Anti-AGEing defences against Alzheimer's disease. Biochem Soc Trans. 2003 Dec;31(Pt 6):1397–1399.
57. Moreira PI, Nunomura A, Nakamura M, Takeda A, Shenk JC, Aliev G, Smith MA, Perry G. Nucleic acid oxidation in Alzheimer disease. Free Radic Biol Med. 2008 Apr 15;44(8):1493–1505.
58. Yang JL, Weissman L, Bohr VA, Mattson MP. Mitochondrial DNA damage and repair in neurodegenerative disorders. DNA Repair (Amst). 2008 Jul 1;7(7):1110–1120.
59. Weissman L, Jo DG, Sorensen MM, de Souza-Pinto NC, Markesbery WR, Mattson MP, Bohr VA. Defective DNA base excision repair in brain from individuals with Alzheimer's disease and amnestic mild cognitive impairment. Nucleic Acids Res. 2007;35(16):5545–5555.

60. Lu T, Pan Y, Kao SY, Li C, Kohane I, Chan J, Yankner BA. Gene regulation and DNA damage in the ageing human brain. Nature. 2004 Jun 24;429(6994):883–891.
61. Kyng KJ, Bohr VA. Gene expression and DNA repair in progeroid syndromes and human aging. Ageing Res Rev. 2005 Nov;4(4):579–602.
62. Boothby LA, Doering PL. Vitamin C and vitamin E for Alzheimer's disease. Ann Pharmacother. 2005 Dec;39(12):2073–2080.
63. Evans J. Antioxidant supplements to prevent or slow down the progression of AMD: a systematic review and meta-analysis. Eye. 2008 Jun;22(6):751–760.
64. Mattson MP, Cheng A. Neurohormetic phytochemicals: low-dose toxins that induce adaptive neuronal stress responses. Trends Neurosci. 2006 Nov;29(11):632–639.
65. Arumugam TV, Gleichmann M, Tang SC, Mattson MP. Hormesis/preconditioning mechanisms, the nervous system and aging. Ageing Res Rev. 2006 May;5(2):165–178.
66. Martin B, Mattson MP, Maudsley S. Caloric restriction and intermittent fasting: two potential diets for successful brain aging. Ageing Res Rev. 2006 Aug;5(3):332–353.
67. Halagappa VK, Guo Z, Pearson M, Matsuoka Y, Cutler RG, Laferla FM, Mattson MP. Intermittent fasting and caloric restriction ameliorate age-related behavioral deficits in the triple-transgenic mouse model of Alzheimer's disease. Neurobiol Dis. 2007 Apr;26(1):212–220.
68. Maswood N, Young J, Tilmont E, Zhang Z, Gash DM, Gerhardt GA, Grondin R, Roth GS, Mattison J, Lane MA, Carson RE, Cohen RM, Mouton PR, Quigley C, Mattson MP, Ingram DK. Caloric restriction increases neurotrophic factor levels and attenuates neurochemical and behavioral deficits in a primate model of Parkinson's disease. Proc Natl Acad Sci U S A. 2004 Dec 28;101(52):18171–18176.
69. Xu X, Zhan M, Duan W, Prabhu V, Brenneman R, Wood W, Firman J, Li H, Zhang P, Ibe C, Zonderman AB, Longo DL, Poosala S, Becker KG, Mattson MP. Gene expression atlas of the mouse central nervous system: impact and interactions of age, energy intake and gender. Genome Biol. 2007;8(11):R234.
70. Radak Z, Kumagai S, Taylor AW, Naito H, Goto S. Effects of exercise on brain function: role of free radicals. Appl Physiol Nutr Metab. 2007 Oct;32(5):942–946.
71. Stranahan AM, Lee K, Pistell PJ, Nelson CM, Readal N, Miller MG, Spangler EL, Ingram DK, Mattson MP. Accelerated cognitive aging in diabetic rats is prevented by lowering corticoserone levels. Neurobiol Learn Mem. 2008;90(2):479–483.
72. Stranahan AM, Arumugam TV, Cutler RG, Lee K, Egan JM, Mattson MP. Diabetes impairs hippocampal function through glucocorticoid-mediated effects on new and mature neurons. Nat Neurosci. 2008 Mar;11(3):309–317.

Ischemia-Reperfusion Induces ROS Production from Three Distinct Sources

Rosemary H. Milton and Andrey Y. Abramov

Abstract Most neurodegenerative disorders are insidious with substantial neural injury present well before symptoms are revealed. Thus, at the time of evident disability or impairment, secondary injuries may obscure the inciting neuropathology. This is not true of stroke, where the onset of injury and symptomatology typically coincide. Moreover, there is a distinct and relatively brief temporal pattern of injury. Therefore, models of ischemia/reperfusion can provide significant insight into early and late injuries and the specific mechanisms of each component of injury. Modelling hypoxia and reoxygenation in primary cultures of hippocampal and cortical neurons to examine temporally the source of reactive oxygen species has identified three phases of reactive oxygen species production in hypoxia/reoxygenation. Mitochondria respond first but are quickly limited by the insufficient oxygen. At this point in ischemia, xanthine oxidase becomes an important source of superoxide, whereas during reperfusion NADPH oxidase is a major source of superoxide. This work highlights the value of a model of early and temporally distinct phases of injury and supports the concept that multitarget approaches will be necessary to effectively prevent or reduce neural injury of stroke.

Keywords Cerebrovascular accident · Mitochondria · Reactive oxygen species · Xanthine oxidase · NADPH oxidase · Ischemia · Hypoxia · Reperfusion · Cortical neurons · Hippocampal neurons

A.Y. Abramov (✉)
Department of Physiology, University College London, London, UK
e-mail: a.abramov@ucl.ac.uk

S.C. Veasey (ed.), *Oxidative Neural Injury*,
Contemporary Clinical Neuroscience, DOI 10.1007/978-1-60327-342-8_6,
© Humana Press, a part of Springer Science+Business Media, LLC 2009

1 Introduction

1.1 The Clinical Significance of Stroke

Stroke remains a devastating and debilitating event – one associated with high mortality and a high emotional cost and financial burden to patients, families, and society. Stroke is the third most common cause of death worldwide, and of the patients who survive, 20% will require institutionalization within 3 months [1]. It is, therefore, of paramount importance to understand more completely the mechanisms by which cerebral ischemic events injure neurons so that we may develop therapies to minimize neural injury in stroke.

1.2 Stroke Pathophysiology

During a stroke, or cerebrovascular accident, the blood supply to a part of the brain becomes restricted and the brain becomes ischemic. The blood supply can be interrupted by thrombosis, embolism, or systemic hypoperfusion (e.g., during cardiac arrest). Alternatively, in a hemorrhagic stroke a ruptured blood vessel may cause a hematoma, affecting perfusion of other brain areas. In all such events, affected cells will experience hypoxia (low oxygen levels) or anoxia (a total absence of oxygen). In addition, birth asphyxia represents a serious challenge to the infant brain and a major cause of brain injury.

Cell injury during periods of ischemia and subsequent reperfusion is caused by the loss of the energy supply due to deprivation of oxygen and glucose, but the area of damage is extended by the effects of excessive accumulation of neurotransmitters. Oxygen and glucose deprivation results in reduced production and depletion of ATP. A pivotal event is the loss of ATP-dependent Na^+–K^+-ATPase function, resulting in profound membrane depolarization and sodium and calcium dyshomeostasis. It has been proposed that increased intracellular calcium causes increased release of glutamate [2], and that ATP depletion causes changes in ionic balance that drive ion-dependent glutamate transport in reverse, leading to further extracellular glutamate accumulation [3]. Glutamate release contributes to a vicious cycle, as the activation of glutamate receptors further increases intracellular calcium concentration, causing calcium overload which can trigger cell death (for review see [3]).

An important aspect of cellular damage, causing severe protein and lipid modification, is the overproduction of reactive oxygen species (ROS). Oxidative stress occurs when the levels of production of ROS overwhelm cellular antioxidant defenses. The exacerbation of hypoxic injury by reoxygenation (the oxygen paradox) represents a major clinical problem in treating episodes of cerebral ischemia; it is obviously important to ensure the return of blood supply, yet this restoration of oxygen provision can itself be an important cause of cellular injury throughout a wide area surrounding the initial pathology.

1.3 Reactive Oxygen Species Implicated in Neuronal Injury in Stroke

Free radicals have been implicated in cerebral ischemic damage since the 1970s, yet the full complexity of their role is still not understood. It is worth noting at this stage that one of the first papers to describe a role for oxidative damage in stroke [4] refers to the species of interest as free radicals, and identified their role through diminution of antioxidant defenses in cat brain following focal ische- mia. This highlights some interesting and enduringly relevant issues. Most literature now refers to these damaging species as ROS, but not all ROS are free radicals. Free radicals have an unpaired electron. Electrons within atoms and molecules can be said to occupy orbitals, each orbital holding a maximum of two electrons. If an orbital contains only one electron, that electron is said to be unpaired. A free radical species is capable of independent existence and contains one or more unpaired electrons. ROS are often radicals or easily converted to radicals, but are defined by their ability to oxidize. For instance, hydrogen peroxide (H_2O_2) is a ROS, but not a free radical.

Furthermore, the use of the blanket term "free radicals" highlights how little it is possible to know about precisely which ROS are implicated, or from whence they came. Even if a certain reaction produces a certain species, the inherent reactivity of this class of molecules results in often very rapid changes into other forms of ROS, in addition to reactive nitrogen- and chlorine-based species. The primary ROS produced at source in mammalian cells is usually the superoxide anion $O^{\bullet-}_2$, but it is rapidly subject to further reactions, resulting in the downstream production of other ROS, such as H_2O_2, $OH^{\bullet}$, and so on. The authors of this original paper [4] also implicated free radicals through having measured levels of antioxidant defenses and found them to be depleted. This emphasizes the dynamic equilibrium not only between different classes of ROS, but also between ROS and the scavengers designed to remove them.

The importance of ROS in instigating neuronal injury has been highlighted more recently through transgenic animal studies. Major enzymes involved in the clearance of ROS include superoxide dismutase, catalase, and glutathione peroxidase. Superoxide dismutase converts superoxide to hydrogen peroxide, which may then be decomposed by either catalase or glutathione peroxidase. One of the three superoxide dismutase (SOD) isoforms, copper/zinc SOD, or SOD1, plays a critical role in minimizing neural injury in cerebral ischemia. SOD1-overexpressing transgenic mice have approximately 35% smaller infarct volume after focal ischemia [5]. In a model of global ischemia, SOD1 over- expression resulted in a 50% reduction in hippocampal CA1 neuron loss. In contrast, SOD1 knockout mice have increased neuronal loss in models of focal or global ischemia [6, 7]. Mice with transgenic absence of both SOD1 and glutathione peroxidase show even larger infarcts, suggesting that glutathione peroxidase is also important in reducing neural injury in ischemia [8]. Reactive nitrogen species may also contribute to neural injury in ischemia, in that mice with transgenic absence of neuronal nitric oxygen synthase, a major source of neuronal nitric oxide, are substantially protected from cerebral ischemia [9, 10]. In summary, reactive oxygen species and nitrogen species play important roles in neuronal injury and loss in animal models of stroke.

2 Sources of Reactive Oxygen Species in Stroke

Since these original observations, we have come some way in toward understanding how, where, and whence particular ROS cause cell damage in ischemia-reperfusion injury. We have some appreciation of causality, and the relative extent and sequence of different contributors. Genetic and pharmacological studies have provided information about enzymatic ROS sources, complimenting each other with their specificity and acuteness, respectively. It is now clear that different sources of ROS combine in complex ways to produce the final damage we observe and shown us that different manipulations must be studied at different time points. It is unlikely that inhibition of any one source of ROS would be sufficient to totally protect neurons from death, yet some inhibitors have proved encouraging in trials, and more general antioxidant therapy offers some hope. If the oxidative damage induced in ischemia-reperfusion can be reduced, the outcome for the patient can be improved. The major sources of ROS in the brain are the mitochondrial respiratory chain, xanthine oxidase (XO), and NADPH oxidase (NOX).

2.1 Mitochondrial Sources of ROS

The first observation of ROS production in mitochondrial fragments was reported by Jensen (1966) [11], while subsequent studies by Britton Chance's group established that intact mitochondria generate ROS [12]. It is now understood that healthy mitochondria convert 1–4% of consumed oxygen into superoxide. Within the mitochondrial respiratory chain, there are several centers which can in principle provide the single electron required to produce superoxide ($O_2^{\cdot -}$) from molecular oxygen (O_2). During hypoxia, the electron transport chain is generally inhibited by lack of substrate and oxygen, and it actually harbors electrons, facilitating the donation of electrons to any available oxygen to form superoxide. Furthermore, mitochondrial ROS production is thought to be subject to positive feedback, a so-called ROS-induced ROS release. This process involves mitochondrial ROS production causing oxidative stress and collapse of the mitochondrial membrane potential, thereby increasing mitochondrial ROS production and the release of ROS to neighboring mitochondria.

There is growing evidence that most of the superoxide anion generated by mammalian mitochondria in vitro is produced by complex I [13, 14], although the exact site of production is not known, with the flavin-containing subunit, the iron–sulfur center, and the semiquinone all being implicated. This superoxide production occurs primarily on the matrix side of the inner mitochondrial membrane, i.e., within the mitochondria, although it is important to bear in mind, when speaking of ROS localization, that superoxide itself has a very short half-life and is rapidly converted to H_2O_2 which can cross membranes. Interestingly, ROS production at complex I is very sensitive to depolarization of the mitochondrial membrane potential, and several authors have therefore suggested that increasing the membrane permeability to protons (mild uncoupling) could be protective. However, it has recently been shown that this intervention is not neuroprotective and does not, in fact, lower superoxide levels within the mitochondrial matrix in cerebellar granule neurons [15]. Since different experiments have been performed in different systems using different techniques, it seems the different results have not provided a clear indication of the effects of modulating the mitochondrial membrane potential on ROS production.

Complex III is also regarded as an important site of superoxide production; ROS produced at this site can spread to either side of the inner membrane [16]. Within the mitochondrial respiratory chain, ubiquinone connects complexes I and II with complex III, and its reduction products are regarded as major contributors to the formation of ROS by complex III [14, 17]. Under hypoxic conditions, the lifetime of the ubisemiquinone radical is likely to be increased, increasing the chances of superoxide formation at complex III [18]. Examination of mitochondria within intact nerve terminals suggests, however, that massive inhibition of complex III (and indeed complex IV) is required for increases in ROS production, whereas complex I is more sensitive and therefore has a lower threshold for induction of ROS production, suggesting it might represent a more important source of ROS in isolated nerve terminals under normoxic conditions [19]. However, the total ROS production capacity of complex III is greater than that of complex I.

Recently, it has been suggested that mitochondrial components outside the respiratory chain also contribute to the production of ROS. A key Krebs cycle enzyme, alpha-ketoglutarate dehydrogenase (alpha-KGDH), is also able to produce ROS and is itself sensitive to oxidative stress [20]. Although this cellular source of ROS has not been studied with respect to ischemia, its newly described properties highlight the complexity of ROS production, the variety of different sources, and the possibility of finding other (albeit minor) contributors.

2.2 Xanthine Oxidase (XO)

XO has the distinction of being expressed in cows' milk, which has made it accessible to investigation for many decades. This complex molybdoflavoenzyme is also widely expressed in the human body and particularly in the brain.

XO is the terminal enzyme of purine catabolism in mammalian cells, catalyzing the hydroxylation of hypoxanthine to xanthine and of xanthine to urate. This reaction also produces superoxide anions and H_2O_2. Interestingly, XO exists in two forms, one of which is xanthine oxidase, the other, and more prevalent and innocuous of which, is known as xanthine dehydrogenase. The conversion of XDH to XO, which is therefore pro-oxidative, is thought to be dependent on calcium and proteolysis. Although there is some suggestion that this conversion is not required for the production of ROS, it is nevertheless generally thought that during ischemia, XO becomes active, while its activity is boosted further through simultaneously increasing levels of its substrate, as ATP is catabolized to hypoxanthine.

Inhibiting XO with allopurinol has, in human trials, been shown to improve survival in severely asphyxiated infants, which is obviously encouraging [21]. Nevertheless, identifying a precise role for XO has been complicated by the pharmacology of its inhibitors, since allopurinol and oxypurinol have been shown to have direct antioxidant capacities themselves, which could prove neuroprotective through means unrelated to XO. It should also be mentioned that oxypurinol and allopurinol can have slightly different effects, since some studies have failed to find protective effects of oxypurinol [22, 23].

2.3 NAPDH Oxidase

The physiological function of NOX enzymes is to generate reactive oxygen species, which may differentiate this enzyme system from others whose ROS production abilities are often considered epiphenomenal. NOX enzymes transfer an electron across a membrane to molecular oxygen, forming superoxide which is then released and can undergo further reactions to produce other ROS. In the brain, two main isoforms of NOX are expressed, namely NOX2 and, to a much lesser degree, NOX4 and NOX5.

NOX2 is the prototype NADPH oxidase, expressed by circulating immune cells (neutrophils and macrophages) but also by microglia, astrocytes, and neurons in the brain. NOX2 is the main subunit of the enzyme complex and is also known as gp91^{phox}. NOX2 is constitutively associated with p22^{phox} within both intracellular and plasma membranes. To activate the enzyme to begin superoxide production, two cytosolic components must translocate to the membrane. It is thought that phosphorylation of p47^{phox} enables it to interact with p22^{phox} and bring the p67^{phox} subunit into contact with NOX2. The p67 is closely linked to a p40 subunit, which binds to phosphotadiylinositides in the membrane. Rac1 then interacts with NOX2 and p67^{phox}. The assembly and phosphorylation of the complex marks the start of the activation of NADPH oxidase activity.

It should be noted that much work on NADPH oxidase has focused on the enzyme in neutrophils, and, while the principles are likely to remain similar, the details may vary between cells types. The regulation of NOX activity has been linked to the Vav family of Rho GTPase guanine exchange factors, phospholipase C, protein kinase C, DAG and calcium. In terms of ischemia, the stimulus activating NADPH oxidase is not clear and could include calcium influx mediated by neurotransmitter receptors or substances released by dying neurons.

NOX has been implicated in ischemia-reperfusion injury by studies in which its contribution has been minimized and the outcome thereby improved. Mice lacking a functional NOX2 and subjected to acute middle cerebral artery occlusion (MCAO) and reperfusion have a significantly reduced infarct size compared to controls [24, 25]. In gerbils, the effects of occlusion of the common carotid arteries were reduced through administration of the NOX inhibitor apocynin [26]. Corroborating evidence of an important role for NOX is derived from the increase in NOX expression observed following ischemia-reperfusion injury. The role of NOX seems mainly linked to reperfusion; microglial ROS production was reduced during hypoxia and stimulated by reoxygenation [27]. This is consistent with a role for NOX in ROS production during reperfusion, therefore, rather than ischemia, and highlights the importance of NOX inhibitors during this later stage during which the patient is more likely to be under medical care. Overexpression of copper–zinc–superoxide disumtase (CuZnSOD) is known to reduce the infarct size following ischemia-reperfusion, but not permanent global ischemia without reperfusion.

There is the potential for NOX enzymes hosted by different cell types to contribute to the total damage. Mice lacking a functional NADPH oxidase, as described, show reduced infarct sizes, yet mice lacking NOX function in neutrophils only were not protected to the same extent, suggesting that NOX in other cell types contributes significantly to ROS production in ischemia-reperfusion [25]. As mentioned, microglia express NOX2 and respond more vigorously to NOX-activating stimuli after hypoxia–reoxygenation [27]. Endothelial cells also express NOX2, and the enzyme has been implicated in ischemic damage to the blood–brain barrier [24]. As we discuss below, NOX expressed in neurons and astrocytes is also required for ROS production during reperfusion. It should be remembered that a variety of isoforms of NOX could be expressed in the brain, for instance, NOX5 in endothelial cells and NOX4 in neurons, and more so in neurons exposed to ischemia-reperfusion [28]. While NOX enzymes have been implicated in ROS production for some time, therefore, the specific cell types and enzyme isoforms which contribute to the damage are not clearly identifiable at this stage.

Interestingly, NOX also seems to play a role in ischemic tolerance (or ischemic preconditioning), the phenomenon wherein a noxious stimulus administered to the brain affords it some protection from subsequent ischemic injury. Administration of lipopolysaccharide to mice before MCAO resulted in a reduced infarct size, and there was only a benefit to mice expressing normal NOX2 which was not seen in mice lacking a functional NOX2 subunit [29]. At this stage, exploitation of the ischemic preconditioning response has remained minimal, but it offers some insight into prevention of damage in stroke and perhaps hope for preemptive treatments in high-risk patients.

3 A More Complete Picture of ROS Production in Ischemia and Reperfusion

3.1 Temporally Distinct Sources of ROS in Ischemia/Reperfusion

Recently, we have examined the kinetics of ROS production during hypoxia and reoxygenation in primary cultures of neurons from rat hippocampus and cortex [30]. The nature and time course of these changes were somewhat unexpected; three distinct phases of ROS generation were identified, occurring with a specific temporal relationship to metabolic events taking place within the cells. We found that these three phases of ROS production in neurons were attributable to three distinguishable sources as we have described above, namely mitochondria, xanthine oxidase, and NADPH oxidase. Most interestingly, we found a role for all these three sources at different time points during ischemia and reperfusion, and demonstrated that they all, having been previously implicated independently, have an important and interactive role to play in generating ROS during hypoxia–reoxygenation.

Mitochondria increase their level of ROS generation during the first few minutes of hypoxia (Fig. 1A–B). Their ROS production ability is limited primarily by changing mitochondrial membrane potential and the availability of oxygen for ROS production. When cells were incubated with the uncoupler FCCP, which dissipates the mitochondrial membrane potential, then the initial burst of ROS production normally seen following the removal of oxygen and glucose was not observed. We examined ROS production not only throughout the cells, but also specifically in mitochondria using a mitochondrially localized ROS-sensitive fluorescent indicator (MitoSOX).

Mitochondrial ROS production is high during the initial minutes of hypoxia, and also, to a certain extent, during reperfusion. During these times, sufficient oxygen will be present and available for reduction to superoxide. During the first phase of mitochondrial ROS production following the onset of hypoxia, we envisage that ROS are generated from complexes I and III, since only complex III can release ROS to the cytosol. Using mitochondrial-specific ROS probes led to the detection mostly of ROS from complex I. In light of this, using only mitochondrially located antioxidants, such as mito-Q, is unlikely to prove sufficient to protect cells at this stage of ischemia-reperfusion injury, and other general antioxidants should also be present. The use of uncouplers is also problematic, since they have effects on energetic metabolism (Fig. 1C). It is also worth noting that, despite constant claims that mitochondria are the prevalent source of ROS in cells and contribute massively to ischemia-reperfusion, we found the contribution of the mitochondrial respiratory chain

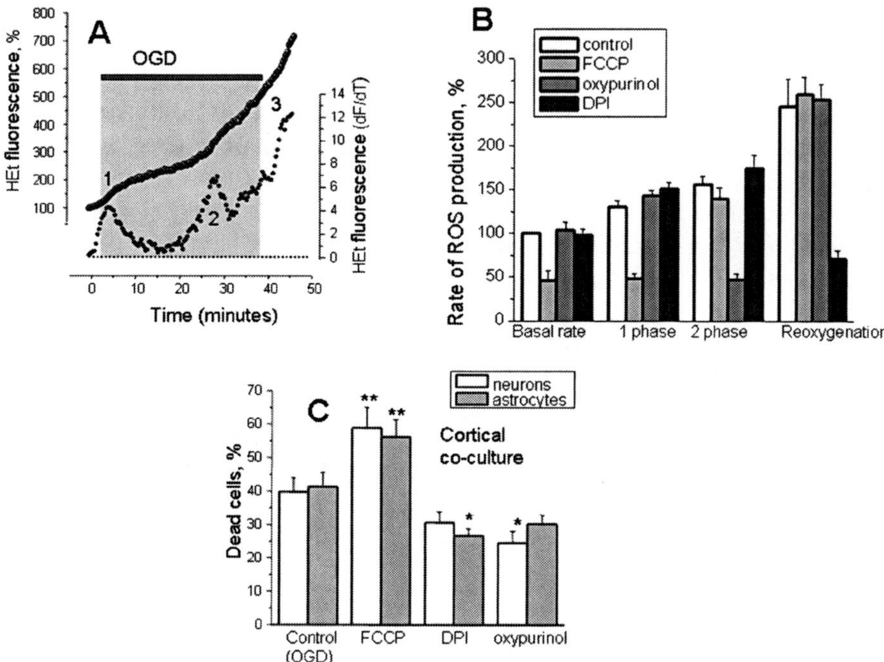

Fig. 1 A triphasic reactive oxygen species response to oxygen glucose deprivation. (A) Cultured primary cortical and hippocampal neurons with glia were exposed to hypoxia (<1mMHg) and medium without glucose at time zero and followed over time for changes in effect on reactive oxygen species generation, measured as dihydroethidium (HEt) fluorescence. *Gray* spheres show the time-dependent increase in HEt across the period of oxygen/ glucose deprivation (*gray shaded box*) and continuing across reperfusion. The *lower line* (•) shows the change in HEt per unit time, revealing three temporally distinct increases and decreases in reactive oxygen species generation. This pattern in the rate of reactive oxygen species generation strongly supports three temporally distinct components in reactive oxygen species production. (B) The triphasic pattern in rates may be altered by inhibiting the production of ROS in mitochondria with *p*-trifluoromethoxy carbonyl cyanide phenyl hydrazone (FCCP) or inhibiting xanthine oxygen ROS production with a xanthine oxidase inhibitor, oxypurinol, or inhibiting NADPH oxidase ROS production with DPI, a NADPH oxidase inhibitor, but each inhibitor only at one specific time point. (C) The effect of blocking ROS production on cell viability. Inhibiting mitochondria (FCCP) as expected increases cell death for both neurons and glia. In contrast, NADPH oxidase inhibition improves survival of glia and xanthine oxidase inhibition improves neuronal viability across oxygen glucose deprivation. Error bars represent standard error of the mean; $*p<0.05$; $**p<0.001$

as a ROS generator to be the smallest and most evasive in this study, highlighting the need for careful and quantitative experiments conducted under relevant physiological (or pathophysiological) conditions.

Xanthine oxidase, most interestingly, is able to generate ROS in the middle phase of hypoxia during which, obviously, the availability of oxygen for ROS-producing reactions is negligible. XO has long been ascribed a major role in ischemia-reperfusion injury, but only more recently has attention been focused on its role during ischemia rather than during reperfusion. We have recently demonstrated that XO is able to produce ROS during ischemia because it is activated by downstream products of ATP degradation and can in fact be inhibited following reoxygenation due to reinstatement of ATP production. An inhibitor of XO, oxypurinol, abolishes this second phase of ROS production (Fig. 1A–B). Most interestingly, oxypurinol is neuroprotective (Fig. 1C).

The third source of ROS, playing an instrumental role during reperfusion, is NADPH oxidase. The NOX enzyme seems to be activated by the calcium signal in neurons initiated by activation of glutamate receptors [30, 31]. The role of NOX during reoxygenation is consistent with its requirement for the presence of oxygen to act as an electron receiver; the enzyme requires 15 mmHg of oxygen to become functional. The oxidative stress induced by NADPH oxidase in particular, as described by ourselves and many other groups, is the most destructive, thus administration of NOX inhibitors is likely to be the most potent means of countering oxidative stress in ischemia-reperfusion injury.

4 Conclusions and Future Directions

Oxygen glucose deprivation rapidly causes a large increase in reactive oxygen species (ROS) in cultured cortical and hippocampal neurons and glia. There are three phases to the generation of ROS: mitochondrial (early deprivation), xanthine oxidase (mid-later deprivation), and NADPH oxidase (reperfusion). It therefore seems evident that several sources contribute to ROS production in brain cells during ischemia-reperfusion injury. In light of these observations, it is most probable that several classes of inhibitors and specifically localized antioxidants will be required to optimally protect brain cells in stroke. Future trials of xanthine oxidase and/or NADPH oxidase inhibition across careful time courses should be compared for effect on stroke in animal models as a translational step toward the development of effective therapies to minimize neural injury in stroke.

References

1. Feigin VL. Stroke epidemiology in the developing world. Lancet. 2005 Jun 25–Jul 1;365(9478):2160–2161.
2. Martin RL, Lloyd HG, Cowan AI. The early events of oxygen and glucose deprivation: setting the scene for neuronal death? Trends Neurosci. 1994 Jun;17(6):251–257.
3. Rossi DJ, Brady JD, Mohr C. Astrocyte metabolism and signaling during brain ischemia. Nat Neurosci. 2007 Nov;10(11):1377–1386.
4. Flamm ES, Demopoulos HB, Seligman ML, Poser RG, Ransohoff J. Free radicals in cerebral ischemia. Stroke. 1978 Sep–Oct;9(5):445–447.
5. Kinouchi H, Epstein CJ, Mizui T, Carlson E, Chen SF, Chan PH. Attenuation of focal cerebral ischemic injury in transgenic mice overexpressing CuZn superoxide dismutase. Proc Natl Acad Sci U S A. 1991 Dec 15;88(24):11158–11162.

6. Kawase M, Murakami K, Fujimura M, Morita-Fujimura Y, Gasche Y, Kondo T, Scott RW, Chan PH. Exacerbation of delayed cell injury after transient global ischemia in mutant mice with CuZn superoxide dismutase deficiency. Stroke. 1999 Sep;30(9):1962–1968.

7. Kondo T, Reaume AG, Huang TT, Carlson E, Murakami K, Chen SF, Hoffman EK, Scott RW, Epstein CJ, Chan PH. Reduction of CuZn-superoxide dismutase activity exacerbates neuronal cell injury and edema formation after transient focal cerebral ischemia. J Neurosci. 1997 Jun 1;17(11):4180–4189.

8. Crack PJ, Taylor JM, de Haan JB, Kola I, Hertzog P, Iannello RC. Glutathione peroxidase-1 contributes to the neuroprotection seen in the superoxide dismutase-1 transgenic mouse in response to ischemia/reperfusion injury. J Cereb Blood Flow Metab. 2003 Jan;23(1):19–22.

9. Eliasson MJ, Huang Z, Ferrante RJ, Sasamata M, Molliver ME, Snyder SH, Moskowitz MA. Neuronal nitric oxide synthase activation and peroxynitrite formation in ischemic stroke linked to neural damage. J Neurosci. 1999 Jul 15;19(14):5910–5918.

10. Hara H, Huang PL, Panahian N, Fishman MC, Moskowitz MA. Reduced brain edema and infarction volume in mice lacking the neuronal isoform of nitric oxide synthase after transient MCA occlusion. J Cereb Blood Flow Metab. 1996 Jul;16(4):605–611.

11. Jensen PK. Antimycin-insensitive oxidation of succinate and reduced nicotinamide-adenine dinucleotide in electron-transport particles. I. pH dependency and hydrogen peroxide formation. Biochim Biophys Acta. 1966 Aug 10;122(2):157–166.

12. Loschen G, Flohe L, Chance B. Respiratory chain linked H(2)O(2) production in pigeon heart mitochondria. FEBS Lett. 1971 Nov 1;18(2):261–264.

13. Smeitink J, van den Heuvel L. Human mitochondrial complex I in health and disease. Am J Hum Genet. 1999 Jun;64(6):1505–1510.

14. Turrens JF. Mitochondrial formation of reactive oxygen species. J Physiol. 2003 Oct 15;552(Pt 2):335–344.

15. Johnson-Cadwell LI, Jekabsons MB, Wang A, Polster BM, Nicholls DG. 'Mild Uncoupling' does not decrease mitochondrial superoxide levels in cultured cerebellar granule neurons but decreases spare respiratory capacity and increases toxicity to glutamate and oxidative stress. J Neurochem. 2007 Jun;101(6):1619–1631.

16. Muller FL, Liu Y, Van Remmen H. Complex III releases superoxide to both sides of the inner mitochondrial membrane. J Biol Chem. 2004 Nov 19;279(47):49064–49073.

17. Andreyev AY, Kushnareva YE, Starkov AA. Mitochondrial metabolism of reactive oxygen species. Biochemistry (Mosc). 2005 Feb;70(2):200–214.

18. Guzy RD, Schumacker PT. Oxygen sensing by mitochondria at complex III: the paradox of increased reactive oxygen species during hypoxia. Exp Physiol. 2006 Sep;91(5):807–819.

19. Sipos I, Tretter L, Adam-Vizi V. The production of reactive oxygen species in intact isolated nerve terminals is independent of the mitochondrial membrane potential. Neurochem Res. 2003 Oct;28(10):1575–1581.

20. Tretter L, Adam-Vizi V. Generation of reactive oxygen species in the reaction catalyzed by alpha-ketoglutarate dehydrogenase. J Neurosci. 2004 Sep 8;24(36):7771–7778.

21. Van Bel F, Shadid M, Moison RM, Dorrepaal CA, Fontijn J, Monteiro L, Van De Bor M, Berger HM. Effect of allopurinol on postasphyxial free radical formation, cerebral hemodynamics, and electrical brain activity. Pediatrics. 1998 Feb;101(2):185–193.

22. Arai T, Takeyama N, Tanaka T. Glutathione monoethyl ester and inhibition of the oxyhemoglobin-induced increase in cytosolic calcium in cultured smooth-muscle cells. J Neurosurg. 1999 Mar;90(3):527–532.

23. Nakashima M, Niwa M, Iwai T, Uematsu T. Involvement of free radicals in cerebral vascular reperfusion injury evaluated in a transient focal cerebral ischemia model of rat. Free Radic Biol Med. 1999 Mar;26(5–6):722–729.

24. Kahles T, Luedike P, Endres M, Galla HJ, Steinmetz H, Busse R, Neumann-Haefelin T, Brandes RP. NADPH oxidase plays a central role in blood-brain barrier damage in experimental stroke. Stroke. 2007 Nov;38(11):3000–3006.

25. Walder CE, Green SP, Darbonne WC, Mathias J, Rae J, Dinauer MC, Curnutte JT, Thomas GR. Ischemic stroke injury is reduced in mice lacking a functional NADPH oxidase. Stroke. 1997 Nov;28(11):2252–2258.
26. Wang Q, Tompkins KD, Simonyi A, Korthuis RJ, Sun AY, Sun GY. Apocynin protects against global cerebral ischemia-reperfusion-induced oxidative stress and injury in the gerbil hippocampus. Brain Res. 2006 May 23;1090(1):182–189.
27. Spranger M, Kiprianova I, Krempien S, Schwab S. Reoxygenation increases the release of reactive oxygen intermediates in murine microglia. J Cereb Blood Flow Metab. 1998 Jun;18(6):670–674.
28. Vallet P, Charnay Y, Steger K, Ogier-Denis E, Kovari E, Herrmann F, Michel JP, Szanto I. Neuronal expression of the NADPH oxidase NOX4, and its regulation in mouse experimental brain ischemia. Neuroscience. 2005;132(2):233–238.
29. Kunz A, Park L, Abe T, Gallo EF, Anrather J, Zhou P, Iadecola C. Neurovascular protection by ischemic tolerance: role of nitric oxide and reactive oxygen species. J Neurosci. 2007 Jul 4;27(27):7083–7093.
30. Abramov AY, Scorziello A, Duchen MR. Three distinct mechanisms generate oxygen free radicals in neurons and contribute to cell death during anoxia and reoxygenation. J Neurosci. 2007 Jan 31;27(5):1129–1138.
31. Abramov AY, Jacobson J, Wientjes F, Hothersall J, Canevari L, Duchen MR. Expression and modulation of an NADPH oxidase in mammalian astrocytes. J Neurosci. 2005 Oct 5;25(40):9176–9184.

Alzheimer Disease: Oxidative Stress and Compensatory Responses

Paula I. Moreira, Akihiko Nunomura, Xiongwei Zhu, Hyoung-Gon Lee, Gjumrakch Aliev, Mark A. Smith, and George Perry

Abstract Oxidative stress occurs early in the progression of Alzheimer disease, significantly before the development of the pathologic hallmarks, neurofibrillary tangles, and senile plaques. All classes of macromolecules are affected by oxidative stress leading, inevitably, to neuronal dysfunction. Extensive data from the literature support the notion that mitochondrial and metal abnormalities are key sources of oxidative stress in Alzheimer disease. Furthermore, it has been suggested that in the first stage of the development of Alzheimer disease, amyloid-β deposition and hyperphosphorylated tau function as compensatory responses to ensure that neuronal cells do not succumb to oxidative damage. However, during the progression of the disease, the antioxidant activity of both agents evolves into prooxidant activity, resulting in the exacerbation of reactive oxygen species production.

Keywords Alzheimer disease · amyloid β-protein · hyperphosphorylated tau protein · metals · mitochondria · oxidative stress

1 Introduction

Alzheimer disease (AD) is a progressive, degenerative brain disorder resulting in cognitive and behavioral decline and is the leading cause of dementia in the Western world. Two pathological hallmarks are observed in the brains of AD patients at autopsy: intracellular neurofibrillary tangles and extracellular senile plaques in the neocortex, hippocampus, and other subcortical regions essential for cognitive function [1]. Neurofibrillary tangles are formed from paired helical filaments composed of neurofilaments and hyperphosphorylated tau protein [2]. In turn, plaque cores are formed mostly from deposition of amyloid-β (Aβ) peptide that results from the cleavage of the amyloid-β-protein precursor (AβPP).

G. Perry (✉)
Department of Pathology, Case Western Reserve University, Cleveland, OH 44106, USA
e-mail: george.perry@utsa.edu

S.C. Veasey (ed.), *Oxidative Neural Injury*,
Contemporary Clinical Neuroscience, DOI 10.1007/978-1-60327-342-8_7,

Oxidative stress is an important issue in understanding the pathogenesis of AD. Indeed, there is accumulating evidence suggesting that oxidative stress occurs prior to the onset of symptoms in AD and oxidative damage is found not only in the vulnerable regions of the brain affected in disease [3–5] but also peripherally [6–9]. Indeed, oxidatively modified products of nucleic acids [e.g., 8-hydroxydeoxyguanosine, 8-hydroxyguanosine (8OHG)] and proteins (e.g., 3-nitrotyrosine, protein carbonyls), as well as products of lipid peroxidation [e.g., 4-hydroxynonenal (HNE), F2-isoprostane, malondialdehyde] and glycoxidation (e.g., carboxymethyllysine, pentosidine) are studied as markers for oxidative damage. Among these, many have been demonstrated in the affected lesions in postmortem brain tissue or premortem cerebrospinal fluid (CSF), plasma, serum, and urine from the patients with these diseases [10, 11]. Moreover, oxidative damage occurs before Aβ plaque formation [3]. Most recently, there have been multiple studies showing that lipid peroxidation, protein oxidation, and nucleic acid oxidation occur in mild cognitive impairment (MCI), which possibly represents a prodromal stage of AD [12–16]. The increased levels of oxidative damage in neurodegenerative conditions are often accompanied by reduced levels of antioxidant defense mechanisms in the subjects [17, 18]. Remarkably, a number of known genetic and environmental factors for neurodegenerative diseases, namely disease-specific gene mutations, risk-modifying gene polymorphisms, and risk-modifying lifestyle factors, are closely associated with oxidative damage [11, 17, 19], which implicates a pathogenic role of oxidative damage in the process of neurodegeneration.

The complex nature and genesis of oxidative damage in AD can now be partly answered by mitochondrial abnormalities and alterations in metal homeostasis that can initiate oxidative stress. Here we provide evidence that mitochondrial and metal abnormalities are key sources of oxidative stress in AD. Finally, the role of AD-specific lesions, Aβ and neurofibrillary tangles, in disease development and progression will be discussed.

2 Main Sources of Oxidative Stress in Alzheimer Disease

2.1 Mitochondria

Mitochondria are essential organelles for neuronal function because the limited glycolytic capacity of these cells makes them highly dependent on aerobic oxidative phosphorylation for their energetic needs. However, oxidative phosphorylation is a major source of endogenous toxic free radicals, including hydrogen peroxide (H_2O_2), hydroxyl ($^\bullet OH$), and superoxide ($O_2^{-\bullet}$), which are products of normal cellular respiration [20]. With inhibition of the electron transport chain, electrons accumulate in complex I and coenzyme Q, where they can be donated directly to molecular oxygen to give $O_2^{-\bullet}$ that can be detoxified by the mitochondrial manganese superoxide dismutase (MnSOD) to give H_2O_2, which, in turn, can be converted to H_2O by glutathione peroxidase (GPx). However, $O_2^{-\bullet}$ in the presence of (nitric oxide) $NO^{\bullet}$, formed during the conversion of arginine to citrulline by nitric oxide synthase (NOS), can form peroxynitrite ($ONOO^-$). Furthermore, H_2O_2 in the presence of reduced transition metals can be converted to toxic $^{\bullet}OH$ via Fenton and/or Haber Weiss reactions, a process that we have specifically localized to neurofibrillary pathology in AD [21]. Inevitably, if the amount of free radical species overwhelms the capacity of neurons to counteract these harmful species, oxidative stress occurs, followed by mitochondrial dysfunction and neuronal damage. Reactive oxygen species (ROS) generated by mitochondria have several cellular targets including mitochondrial components themselves (lipids, proteins, and DNA).

Besides the key role of mitochondria in the maintenance of cell energy and generation of free radicals, these organelles are also involved in cell death pathways, namely apoptosis. There are three main apoptotic pathways leading to the activation of caspases, which converge onto mitochondria and are mediated through members of Bcl-2 family such as Bid, Bax, and Bad [22]. The end result of each pathway is the cleavage of specific cellular substrates, resulting in the morphological and biochemical changes associated with the apoptotic phenotype. The first of these depends on the participation of mitochondria (mitochondrial pathway), the second involves the interaction of a death receptor with its ligand (death receptor pathway), and the third is triggered under conditions of endoplasmic-reticulum (ER) stress (ER-specific pathway) [23].

Previous studies from our laboratory show that the neurons showing increased oxidative damage in AD also possess a striking and significant increase in mtDNA, cytochrome oxidase, and lipoic acid [5, 24]. Surprisingly, much of the mtDNA, cytochrome oxidase, and lipoic acid are found in the neuronal cytoplasm and in the case of mtDNA and lipoic acid, in vacuoles associated with lipofuscin. We also observed an overall reduction in microtubules in AD compared to controls [25]. Altogether, these data indicated that the abnormal mitochondrial turnover, as indicated by increased perikaryal mtDNA and mitochondrial protein accumulation in the face of reduced numbers of mitochondria, could be due to a defective microtubule system resulting in deficient mitochondrial transport.

Furthermore, we analyzed the ultrastructural features of vascular lesions and mitochondria in brain vascular wall cells from human AD, yeast artificial chromosome (YAC), or C57B6/SJL transgenic-positive Tg(+) mice overexpressing AβPP. We observed a higher degree of amyloid deposition, overexpression of oxidative stress markers, mtDNA deletion, and mitochondrial structural abnormalities in the vascular walls in human AD and in both YAC and C57B6/SJL Tg(+) mice when compared to the respective controls. All the abnormalities observed occur before neuronal degeneration and amyloid deposition [26, 27].

These results indicate a clear involvement of oxidative stress, mitochondria dysfunction, and neuronal damage/death during AD evolution. In fact, an intricate interorganelle cross-talk was previously suggested by Ferri and Kroemer [22], who reviewed the participation of distinct organelles, namely the nuclei, lysosomes, ER, and Golgi, in the release of death signals that converged in mitochondria, the central executioner. To obtain more information on the involvement of mitochondria in AD pathophysiology, please see review articles by Moreira et al. [28, 29] and Zhu et al. [30].

2.2 Redox-Active Metals

The loss of homeostasis of iron and copper in the brain is accompanied by severe neurological consequences. In AD patients, overaccumulation of iron in the hippocampus, cerebral cortex, and basal nucleus of Meynert colocalizes with AD lesions, senile plaques, and neurofibrillary tangles [31]. Iron is an important cause of oxidative stress in AD because it is found in considerable amounts in the AD brain [32] and, as a transition metal, is involved in the formation of $^{\bullet}OH$ via Fenton reaction. Furthermore, it has been reported that $A\beta$ itself is a substrate for $^{\bullet}OH$ [31]. $A\beta$ extracted from postmortem AD brains presents oxidative modifications such as carbonyl adduct formation, histidine loss, and dityrosine cross-linking, making this protein less water-soluble and less susceptible to degradation by the proteases [31]. Furthermore, it has been reported that $A\beta$ deposition and $A\beta PP$ cleavage and synthesis are promoted by the presence of iron [33]. Huang and collaborators [34] presented in vitro evidence that trace levels of zinc, copper, and iron are initiators of $A\beta_{1-42}$-mediated seeding process and $A\beta$ oligomerization and these effects were abolished by chelation of trace metals. Recently, we reported that rRNA provides a binding site for redox-active iron and serves as a redox center within the cytoplasm of vulnerable neurons in AD in advance of the appearance of morphological change indicating neurodegeneration [4].

Data from our laboratory and others indicate that heme oxygenase-1 (HO-1) is induced in the brain during the development of AD [35, 36]. HO-1 catalyzes the conversion of heme to biliverdin and iron, and biliverdin, in turn, is reduced to bilirubin, an antioxidant. Since HO-1 is induced in proportion to the level of heme [37], the induction of HO-1 suggests that there may be abnormal turnover of heme in AD. This idea is consistent with the mitochondrial abnormalities associated with AD since it is well known that many heme-containing enzymes are found in mitochondria. As previously discussed, our ultrastructural studies suggest a high rate of mitochondrial turnover and accumulation of iron in the residual body of lysosomes [5, 24].

Furthermore, we observed that oxidized nucleic acids are commonly observed in the cytoplasm of the neurons that are particularly vulnerable to degeneration in AD [3]. 8OHG, a marker of nucleic acid oxidation, is likely to form at the site of $^{\bullet}OH$ production, a process dependent on redox-active, metal-catalyzed reduction of H_2O_2 together with cellular reductants such as ascorbate or $O_2^{-\bullet}$. Interestingly, the levels of 8OHG are inversely related to the extent of Aβ deposits although this oxidative marker is found distant from the Aβ deposits [3], suggesting a complex interplay between Aβ and redox metal activity that may be critical to metal dynamics within the neuronal cytoplasm. A possible key element to these dynamics is mitochondria in the neuronal cell body.

Copper can also participate in the Fenton reaction to generate ROS [38, 39]. Although conflicting results exist concerning the amount of copper and the formation of senile plaques [40, 41], there is accumulating evidence that both iron and copper in their redox-competent states are bound to neurofibrillary tangles and Aβ deposits [21, 32, 42]. However, a recent study reported that cognitive decline correlates with low plasma concentrations of copper in patients with mild-to-moderate AD [43]. To obtain more information on the involvement of metals in AD pathophysiology, please see the review articles by Adlard and Bush [44] and Zhu et al. [30].

3 Alzheimer Disease Lesions: Compensatory Responses?

3.1 Amyloid-β peptide

It is now over two decades since Aβ was first sequenced and recognized as a potential marker of AD [45]. Soon after this, a 39–43 amino-acid peptide was identified as the major component of the senile plaque, one of the hallmarks of AD. Since then a wealth of academic and commercial research has been aimed at understanding where this peptide comes from because many believe that if its production is stopped, the development of AD may be prevented. Yet, it has been shown that this peptide is present in the CSF and plasma of healthy individuals throughout life [46, 47]. Kamenetz and collaborators [48] reported that Aβ is secreted from healthy neurons in response to activity and that Aβ, in turn, downregulates excitatory synaptic transmission. This negative feedback loop provides a physiological homeostatic mechanism aimed to maintain normal levels of neuronal activity. More recently, Lesne and collaborators [49] reported that the specific stimulation of NMDA receptors upregulates AβPP, inhibits α-secretase activity, and promotes Aβ production. Together these studies argue strongly that AβPP processing and the presence of Aβ itself are closely associated with synaptic activity and may serve to provide physiological control of activity, guarding against excessive glutamate release. Previous studies also suggest that Aβ might act as a regulator of ion channel function in neurons [50, 51]. Furthermore, it has been shown that neurons respond to oxidative stress, both in vitro and in vivo, by increasing Aβ production [52–54].

An antioxidant role for Aβ in vivo is in agreement with recent data on the distribution of oxidative damage to AD neurons. As previously discussed, 8OHG markedly accumulates in the cytoplasm of cerebral neurons in AD. Unexpectedly, an increase in Aβ deposition in AD cortex is associated with a decrease in neuronal levels of 8OHG, i.e., with decreased oxidative damage [55]. Similar negative correlation between Aβ deposition and oxidative damage is found in patients with Down syndrome [56]. Aβ deposits observed in both studies mainly consist of early diffuse plaques, meaning that these diffuse amyloid plaques may be considered as a compensatory response that reduces oxidative stress [57]. The strong chelating properties of Aβ for zinc, iron, and copper explain the reported enrichment of these metals in amyloid plaques in AD [32] and suggest that one function of Aβ is to sequester these metal ions. Chelation of transition metals in a redox-inactive form may theoretically serve to inhibit metal-catalyzed oxidation of biomolecules. Methionine at residue 35 on Aβ sequence can both scavenge free radicals [58–60] and reduce metals to their high-active, low-valence form [61], thereby possessing both anti- and prooxidative properties. As discussed above, reduced metal ions are highly active oxidants and can catalyze further oxidation of biomolecules. These data indicate that Aβ is a lipophilic metal chelator with a metal-reducing activity.

The redox properties of Aβ indicate that it could function as both an antioxidant and a prooxidant under specific conditions. However, the conditions under which Aβ ceases to act as an antioxidant and functions as a pro-oxidant are not clearly understood, although several lines of evidence indicate that the activity is dependent on the concentration of the peptide. The concentrations of Aβ required to induce toxicity in vitro are in the micromolar range [62]. Indeed, it has been shown that Aβ induces peroxidation of membrane lipids [63], generates H_2O_2 [64] and HNE [65] in neurons, damages DNA [66], and inactivates transport enzymes [67]. In CSF where Aβ has been reported to act as an antioxidant, the concentration of the peptide is between 0.1 and 1 nM, while at higher concentrations this activity is ablated [68]. Similarly, antiapoptotic activity is observed only at nanomolar levels and conversely at higher concentrations it is toxic [69].

The presence of transition metals and methionine on residue 35 is necessary for Aβ to induce oxidation fibrillation. The presence of transition metals is a requisite for Aβ aggregation and its prooxidative activity [70, 71]. The toxicity of Aβ is likely to be mediated by a direct interaction between this peptide and transition metals with subsequent generation of ROS [38]. Another factor essential for the prooxidative activity of Aβ seems to be the presence of methionine on residue 35. It has been demonstrated that the substitution of this residue by another amino acid abrogates or significantly diminishes the prooxidant action of Aβ [72, 73]. Methionine 35 can scavenge free radicals [74] and reduce transition metals to their high-active, low-valence form [75], thereby exhibiting both anti- and prooxidative properties.

3.2 Hyperphosphorylated tau protein

In the adult human brain, tau proteins are found essentially in neurons. They bind microtubules through the microtubule-binding domains and this assembly depends partially upon the degree of phosphorylation, since hyperphosphory-lated tau is less effective than hypophosphorylated tau on microtubule poly-merization [76]. Besides the role in microtubule stabilization, tau has other functions such as membrane interactions or anchoring of enzymes [77, 78]. Among the 80 Ser/Thr residues on tau, at least 30 phosphorylation sites have been described, most of which occur on Ser-Pro and Thr-Pro motives. In fact, phosphorylation of Ser262, located in the first microtubule-binding domain, dramatically reduces the affinity of tau for microtubules in vitro [79]. Never-theless, this site alone is insufficient to abolish tau binding to microtubules. Thus, phosphorylation outside the microtubule-binding domains may also strongly influence tubulin assembly by modifying the affinity between tau and microtubules. By regulating microtubule assembly, tau has a role in modulating the functional organization of the neuron, particularly in axonal morphology, growth, and polarity [76]. In AD, hyperphosphorylated tau accumulates in neurons, aggregates into paired helical filaments, and loses its microtubule-binding and stabilizing function, leading to neuronal degeneration [80]. The abnormal phosphorylation of tau associated with AD may be related to either an increase in kinase activity (glycogen synthase kinase 3β, cyclin-dependent kinase-5, p42/44 MAP kinase, p38 MAPK, stress-activated protein kinases, mitotic protein kinases) or a decrease in phosphatase activity (protein phos-phatases 1, 2α, 2β) [76, 79, 81–83]. However, there is evidence indicating that hyperphosphorylated tau exerts protective functions. It has been shown that oxidative stress and the modification of tau by products of oxidative stress [84] lead to protein aggregation (formation of neurofibrillary tangles) and enable neurons to survive for decades [85]. Intriguingly, although neurofilament heavy subunit has a long half-life, the same extent of carbonyl modification is found throughout the normal aging process as well as along the length of the axon [86], suggesting that oxidative stress-modified molecules are under tight regulation. Neurofilament and tau proteins appear adapted to oxidative stress due to their high content of lysine–serine–proline (KSP) domains [86]. So, these molecules may work as a buffer by absorbing lipoxidation-derived and glycoxidation-derived aldehydes. Furthermore, transfection studies of neuroblastoma cells showed that HO-1 expression is coordinated with tau phosphorylation suggest-ing a physiological interaction between cytoskeleton reorganization and oxida-tion [84]. In addition, the coordination between the enzymatic activity of HO-1 to produce iron and tau to bind iron [42] seems to be related to iron home-ostasis. These results clearly implicate a regulated process for modifications and response. Changes such as MAP kinase and HO-1 may be a few of many responses that interrelate to lipid peroxidative modification [87, 88]. While more studies are required to understand the role of these oxidative modifica-tions in neuronal homeostasis, it is tempting to consider them as augmentations to the neuronal defenses important in protecting the axon. The slow turnover rate of proteins in the axon, which can take years, may necessitate this protection.

4 Conclusions

Oxidative stress plays a major role in the development and progression of AD. Extensive evidence suggests ROS-mediated oxidative damage to proteins, lipids, nucleic acids, and sugars in AD and this damage results from extensive mitochondrial and metal abnormalities. Data from the literature support the notion that the oxidative modifications that occur in AD may elicit compensatory mechanisms, such as Aβ deposition and hyperphosphorylated tau that try to restore the redox balance in an attempt to avoid neuronal death. However, with the progression of AD and the consequent increase of ROS, efficient removal of Aβ–metal complexes and hyperphosphorylated tau would be overtaken by their disproportionably high generation, resulting in an uncontrollable growth of plaques and neurofibrillary tangles and, consequently, an increase in ROS generation.

Acknowledgments This research was supported by the National Institutes of Health, Alzheimer's Association, and Philip Morris USA Inc. and Philip Morris International.

References

1. Mattson MP. Pathways towards and away from Alzheimer's disease. Nature 2004;430:631–639.
2. Perry G, Rizzuto N, Autilio-Gambetti L, Gambetti P. Paired helical filaments from Alzheimer disease patients contain cytoskeletal components. Proc Natl Acad Sci U S A 1985;82:3916–3920.
3. Nunomura A, Perry G, Aliev G, et al. Oxidative damage is the earliest event in Alzheimer disease. J Neuropathol Exp Neurol 2001;60:759–767.
4. Honda K, Smith MA, Zhu X, et al. Ribosomal RNA in Alzheimer disease is oxidized by bound redox-active iron. J Biol Chem 2005;280:20978–20986.
5. Moreira PI, Siedlak SL, Wang X, et al. Autophagocytosis of mitochondria is prominent in Alzheimer disease. J Neuropathol Exp Neurol 2007;66:525–532.
6. Perry G, Castellani RJ, Smith MA, et al. Oxidative damage in the olfactory system in Alzheimer's disease. Acta Neuropathol (Berl) 2003;106:552–556.
7. Ghanbari HA, Ghanbari K, Harris PL, et al. Oxidative damage in cultured human olfactory neurons from Alzheimer's disease patients. Aging Cell 2004; 3: 41–44.
8. Migliore L, Fontana I, Trippi F, et al. Oxidative DNA damage in peripheral leukocytes of mild cognitive impairment and AD patients. Neurobiol Aging 2005;26:567–573.
9. Moreira PI, Harris PL, Zhu X, et al. Lipoic acid and N-acetyl cysteine decrease mitochondrial-related oxidative stress in Alzheimer disease patient fibroblasts. J Alzheimers Dis 2007;12:195–206.
10. Sayre LM, Smith MA, Perry G. Chemistry and biochemistry of oxidative stress in neurodegenerative disease. Curr Med Chem 2001;8:721–738.
11. Nunomura A, Castellani RJ, Zhu X, Moreira PI, Perry G, Smith MA. Involvement of oxidative stress in Alzheimer disease. J Neuropathol Exp Neurol 2006;65:631–641.
12. Butterfield DA, Poon HF, St Clair D, et al. Redox proteomics identification of oxidatively modified hippocampal proteins in mild cognitive impairment: insights into the development of Alzheimer's disease. Neurobiol Dis 2006;22:223–232.

13. Ding Q, Markesbery WR, Chen Q, Li F, Keller JN. Ribosome dysfunction is an early event in Alzheimer's disease. J Neurosci 2005;25:9171–9175.

14. Keller JN, Schmitt FA, Scheff SW, et al. Evidence of increased oxidative damage in subjects with mild cognitive impairment. Neurology 2005;64:1152–1156.

15. Markesbery WR, Kryscio RJ, Lovell MA, Morrow JD. Lipid peroxidation is an early event in the brain in amnestic mild cognitive impairment. Ann Neurol 2005;58:730–735.

16. Wang J, Markesbery WR, Lovell MA. Increased oxidative damage in nuclear and mitochondrial DNA in mild cognitive impairment. J Neurochem 2006;96:825–832.

17. Ischiropoulos H, Beckman JS. Oxidative stress and nitration in neurodegeneration: cause, effect, or association? J Clin Invest 2003;111:163–169.

18. Andersen JK. Oxidative stress in neurodegeneration: cause or consequence? Nat Med 2004;10 Suppl:S18–S25.

19. Barnham KJ, Masters CL, Bush AI. Neurodegenerative diseases and oxidative stress. Nat Rev Drug Discov 2004;3:205–214.

20. Wallace DC. Mitochondrial diseases in man and mouse. Science 1999;283:1482–1488.

21. Smith MA, Harris PL, Sayre LM, Perry G. Iron accumulation in Alzheimer disease is a source of redox-generated free radicals. Proc Natl Acad Sci U S A 1997;94:9866–9868.

22. Ferri KF, Kroemer G. Organelle-specific initiation of cell death pathways. Nat Cell Biol 2001;3:E255–E263.

23. Pereira C, Ferreiro E, Cardoso SM, de Oliveira CR. Cell degeneration induced by amyloid-beta peptides: implications for Alzheimer's disease. J Mol Neurosci 2004;23:97–104.

24. Hirai K, Aliev G, Nunomura A, et al. Mitochondrial abnormalities in Alzheimer's disease. J Neurosci 2001;21:3017–3023.

25. Cash AD, Aliev G, Siedlak SL, et al. Microtubule reduction in Alzheimer's disease and aging is independent of tau filament formation. Am J Pathol 2003;162:1623–1627.

26. Aliev G, Smith MA, de la Torre JC, Perry G. Mitochondria as a primary target for vascular hypoperfusion and oxidative stress in Alzheimer's disease. Mitochondrion 2004;4:649–663.

27. Aliyev A, Chen SG, Seyidova D, et al. Mitochondria DNA deletions in atherosclerotic hypoperfused brain microvessels as a primary target for the development of Alzheimer's disease. J Neurol Sci 2005;229–230:285–292.

28. Moreira PI, Cardoso SM, Santos MS, Oliveira CR. The key role of mitochondria in Alzheimer's disease. J Alzheimers Dis 2006;9:101–110.

29. Moreira PI, Santos MS, Oliveira CR. Alzheimer's disease: a lesson from mitochondrial dysfunction. Antioxid Redox Signal 2007;9:1621–1630.

30. Zhu X, Su B, Wang X, Smith MA, Perry G. Causes of oxidative stress in Alzheimer disease. Cell Mol Life Sci 2007;64:2202–2210.

31. Connor JR, Milward EA, Moalem S, et al. Is hemochromatosis a risk factor for Alzheimer's disease? J Alzheimers Dis 2001;3:471–477.

32. Lovell MA, Robertson JD, Teesdale WJ, Campbell JL, Markesbery WR. Copper, iron and zinc in Alzheimer's disease senile plaques. J Neurol Sci 1998;158:47–52.

33. Rogers JT, Randall JD, Cahill CM, et al. An iron-responsive element type II in the 5′-untranslated region of the Alzheimer's amyloid precursor protein transcript. J Biol Chem 2002;277:45518–45528.

34. Huang X, Moir RD, Tanzi RE, Bush AI, Rogers JT. Redox-active metals, oxidative stress, and Alzheimer's disease pathology. Ann N Y Acad Sci 2004;1012:153–163.

35. Smith MA, Kutty RK, Richey PL, et al. Heme oxygenase-1 is associated with the neurofibrillary pathology of Alzheimer's disease. Am J Pathol 1994;145:42–47.

36. Premkumar DR, Smith MA, Richey PL, et al. Induction of heme oxygenase-1 mRNA and protein in neocortex and cerebral vessels in Alzheimer's disease. J Neurochem 1995;65:1399–1402.

37. Keyse SM, Tyrrell RM. Heme oxygenase is the major 32-kDa stress protein induced in human skin fibroblasts by UVA radiation, hydrogen peroxide, and sodium arsenite. Proc Natl Acad Sci U S A 1989;86:99–103.

38. Huang X, Atwood CS, Hartshorn MA, et al. The A beta peptide of Alzheimer's disease directly produces hydrogen peroxide through metal ion reduction. Biochemistry (Mosc) 1999;38:7609–7616.
39. Finefrock AE, Bush AI, Doraiswamy PM. Current status of metals as therapeutic targets in Alzheimer's disease. J Am Geriatr Soc 2003;51:1143–1148.
40. Bayer TA, Schafer S, Simons A, et al. Dietary Cu stabilizes brain superoxide dismutase 1 activity and reduces amyloid Abeta production in APP23 transgenic mice. Proc Natl Acad Sci U S A 2003;100:14187–14192.
41. Sparks DL, Schreurs BG. Trace amounts of copper in water induce beta-amyloid plaques and learning deficits in a rabbit model of Alzheimer's disease. Proc Natl Acad Sci U S A 2003;100:11065–11069.
42. Sayre LM, Perry G, Harris PL, Liu Y, Schubert KA, Smith MA. In situ oxidative catalysis by neurofibrillary tangles and senile plaques in Alzheimer's disease: a central role for bound transition metals. J Neurochem 2000;74:270–279.
43. Pajonk FG, Kessler H, Supprian T, et al. Cognitive decline correlates with low plasma concentrations of copper in patients with mild to moderate Alzheimer's disease. J Alzheimers Dis 2005;8:23–27.
44. Adlard PA, Bush AI. Metals and Alzheimer's disease. J Alzheimers Dis 2006;10:145–163.
45. Glenner GG, Wong CW. Alzheimer's disease: initial report of the purification and characterization of a novel cerebrovascular amyloid protein. Biochem Biophys Res Commun 1984;120:885–890.
46. Seubert P, Vigo-Pelfrey C, Esch F, et al. Isolation and quantification of soluble Alzheimer's beta-peptide from biological fluids. Nature 1992;359:325–327.
47. Shoji M, Golde TE, Ghiso J, et al. Production of the Alzheimer amyloid beta protein by normal proteolytic processing. Science 1992;258:126–129.
48. Kamenetz F, Tomita T, Hsieh H, et al. APP processing and synaptic function. Neuron 2003;37:925–937.
49. Lesne S, Ali C, Gabriel C, et al. NMDA receptor activation inhibits alpha-secretase and promotes neuronal amyloid-beta production. J Neurosci 2005;25:9367–9377.
50. Ramsden M, Henderson Z, Pearson HA. Modulation of Ca2+ channel currents in primary cultures of rat cortical neurones by amyloid beta protein (1–40) is dependent on solubility status. Brain Res 2002;956:254–261.
51. Ramsden M, Plant LD, Webster NJ, Vaughan PF, Henderson Z, Pearson HA. Differential effects of unaggregated and aggregated amyloid beta protein (1–40) on K(+) channel currents in primary cultures of rat cerebellar granule and cortical neurones. J Neurochem 2001;79:699–712.
52. Yan SD, Yan SF, Chen X, et al. Non-enzymatically glycated tau in Alzheimer's disease induces neuronal oxidant stress resulting in cytokine gene expression and release of amyloid beta-peptide. Nat Med 1995;1:693–699.
53. Tamagno E, Parola M, Bardini P, et al. Beta-site APP cleaving enzyme up-regulation induced by 4-hydroxynonenal is mediated by stress-activated protein kinases pathways. J Neurochem 2005;92:628–636.
54. van Groen T, Puurunen K, Maki HM, Sivenius J, Jolkkonen J. Transformation of diffuse beta-amyloid precursor protein and beta-amyloid deposits to plaques in the thalamus after transient occlusion of the middle cerebral artery in rats. Stroke 2005;36:1551–1556.
55. Nunomura A, Perry G, Pappolla MA, et al. RNA oxidation is a prominent feature of vulnerable neurons in Alzheimer's disease. J Neurosci 1999;19:1959–1964.
56. Nunomura A, Perry G, Pappolla MA, et al. Neuronal oxidative stress precedes amyloid-beta deposition in Down syndrome. J Neuropathol Exp Neurol 2000;59:1011–1017.
57. Smith MA, Rottkamp CA, Nunomura A, Raina AK, Perry G. Oxidative stress in Alzheimer's disease. Biochim Biophys Acta 2000;1502:139–144.
58. Unnikrishnan MK, Rao MN. Antiinflammatory activity of methionine, methionine sulfoxide and methionine sulfone. Agents Actions 1990;31:110–112.

59. Nakamura M, Shishido N, Nunomura A, et al. Three histidine residues of amyloid-beta peptide control the redox activity of copper and iron. Biochemistry (Mosc) 2007;46:12737–12743.
60. Hayashi T, Shishido N, Nakayama K, et al. Lipid peroxidation and 4-hydroxy-2-nonenal formation by copper ion bound to amyloid-beta peptide. Free Radic Biol Med 2007;43:1552–1559.
61. Hiller KO, Asmus KD. Tl2+ and Ag2+ metal-ion-induced oxidation of methionine in aqueous solution. A pulse radiolysis study. Int J Radiat Biol Relat Stud Phys Chem Med 1981;40:597–604.
62. Simmons MA, Schneider CR. Amyloid beta peptides act directly on single neurons. Neurosci Lett 1993;150:133–136.
63. Varadarajan S, Yatin S, Aksenova M, Butterfield DA. Review: Alzheimer's amyloid beta-peptide-associated free radical oxidative stress and neurotoxicity. J Struct Biol 2000;130:184–208.
64. Behl C, Davis JB, Lesley R, Schubert D. Hydrogen peroxide mediates amyloid beta protein toxicity. Cell 1994;77:817–827.
65. Mark RJ, Lovell MA, Markesbery WR, Uchida K, Mattson MP. A role for 4-hydroxynonenal, an aldehydic product of lipid peroxidation, in disruption of ion homeostasis and neuronal death induced by amyloid beta-peptide. J Neurochem 1997;68:255–264.
66. Xu J, Chen S, Ahmed SH, et al. Amyloid-beta peptides are cytotoxic to oligodendrocytes. J Neurosci 2001;21:RC118.
67. Mark RJ, Pang Z, Geddes JW, Uchida K, Mattson MP. Amyloid beta-peptide impairs glucose transport in hippocampal and cortical neurons: involvement of membrane lipid peroxidation. J Neurosci 1997;17:1046–1054.
68. Kontush A, Berndt C, Weber W, et al. Amyloid-beta is an antioxidant for lipoproteins in cerebrospinal fluid and plasma. Free Radic Biol Med 2001;30:119–128.
69. Chan C-W, Dharmarajan A, Atwood CS, et al. Anti-apoptotic action of Alzheimer Abeta. Alzheimers Rep 1999;2:1–6.
70. Bondy SC, Guo-Ross SX, Truong AT. Promotion of transition metal-induced reactive oxygen species formation by beta-amyloid. Brain Res 1998;799:91–96.
71. Schubert D, Chevion M. The role of iron in beta amyloid toxicity. Biochem Biophys Res Commun 1995;216:702–707.
72. Walter MF, Mason PE, Mason RP. Alzheimer's disease amyloid beta peptide 25–35 inhibits lipid peroxidation as a result of its membrane interactions. Biochem Biophys Res Commun 1997;233:760–764.
73. Butterfield DA, Bush AI. Alzheimer's amyloid beta-peptide (1–42): involvement of methionine residue 35 in the oxidative stress and neurotoxicity properties of this peptide. Neurobiol Aging 2004;25:563–568.
74. Soriani M, Pietraforte D, Minetti M. Antioxidant potential of anaerobic human plasma: role of serum albumin and thiols as scavengers of carbon radicals. Arch Biochem Biophys 1994;312:180–188.
75. Lynch SM, Frei B. Physiological thiol compounds exert pro- and anti-oxidant effects, respectively, on iron- and copper-dependent oxidation of human low-density lipoprotein. Biochim Biophys Acta 1997;1345:215–221.
76. Buee L, Bussiere T, Buee-Scherrer V, Delacourte A, Hof PR. Tau protein isoforms, phosphorylation and role in neurodegenerative disorders. Brain Res Brain Res Rev 2000;33:95–130.
77. Brandt R, Leger J, Lee G. Interaction of tau with the neural plasma membrane mediated by tau's amino-terminal projection domain. J Cell Biol 1995;131:1327–1340.
78. Sontag E, Nunbhakdi-Craig V, Lee G, et al. Molecular interactions among protein phosphatase 2A, tau, and microtubules. Implications for the regulation of tau phosphorylation and the development of tauopathies. J Biol Chem 1999;274:25490–25498.

79. Hamdane M, Delobel P, Sambo AV, et al. Neurofibrillary degeneration of the Alzheimer-type: an alternate pathway to neuronal apoptosis? Biochem Pharmacol 2003;66:1619–1625.
80. Garcia ML, Cleveland DW. Going new places using an old MAP: tau, microtubules and human neurodegenerative disease. Curr Opin Cell Biol 2001;13:41–48.
81. Brion JP, Anderton BH, Authelet M, et al. Neurofibrillary tangles and tau phosphorylation. Biochem Soc Symp 2001;67:81–88.
82. Geschwind DH. Tau phosphorylation, tangles, and neurodegeneration: the chicken or the egg? Neuron 2003;40:457–460.
83. Liu F, Liang Z, Gong CX. Hyperphosphorylation of tau and protein phosphatases in Alzheimer disease. Panminerva Med 2006;48:97–108.
84. Takeda A, Smith MA, Avila J, et al. In Alzheimer's disease, heme oxygenase is coincident with Alz50, an epitope of tau induced by 4-hydroxy-2-nonenal modification. J Neurochem 2000;75:1234–1241.
85. Morsch R, Simon W, Coleman PD. Neurons may live for decades with neurofibrillary tangles. J Neuropathol Exp Neurol 1999;58:188–197.
86. Wataya T, Nunomura A, Smith MA, et al. High molecular weight neurofilament proteins are physiological substrates of adduction by the lipid peroxidation product hydroxynonenal. J Biol Chem 2002;277:4644–4648.
87. Zhu X, Rottkamp CA, Boux H, Takeda A, Perry G, Smith MA. Activation of p38 kinase links tau phosphorylation, oxidative stress, and cell cycle-related events in Alzheimer disease. J Neuropathol Exp Neurol 2000;59:880–888.
88. Zhu X, Rottkamp CA, Hartzler A, et al. Activation of MKK6, an upstream activator of p38, in Alzheimer's disease. J Neurochem 2001;79:311–318.

Oxidative Stress Associated Signal Transduction Cascades in Alzheimer Disease

Robert B. Petersen, Akihiko Nunomura, Hyoung-gon Lee, Gemma Casadesus, George Perry, Mark A. Smith, and Xiongwei Zhu

Abstract An unfortunate consequence of high metabolism in the brain is the age-related increase in oxidative stress observed. Multiple lines of evidence indicate that oxidative stress is one of the earliest events in Alzheimer disease, occurring before the development of plaques and tangles. The large number of metabolic signs of oxidative stress and markers of oxidative damage suggest that oxidative stress likely plays a key pathogenic role in the disease and is clearly involved in the cell loss and other neuropathology associated with Alzheimer disease. However, although long-lived markers of oxidative damage persist throughout the disease, the levels of rapidly turned over markers of oxidative damage, i.e., oxidized nucleic acids, which are initially elevated, decrease as the disease progresses to advanced Alzheimer disease indicating that oxidative stress decreases with disease progression. Thus, the initial burst of reactive oxygen species not only results in damage to cellular structures but also engenders a cellular response(s), i.e., the compensatory up-regulation of antioxidant enzymes found in vulnerable neurons in Alzheimer disease. In addition, oxidative stress also stimulates the stress-activated protein kinase pathways, which are extensively activated during Alzheimer disease. In this chapter, we review the evidence of oxidative stress and compensatory responses in Alzheimer disease and conclude with a focus mechanism of activation of stress-activated protein kinase pathways and the role of this pathway in the disease process.

Keywords Alzheimer disease · Heme oxygenase · Mitochondria · Oxidative stress · Stress-activated protein kinase (SAPK) · Transition metals

R.B. Petersen (✉)
Department of Pathology, Case Western Reserve University,
Cleveland, OH 44106, USA
e-mail: robert.petersen@case.edu

S.C. Veasey (ed.), *Oxidative Neural Injury*,
Contemporary Clinical Neuroscience, DOI 10.1007/978-1-60327-342-8_8,
© Humana Press, a part of Springer Science+Business Media, LLC 2009

1 Introduction

All aerobic organisms produce free radicals, as a direct consequence of energy production, predominantly in the form of superoxide, a side product formed during the reduction of molecular oxygen by mitochondria. The average cell utilizes 10^{13} O_2 molecules per day, and of these 10^{13} O_2 molecules, approximately 1% will form O_2^-. As a result, the average cell produces 10^{11} free radical species in a day. Neuronal cells, due to their heightened oxygen metabolism, undoubtedly generate more. This suggestion is supported by the observation that although the brain constitutes only 2–3% of total body mass, 20% of the basal oxygen supplied to the body is utilized in the brain. Normally mitochondria sequester most of the radicals produced; however, the likelihood of untoward oxidative damage is exacerbated by age and metabolic demand ultimately contributing to disease conditions such as Alzheimer disease (AD).

Cerebral metabolism is reduced in AD and impaired mitochondrial function has been implicated as the cause [1–3]. The reduced activity of specific mitochondrial enzyme complexes in AD such as cytochrome oxidase (COX), the pyruvate dehydrogenase complex (PDHC), and the α-ketoglutarate dehydrogenase complex (KGDHC) are reflected in the decrease in metabolism [2, 4–9]. The reduction in COX activity was demonstrated to be the result of the complex losing a kinetically identifiable site for reduced cytochrome c [9]. Most importantly, the alteration of these key enzymes may favor the increased production of reactive oxygen species (ROS). Loss of COX activity, the component of the mitochondrial electron transport chain that directly interacts with molecular oxygen, may lead to increased side-production of superoxide and that could in turn back up electrons at the complex III site resulting in elevated formation of ROS in mitochondria. Further, detoxifying ROS requires electrons provided by a functional Krebs tricarboxylic acid cycle for their chemical resolution. Thus, reduced efficiency of the Krebs cycle as the result of the deficiency in PDHC and KGDHC favors the production of ROS. Although mitochondria are significant producers of ROS they do not themselves exhibit striking evidence of oxidative damage [10]. For example, oxidative damage to nucleic acids is primarily limited to the cytoplasm of susceptible neurons in AD using 8OHG as a marker for nucleic acid oxidation [10, 11]. Thus, mitochondria in AD likely supply a key reactant that, once in the cytoplasm, produces free radicals resulting in cellular damage. H_2O_2, which can freely diffuse across the outer membrane of the mitochondria, is a likely potential candidate. H_2O_2 in the cytoplasm or other cellular compartments, which have a lower ability to resolve free radicals than the mitochondria, provides the potential for increased free radical damage [12]. Although H_2O_2 itself does not pose a significant threat to the cell, because it is not highly reactive, it is toxic at high levels. Of relevance to neurodegeneration, H_2O_2 can participate in chemical reactions producing highly reactive radicals that are able to modify cellular macromolecules.

The most significant reaction involving H_2O_2 in the cytoplasm is that with redox-active iron that results in the production of the highly damaging hydroxyl radical, •OH, via the Fenton reaction. In AD there are conspicuous changes in iron distribution localized to the cytoplasm of vulnerable neurons in AD; in addition, Aβ deposits and neurofibrillary tangles (NFT) contain redox-active iron, which mirrors the site of oxidative damage [13]. Dysregulation of cellular iron metabolism in AD results in impaired iron homeostasis. IRP-1 and IRP-2 are the two major regulatory proteins involved in iron metabolism in the central nervous system. IRP-2 is increased in AD and is found in association with the pathologic hallmarks of AD [14]. Ferritin is decreased in the AD brain while the iron concentration is increased, which would lead to an increase in free iron [15, 16]. Ferritin expression is regulated by the binding of IRPs to a conserved RNA structure, termed the iron-responsive element (IRE), found in the ferritin mRNA. Binding to the mRNA can repress translation, 5' end of the message, or enhance mRNA stability, 3' end of the message, thereby regulating the amount of ferritin produced in the cell. The disruption observed in iron homeostasis in AD seems to be linked to alterations in the IRP/IRE interaction [14, 17]. Another source of increased ROS production is provided by the alteration in copper homeostasis found in AD [18], but in this case the alteration is specifically associated with Aβ deposits.

A final category of ROS producers is the lesion-associated proteins, such as Aβ, found in AD. Aβ-mediated neuronal damage appears to be caused by free radical damage to membranes and, as such, can be attenuated using antioxidants such as vitamin E [19, 20] or catalase [21]. Aβ-induced lipid peroxidation results in the production of several reactive aldehydes including HNE that are capable of modifying membrane proteins and changing their function which in turn leads to deleterious consequences to neuronal cells. For example, Aβ-induced HNE production impairs the glutamate transporter, the glucose transporter, mitochondrial function in synaptosomes, and disrupts ion homeostasis [22–26]. Fibrillar Aβ generates ROS by binding to the receptor for advanced glycation end products (RAGE) which initiates an oxidative inflammatory response [27]. Senile plaques activate microglia, the macrophages of the brain, which produce free radicals [28]. Lastly, NFT and senile plaques contain abundant iron [29], which, as described above, is critical to the initiation of free radical formation. Indeed, the cell free generation of H_2O_2 by Aβ resulting in its neurotoxicity is mediated by the peptide-binding catalytic amounts of Cu^{2+} [30] or by its direct interaction with iron [31]. The latter observation led us to consider Aβ as redox-active with both prooxidant (in vitro) and antioxidant (in vivo) activities [10, 32, 33].

2 Evidence of Oxidative Stress in Alzheimer Disease

As detailed above, free radical generation is an inherent property of the high metabolism found in neurons and the potential for oxidative damage is increased with age. Free radicals produced during oxidative stress have been postulated to be pathologically important in the development of Alzheimer disease (AD) and other neurodegenerative diseases [34, 35]. Oxidative modification is observed in virtually all classes of biomacromolecules in the susceptible neurons of AD. First, DNA and RNA are oxidized as shown by the increased levels of 8-hydroxyl-2'-deoxyguanosine [8OHdG) and 8-hydroxyguanosine [8OHG) found in vulnerable neurons. Higher levels of DNA breaks, DNA nicking, and fragmentation are observed in AD patients indicating a deficiency in DNA repair [10, 11, 32, 36, 37]. Oxidative modification of proteins is indicated by the significantly elevated levels of protein carbonyls and the widespread nitration of tyrosine residues [38–40]. Proteomic analysis has been used to identify proteins that are specifically oxidized [41–43]; many are enzymes related to ATP generation or enzymes involved in glycolysis, for example, creatine kinase BB, an enzyme involved in production of high-energy phosphate used for ATP synthesis. Oxidative modification of these proteins may in turn lead to the metabolic impairment evident in AD [44–46]. Protein crosslinking by oxidation may render the proteins found in lesions unavailable for intracellular or extracellular degradation even though they are extensively ubiquitinated [47–49]. Lipid peroxidation is marked by higher levels of thiobarbituric acid reactive substances (TBARS), malondialdehyde (MDA), 4-hydroxy-2-transnonenal (HNE), isoprostane, and the altered phospholipid composition of membranes [26, 50]. Lastly, modification to sugars is marked by increased glycation and glycoxidation products [34, 51–53].

Oxidative stress represents a very early contributor to the disease process. The very earliest neuronal and pathological changes characteristic of AD show evidence of oxidative damage [10, 32, 54, 55]. A systematic examination of the spatiotemporal relationship between oxidative modification and the presence of hallmark AD lesions at early AD stages in our laboratory suggests that oxidative damage is present in susceptible neurons even before they exhibit neurofibrillary pathology [11, 56]. In addition, analysis of the cytoplasm of cerebral neurons from Down syndrome cases showed a marked accumulation of 8OHG, as well as advanced glycation end products and nitrotyrosine, at ages preceding amyloid-β (Aβ) deposition by decades [32, 33, 57]. The primacy of oxidative damage has been recapitulated in a transgenic mouse model of AD, Tg2576, in which lipid peroxidation was significantly increased in 7–8-month-old animals, which precedes apparent Aβ deposition and increases in Aβ levels by several months [55].

Neurites with membrane abnormalities, i.e., extensive lipid peroxidation, more frequently contain paired helical filaments (PHF) suggesting that oxidative stress may play a role in the development of neuritic abnormalities [58]. Further linkage between oxidative stress and dystrophic neurites is provided by a vitamin E-deficient rat model, which undergo continuous tissue-selective oxidative stress. These rats contain dystrophic neurites analogous to those associated with the senile plaques of AD [59].

3 Cellular Responses to Oxidative Stress

Neurons contain a number of systems to regulate oxidative balance. Consequently, utilization of compensatory mechanisms in the face of oxidative stress would be expected to occur in AD. The finding that the pentose phosphate pathway is induced in AD provided one of the first examples of oxidative stress in AD [60]. Oxidant defense in cells is accomplished by the accumulation of reducing equivalents in the form of NADPH. Heme oxygenase-1 (HO-1], an NADPH-requiring enzyme and the rate-limiting enzyme for conversion of the prooxidant heme to the antioxidant bilirubin, is induced in AD, substantiating the presence of an active antioxidant defense in AD [61, 62]. HO-1 induction is synchronous with tau accumulation in neurons [63, 64], so the reduced oxidative damage in neurons with tau accumulation may reflect the induction of an antioxidant response [56]. Generation of NADPH results in a concomitant increase in glutathione and free sulfhydryls. Vulnerable neurons show a specific increase in free sulfhydryls [65]; sulfhydryls are a major component of the cellular antioxidant defenses to ROS and secondary metabolites. Additional evidence that the vulnerable neuronal cells are mobilizing antioxidant defense in the face of increased oxidative stress is the induction of Cu/Zn superoxide dismutase, catalase, GSHPx, GSSG-R, peroxiredoxins, and several heat shock proteins and their association with intracellular fibrillary pathology [64, 66–68].

The relationship between oxidative modification and the hallmark AD lesions in a spatiotemporal context would appear to present a paradox. Whereas stable glycation products and HNE-adducts are predominantly associated with NFT and Aβ deposits, reversible or rapidly degraded adduction products such as 8OHG and nitrotyrosine are predominantly found in the cytoplasm of vulnerable neurons lacking cytopathology [10, 11, 39]. Thus, the data support a model in which oxidative stress, with its transient markers such as 8OHG and nitrotyrosine, precedes formation of long-lived lesions. In fact, when short-lived products are analyzed, the damage is restricted to cytosolic compartments, and NFT and Aβ are inversely correlated [32]; i.e., cases of AD with the most extensive Aβ deposits show the lowest 8OHG levels and neurons containing NFTs also have significantly lower levels of 8OHG, despite an obvious history of oxidative damage, i.e., advanced glycation end products (AGEs) or lipid peroxidation. One possible interpretation is that both Aβ production and NFT formation represent cellular responses to increased oxidative stress by serving an antioxidant function. The fact that neurons containing NFTs can actually survive for decades supports this view [69]. HNE modification of tau is phosphorylation dependent [70, 71], which greatly increases its ability to form filaments similar to those found in NFTs in vitro [72] and in cells [72], suggesting that lipid peroxidation and NFT formation may be regulated by signal transduction pathways [73]. An obvious candidate is the stress-activated protein kinase (SAPK) pathways. In addition, oxidative stress upregulates production of both amyloid-β protein precursor (AβPP) and Aβ [74, 75], which depends on the SAPK pathways.

4 Role of Stress-Activated Protein Kinase in Oxidative Stress Signaling

Multiple signaling pathways are induced by cellular stresses such as oxidative stress resulting in alterations in gene expression and enzyme activity. The SAPK pathways are the central mediators propagating stress signals from the membrane to the nucleus and are one such pathway. In neuronal cells, potentially deleterious stimuli such as deprivation of trophic factors, UV irradiation, free radicals, hypoxia, ischemia, heat shock, and cytokines provoke an intracellular stress response that either leads to apoptosis or defensive/protective adaptations. SAPK and its downstream effectors, in conjunction with other signaling pathways, are the major molecules involved in this bipartite response, which can accordingly lead to either neurodegeneration or neuroprotection depending on the cellular and environmental conditions [76]. The two major SAPK pathways are JNK/SAPK and p38/SAPK2. The importance of SAPK as a pathological modulator is being increasingly recognized [77]. SAPK pathways play important roles in cellular processes from gene expression, to inflammation, to cell death, all of which are likely involved in a chronic disease condition such as AD.

The entire JNK/SAPK pathway is altered in AD, as delineated in a number of studies dissecting the signaling cascade involved in AD (reviewed in [78–80]). JNK1 was associated with Hirano bodies in cases of AD and JNK2 and JNK3 were associated with neurofibrillary pathology whereas, in contrast, there was only diffuse staining in the cytoplasm of all neurons in control cases and in non-involved neurons in diseased brain [81]. Significantly, JNK was redistributed from the nucleus to the cytoplasm in AD [81, 82] in a manner that correlates with the progression of the disease; phospho-JNK is exclusively localized in association with neurofibrillar alterations in severe AD cases [81]. Initially, JNK/SAPK activation precedes Aβ deposition [81, 83], suggesting that JNK/SAPK activation is independent of Aβ, although Aβ may provide feedback at a later time enhancing its activation. In early AD, nuclear localization of active JNK/SAPK is almost uniformly detected in most susceptible neurons. This pattern is similar to the oxidative marker 8OHG, which suggests that oxidative stress is a likely activator of the JNK/SAPK pathway in AD. As discussed above, mitochondria are one of the major sources of oxidative stress and likely produce some key effector molecule in the cytoplasm such as H_2O_2. In addition to causing damage to macromolecules via the formation of more reactive free radicals, ROS such as H_2O_2 also serve as a signaling molecule that stimulates protein kinase cascades coupled to inflammatory gene expression, antioxidant responses, or in control of the cell cycle [84]. Therefore, the same compound supplied by mitochondria that leads to the oxidative damage as demonstrated by the production of 8OHG may also lead to the cellular responses that are coordinated by the activation of JNK/SAPK.

The activation of JNK/SAPK by oxidative stress has been linked with consequent apoptosis in several models, however, the actual number of cells dying by JNK/SAPK-mediated apoptosis occurring in the AD brain at any point in time is small [85–89] given the large population of neuronal cells that demonstrate activated JNK/SAPK. Thus, the nuclear localization of active JNK/SAPK [81, 78] may reflect an effect on gene expression, which is supported by the observation that c-Jun is also induced and activated in the same neuronal populations in AD [90–92] (Zhu and colleagues, unpublished results) as an antioxidant response rather than as an initiator of the apoptotic machinery resulting from oxidative stress. Activation of JNK/SAPK pathway may modulate the induction of several antioxidant enzymes that are induced in AD such as HO-1 and superoxide dismutase-1 [61, 66, 93, 94], however, whether the induction of those enzymes is indeed downstream of JNK/SAPK activation in AD merits further investigation. Nevertheless, an increase in the expression of certain mitochondrial genes that are involved in functional compensation has been described in both AD brain and the Tg2576 AβPP transgenic mouse model [95, 96]. Further, a distinctively different pattern of gene expression in susceptible neurons in AD compared to neurons from control cases is observed and many of the differentially expressed genes are regulated by c-Jun [97, 98], supporting the idea of compensatory adaptation via the JNK/c-Jun pathway. As the disease progresses, active JNK/SAPK is redistributed from the nucleus to the cytoplasm [81] suggesting that in late stage AD, JNK/SAPK plays a role in the phosphorylation of the tau protein resulting ultimately in the formation of NFTs. Under conditions of chronic oxidative stress where the cellular antioxidant defenses may be overwhelmed, which is likely to be the case in a chronic neurodegenerative disease such as AD, structural adaptations such as phosphorylation of the tau protein via JNK/SAPK activation and formation of NFTs may serve an antioxidant function in susceptible neurons [10, 56, 99]. Activation of MKK6 and p38 is also well documented in AD and other tauopathies as well as their association with neurofibrillary pathology [84, 78, 100–105]. The nearly identical activation profiles for phospho-JNK and phospho-p38 in AD cases suggest that JNK and p38 are activated in neurons by the same signal [78].

SAPKs may be activated by Aβ given that Aβ plays a key role in the pathogenesis of AD and that an oxidative event mediates Aβ toxicity. Several groups have shown that Aβ induces a two- to threefold activation of JNK in different neuronal cell types and that this activation is directly related to Aβ-induced cell death [106–109]. An in vivo study showing that JNK and p38 are activated in an age-dependent manner in Tg2576/PS1^{P264L} mice and that JNK activation is localized to abnormal neurites within amyloid deposits provides further support [110]. Based on the observation that lipid peroxidation, a marker of oxidative stress, precedes $A\beta$ deposition in Tg2576 mice [55], it is tempting to suggest that oxidative stress activates JNK/SAPK and that subsequently elevated levels of $A\beta$ contributes to further activation of JNK/SAPK. In fact, recent evidence suggests that activation of JNK/SAPK in response to oxidative stress increases BACE activity which in turn results in elevated levels of $A\beta$, suggesting a feed-forward cycle [111–113]. To delineate the primacy of these events, a systematic examination of the temporal relationship between oxidative stress, JNK/SAPK activation, and $A\beta$ deposition should be undertaken in these mice. Moreover, the effect of an increase in $A\beta$ on JNK activation is currently under debate; it has been suggested that JNK activation is mediated by an oxidative stress mechanism. Indeed, causality of $A\beta$-mediated JNK activation is cast in doubt by the observation that some transgenic mice (such as $PS1^{P264L}$ mice) with elevated $A\beta$ levels do not show JNK activation and not all $A\beta$-containing neurons show JNK activation [82], suggesting that some additional element(s) is/are involved. One possibility is provided by the finding that JNK is strongly activated in $A\beta$PP transgenic mice with extensive iron accumulation and oxidative damage, but not in $A\beta$PP transgenic mice with little iron accumulation and oxidative damage [114, 115]. Thus, iron and ROS may play an important role in mediating $A\beta$-induced JNK activation since $A\beta$ plaques are present in both cases. This is further supported by in vitro studies suggesting that ROS, like H_2O_2, are involved in mediating JNK activation induced by $A\beta$ [116, 117].

5 Conclusions

As one of the earliest events in AD pathogenesis, oxidative stress plays a significant role in the initiation of AD pathology. Mitochondrial abnormalities may cause the genesis of oxidative damage and the responses to it in AD by initiating oxidative stress (Fig. 1). As a consequence of releasing excess H_2O_2, dysfunctional mitochondria propagate a series of events involving redox metals and oxidative stress response elements. This may be exacerbated by increases in $A\beta$, which, in the presence of redox-active ions, may also cause H_2O_2 generation. The formation of highly reactive $\bullet OH$ by the interaction of H_2O_2 and metal ions poses a great threat to neuronal cells by damaging important macromolecules, however, compensatory responses provoked by H_2O_2 via the activation of SAPK pathways and downstream effectors such as antioxidant enzymes, tau phosphorylation, and NFT formation may protect neurons from succumbing to the oxidative insults. As may be expected in a chronic disease, a shift in homeostasis may achieve a dynamic balance between oxidative damage and compensatory responses. While this may preserve the life of the neuron it may not maintain proper functioning. The challenge is to determine the extent to which the disease process is driven by oxidative stress and elucidating the relationship between dysfunctional mitochondria and $A\beta$.

Fig. 1 Schematic of proposed oxidative stress pathway in Alzheimer disease. Normally H_2O_2 generated by mitochondria is resolved by catalase to yield H_2O (*top*). In the disease state, H_2O_2 production is increased while catalase is decreased. In the presence of transition metals, the increased H_2O_2 leads to oxidation of biomolecules and activation of the SAPK/Jnk pathway. This leads to induction of antioxidant genes, i.e., HO-1, as well as BACE-1, which results in increased $A\beta$ production potentially feeding back to create more oxidative stress

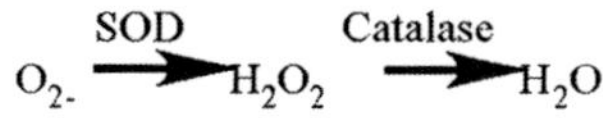

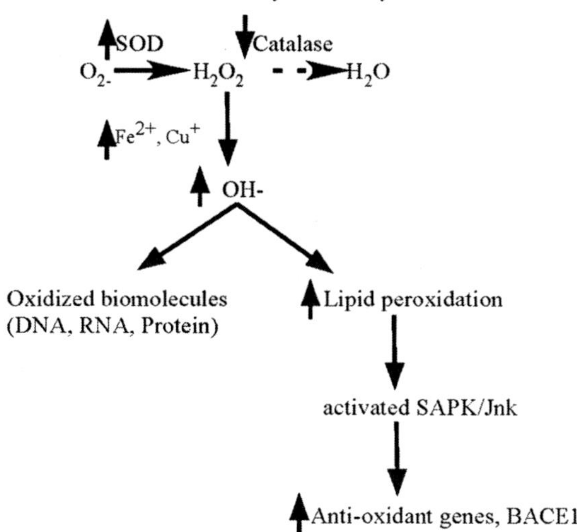

Acknowledgments Work in the authors' laboratory was supported by grants from the National Institutes of Health, the Alzheimer's Association, and by Philip Morris USA Inc. and Philip Morris International.

References

1. Aliev G, Seyidova D, Raina AK, Obrenovich ME, Neal ML, Siedlak SL, Lamb BT, Vinters H, Lamanna JC, Smith MA, Perry G. Vascular hypoperfusion, mitochondria failure and oxidative stress in Alzheimer disease. Proc Indian Natl Sci Acad. 2003;B69(2):209–238.
2. Gibson GE, Sheu KF, Blass JP. Abnormalities of mitochondrial enzymes in Alzheimer disease. J Neural Transm. 1998;105(8–9):855–870.
3. Zhu X, Perry G, Moreira PI, Aliev G, Cash AD, Hirai K, Smith VA. Mitochondrial abnormalities and oxidative imbalance in Alzheimer disease. J Alzheimers Dis. 2006 Jul;9(2):147–153.
4. Cottrell DA, Blakely EL, Johnson MA, Ince PG, Turnbull DM. Mitochondrial enzyme-deficient hippocampal neurons and choroidal cells in AD. Neurology. 2001 Jul 24;57(2):260–264.
5. Maurer I, Zierz S, Moller HJ. A selective defect of cytochrome c oxidase is present in brain of Alzheimer disease patients. Neurobiol Aging. 2000 May–Jun;21(3):455–462.

6. Nagy Z, Esiri MM, LeGris M, Matthews PM. Mitochondrial enzyme expression in the hippocampus in relation to Alzheimer-type pathology. Acta Neuropathol (Berl). 1999 Apr;97(4):346–354.

7. Chandrasekaran K, Giordano T, Brady DR, Stoll J, Martin LJ, Rapoport SI. Impairment in mitochondrial cytochrome oxidase gene expression in Alzheimer disease. Brain Res Mol Brain Res. 1994 Jul;24(1–4):336–340.

8. Parker WD, Jr., Mahr NJ, Filley CM, Parks JK, Hughes D, Young DA, Cullum CM. Reduced platelet cytochrome c oxidase activity in Alzheimer's disease. Neurology. 1994 Jun;44(6):1086–1090.

9. Parker WD, Jr., Parks J, Filley CM, Kleinschmidt-DeMasters BK. Electron transport chain defects in Alzheimer's disease brain. Neurology. 1994 Jun;44(6):1090–1096.

10. Nunomura A, Perry G, Aliev G, Hirai K, Takeda A, Balraj EK, Jones PK, Ghanbari H, Wataya T, Shimohama S, Chiba S, Atwood CS, Petersen RB, Smith MA. Oxidative damage is the earliest event in Alzheimer disease. J Neuropathol Exp Neurol. 2001 Aug;60(8):759–767.

11. Nunomura A, Perry G, Pappolla MA, Wade R, Hirai K, Chiba S, Smith MA. RNA oxidation is a prominent feature of vulnerable neurons in Alzheimer's disease. J Neurosci. 1999 Mar 15;19(6):1959–1964.

12. Marlatt M, Lee HG, Perry G, Smith MA, Zhu X. Sources and mechanisms of cytoplasmic oxidative damage in Alzheimer's disease. Acta Neurobiol Exp (Wars). 2004;64(1):81–87.

13. Perry G, Nunomura A, Hirai K, Zhu X, Perez M, Avila J, Castellani RJ, Atwood CS, Aliev G, Sayre LM, Takeda A, Smith MA. Is oxidative damage the fundamental pathogenic mechanism of Alzheimer's and other neurodegenerative diseases? Free Radic Biol Med. 2002 Dec 1;33(11):1475–1479.

14. Smith MA, Wehr K, Harris PL, Siedlak SL, Connor JR, Perry G. Abnormal localization of iron regulatory protein in Alzheimer's disease. Brain Res. 1998 Mar 30;788(1–2):232–236.

15. Connor JR, Snyder BS, Arosio P, Loeffler DA, LeWitt P. A quantitative analysis of isoferritins in select regions of aged, parkinsonian, and Alzheimer's diseased brains. J Neurochem. 1995 Aug;65(2):717–724.

16. Loeffler DA, Connor JR, Juneau PL, Snyder BS, Kanaley L, DeMaggio AJ, Nguyen H, Brickman CM, LeWitt PA. Transferrin and iron in normal, Alzheimer's disease, and Parkinson's disease brain regions. J Neurochem. 1995 Aug;65(2):710–724.

17. Pinero DJ, Hu J, Connor JR. Alterations in the interaction between iron regulatory proteins and their iron responsive element in normal and Alzheimer's diseased brains. Cell Mol Biol (Noisy-le-grand). 2000 Jun;46(4):761–776.

18. Perry G, Taddeo MA, Petersen RB, Castellani RJ, Harris PL, Siedlak SL, Cash AD, Liu Q, Nunomura A, Atwood CS, Smith MA. Adventiously-bound redox active iron and copper are at the center of oxidative damage in Alzheimer disease. Biometals. 2003 Mar;16(1):77–81.

19. Behl C, Davis JB, Lesley R, Schubert D. Hydrogen peroxide mediates amyloid beta protein toxicity. Cell. 1994;77(6):817–827.

20. Behl C, Davis J, Cole GM, Schubert D. Vitamin E protects nerve cells from amyloid beta protein toxicity. Biochem Biophys Res Commun. 1992;186(2):944–950.

21. Lockhart BP, Benicourt C, Junien JL, Privat A. Inhibitors of free radical formation fail to attenuate direct beta-amyloid25-35 peptide-mediated neurotoxicity in rat hippocampal cultures. J Neurosci Res. 1994;39(4):494–505.

22. Keller JN, Mark RJ, Bruce AJ, Blanc E, Rothstein JD, Uchida K, Waeg G, Mattson MP. 4-Hydroxynonenal, an aldehydic product of membrane lipid peroxidation, impairs glutamate transport and mitochondrial function in synaptosomes. Neuroscience. 1997 Oct;80(3):685–696.

23. Mark RJ, Lovell MA, Markesbery WR, Uchida K, Mattson MP. A role for 4-hydroxynonenal, an aldehydic product of lipid peroxidation, in disruption of ion

homeostasis and neuronal death induced by amyloid beta-peptide. J Neurochem. 1997 Jan;68(1):255–264.

24. Mark RJ, Pang Z, Geddes JW, Uchida K, Mattson MP. Amyloid beta-peptide impairs glucose transport in hippocampal and cortical neurons: involvement of membrane lipid peroxidation. J Neurosci. 1997 Feb 1;17(3):1046–1054.

25. Markesbery WR, Lovell MA. Four-hydroxynonenal, a product of lipid peroxidation, is increased in the brain in Alzheimer's disease. Neurobiol Aging. 1998 Jan–Feb;19(1):33–36.

26. Sayre LM, Zelasko DA, Harris PL, Perry G, Salomon RG, Smith MA. 4-Hydroxynonenal-derived advanced lipid peroxidation end products are increased in Alzheimer's disease. J Neurochem. 1997 May;68(5):2092–2097.

27. Yan SD, Chen X, Fu J, Chen M, Zhu H, Roher A, Slattery T, Zhao L, Nagashima M, Morser J, Migheli A, Nawroth P, Stern D, Schmidt AM. RAGE and amyloid-beta peptide neurotoxicity in Alzheimer's disease. Nature. 1996 Aug 22;382(6593):685–691.

28. Colton CA, Gilbert DL. Production of superoxide anions by a CNS macrophage, the microglia. FEBS Lett. 1987;223(2):284–288.

29. Good PF, Perl DP, Bierer LM, Schmeidler J. Selective accumulation of aluminum and iron in the neurofibrillary tangles of Alzheimer's disease: a laser microprobe (LAMMA) study. Ann Neurol. 1992;31(3):286–292.

30. Huang X, Cuajungco MP, Atwood CS, Hartshorn MA, Tyndall JD, Hanson GR, Stokes KC, Leopold M, Multhaup G, Goldstein LE, Scarpa RC, Saunders AJ, Lim J, Moir RD, Glabe C, Bowden EF, Masters CL, Fairlie DP, Tanzi RE, Bush AI. Cu(II) potentiation of alzheimer abeta neurotoxicity. Correlation with cell-free hydrogen peroxide production and metal reduction. J Biol Chem. 1999;274(52):37111–37116.

31. Rottkamp CA, Raina AK, Zhu X, Gaier E, Bush AI, Atwood CS, Chevion M, Perry G, Smith MA. Redox-active iron mediates amyloid-beta toxicity. Free Radic Biol Med. 2001;30(4):447–450.

32. Nunomura A, Perry G, Pappolla MA, Friedland RP, Hirai K, Chiba S, Smith MA. Neuronal oxidative stress precedes amyloid-beta deposition in Down syndrome. J Neuropathol Exp Neurol. 2000 Nov;59(11):1011–1017.

33. Lee HG, Casadesus G, Zhu X, Takeda A, Perry G, Smith MA. Challenging the amyloid cascade hypothesis: senile plaques and amyloid-beta as protective adaptations to Alzheimer disease. Ann N Y Acad Sci. 2004 Jun;1019:1–4.

34. Smith MA, Sayre LM, Monnier VM, Perry G. Radical AGEing in Alzheimer's disease. Trends Neurosci. 1995 Apr;18(4):172–176.

35. Cross CE, Halliwell B, Borish ET, Pryor WA, Ames BN, Saul RL, McCord JM, Harman D. Oxygen radicals and human disease. Ann Intern Med. 1987;107(4):526–545.

36. Mecocci P, MacGarvey U, Beal MF. Oxidative damage to mitochondrial DNA is increased in Alzheimer's disease. Ann Neurol. 1994 Nov;36(5):747–751.

37. Mecocci P, Beal MF, Cecchetti R, Polidori MC, Cherubini A, Chionne F, Avellini L, Romano G, Senin U. Mitochondrial membrane fluidity and oxidative damage to mitochondrial DNA in aged and AD human brain. Mol Chem Neuropathol. 1997 May;31(1):53–64.

38. Smith MA, Perry G, Richey PL, Sayre LM, Anderson VE, Beal MF, Kowall N. Oxidative damage in Alzheimer's. Nature. 1996 Jul 11;382(6587):120–121.

39. Smith MA, Richey Harris PL, Sayre LM, Beckman JS, Perry G. Widespread peroxynitrite-mediated damage in Alzheimer's disease. J Neurosci. 1997 Apr 15;17(8):2653–2657.

40. Good PF, Werner P, Hsu A, Olanow CW, Perl DP. Evidence of neuronal oxidative damage in Alzheimer's disease. Am J Pathol. 1996 Jul;149(1):21–28.

41. Castegna A, Aksenov M, Thongboonkerd V, Klein JB, Pierce WM, Booze R, Markesbery WR, Butterfield DA. Proteomic identification of oxidatively modified proteins in Alzheimer's disease brain. Part II: dihydropyrimidinase-related protein 2, alpha-enolase and heat shock cognate 71. J Neurochem. 2002 Sep;82(6):1524–1532.

42. Castegna A, Aksenov M, Aksenova M, Thongboonkerd V, Klein JB, Pierce WM, Booze R, Markesbery WR, Butterfield DA. Proteomic identification of oxidatively modified proteins in Alzheimer's disease brain. Part I: creatine kinase BB, glutamine synthase, and ubiquitin carboxy-terminal hydrolase L-1. Free Radic Biol Med. 2002 Aug 15;33(4):562–571.

43. Castegna A, Thongboonkerd V, Klein JB, Lynn B, Markesbery WR, Butterfield DA. Proteomic identification of nitrated proteins in Alzheimer's disease brain. J Neurochem. 2003 Jun;85(6):1394–1401.

44. Hirai K, Aliev G, Nunomura A, Fujioka H, Russell RL, Atwood CS, Johnson AB, Kress Y, Vinters HV, Tabaton M, Shimohama S, Cash AD, Siedlak SL, Harris PL, Jones PK, Petersen RB, Perry G, Smith MA. Mitochondrial abnormalities in Alzheimer's disease. J Neurosci. 2001 May 1;21(9):3017–3023.

45. Perry G, Nunomura A, Raina AK, Aliev G, Siedlak SL, Harris PL, Casadesus G, Petersen RB, Bligh-Glover W, Balraj E, Petot GJ, Smith MA. A metabolic basis for Alzheimer disease. Neurochem Res. 2003 Oct;28(10):1549–1552.

46. Moreira PI, Zhu X, Lee HG, Honda K, Smith MA, Perry G. The (un)balance between metabolic and oxidative abnormalities and cellular compensatory responses in Alzheimer disease. Mech Ageing Dev. 2006 Jun;127(6):501–506.

47. Cras P, Smith MA, Richey PL, Siedlak SL, Mulvihill P, Perry G. Extracellular neurofibrillary tangles reflect neuronal loss and provide further evidence of extensive protein cross-linking in Alzheimer disease. Acta Neuropathol. 1995;89(4):291–295.

48. Smith MA, Sayre LM, Anderson VE, Harris PL, Beal MF, Kowall N, Perry G. Cytochemical demonstration of oxidative damage in Alzheimer disease by immunochemical enhancement of the carbonyl reaction with 2,4-dinitrophenylhydrazine. J Histochem Cytochem. 1998 Jun;46(6):731–735.

49. Smith MA, Siedlak SL, Richey PL, Nagaraj RH, Elhammer A, Perry G. Quantitative solubilization and analysis of insoluble paired helical filaments from Alzheimer disease. Brain Res. 1996 Apr 22;717(1–2):99–108.

50. Butterfield DA, Drake J, Pocernich C, Castegna A. Evidence of oxidative damage in Alzheimer's disease brain: central role for amyloid beta-peptide. Trends Mol Med. 2001 Dec;7(12):548–554.

51. Castellani RJ, Harris PL, Sayre LM, Fujii J, Taniguchi N, Vitek MP, Founds H, Atwood CS, Perry G, Smith MA. Active glycation in neurofibrillary pathology of Alzheimer disease: N(epsilon)-(carboxymethyl) lysine and hexitol-lysine. Free Radic Biol Med. 2001 Jul 15;31(2):175–180.

52. Vitek MP, Bhattacharya K, Glendening JM, Stopa E, Vlassara H, Bucala R, Manogue K, Cerami A. Advanced glycation end products contribute to amyloidosis in Alzheimer disease. Proc Natl Acad Sci U S A. 1994 May 24;91(11):4766–4770.

53. Smith MA, Taneda S, Richey PL, Miyata S, Yan SD, Stern D, Sayre LM, Monnier VM, Perry G. Advanced Maillard reaction end products are associated with Alzheimer disease pathology. Proc Natl Acad Sci U S A. 1994;91(12):5710–5714.

54. Perry G, Smith MA. Is oxidative damage central to the pathogenesis of Alzheimer disease? Acta Neurol Belg. 1998;98(2):175–179.

55. Pratico D, Uryu K, Leight S, Trojanoswki JQ, Lee VM. Increased lipid peroxidation precedes amyloid plaque formation in an animal model of Alzheimer amyloidosis. J Neurosci. 2001 Jun 15;21(12):4183–4187.

56. Lee HG, Perry G, Moreira PI, Garrett MR, Liu Q, Zhu X, Takeda A, Nunomura A, Smith MA. Tau phosphorylation in Alzheimer's disease: pathogen or protector? Trends Mol Med. 2005 Apr;11(4):164–169.

57. Odetti P, Angelini G, Dapino D, Zaccheo D, Garibaldi S, Dagna-Bricarelli F, Piombo G, Perry G, Smith M, Traverso N, Tabaton M. Early glycoxidation damage in brains from Down's syndrome. Biochem Biophys Res Commun. 1998 Feb 24;243(3):849–851.

58. Praprotnik D, Smith MA, Richey PL, Vinters HV, Perry G. Plasma membrane fragility in dystrophic neurites in senile plaques of Alzheimer's disease: an index of oxidative stress. Acta Neuropathol. 1996;91(1):1–5.
59. Heslop KE, Goss-Sampson MA, Muller DP, Curzon G. Serotonin metabolism and release in frontal cortex of rats on a vitamin E-deficient diet. J Neurochem. 1996;66(2):860–864.
60. Martins RN, Harper CG, Stokes GB, Masters CL. Increased cerebral glucose-6-phosphate dehydrogenase activity in Alzheimer's disease may reflect oxidative stress. J Neurochem. 1986 Apr;46(4):1042–1045.
61. Premkumar DR, Smith MA, Richey PL, Petersen RB, Castellani R, Kutty RK, Wiggert B, Perry G, Kalaria RN. Induction of heme oxygenase-1 mRNA and protein in neocortex and cerebral vessels in Alzheimer's disease. J Neurochem. 1995;65(3):1399–1402.
62. Smith MA, Kutty RK, Richey PL, Yan SD, Stern D, Chader GJ, Wiggert B, Petersen RB, Perry G. Heme oxygenase-1 is associated with the neurofibrillary pathology of Alzheimer's disease. Am J Pathol. 1994 Jul;145(1):42–47.
63. Takeda A, Perry G, Abraham NG, Dwyer BE, Kutty RK, Laitinen JT, Petersen RB, Smith MA. Overexpression of heme oxygenase in neuronal cells, the possible interaction with Tau. J Biol Chem. 2000 Feb 25;275(8):5395–5399.
64. Takeda A, Smith MA, Avila J, Nunomura A, Siedlak SL, Zhu X, Perry G, Sayre LM. In Alzheimer's disease, heme oxygenase is coincident with Alz50, an epitope of tau induced by 4-hydroxy-2-nonenal modification. J Neurochem. 2000;75(3):1234–1241.
65. Russell RL, Siedlak SL, Raina AK, Bautista JM, Smith MA, Perry G. Increased neuronal glucose-6-phosphate dehydrogenase and sulfhydryl levels indicate reductive compensation to oxidative stress in Alzheimer disease. Arch Biochem Biophys. 1999 Oct 15;370(2):236–239.
66. Pappolla MA, Omar RA, Kim KS, Robakis NK. Immunohistochemical evidence of oxidative (corrected) stress in Alzheimer's disease. Am J Pathol. 1992;140(3):621–628.
67. Lee SC, Zhao ML, Hirano A, Dickson DW. Inducible nitric oxide synthase immunoreactivity in the Alzheimer disease hippocampus: association with Hirano bodies, neurofibrillary tangles, and senile plaques. J Neuropathol Exp Neurol. 1999;58(11):1163–1169.
68. Aksenov MY, Tucker HM, Nair P, Aksenova MV, Butterfield DA, Estus S, Markesbery WR. The expression of key oxidative stress-handling genes in different brain regions in Alzheimer's disease. J Mol Neurosci. 1998 Oct;11(2):151–164.
69. Morsch R, Simon W, Coleman PD. Neurons may live for decades with neurofibrillary tangles. J Neuropathol Exp Neurol. 1999;58(2):188–197.
70. Takeda A, Smith MA, Avila J, Nunomura A, Siedlak SL, Zhu X, Perry G, Sayre LM. In Alzheimer's disease, heme oxygenase is coincident with Alz50, an epitope of tau induced by 4-hydroxy-2-nonenal modification. J Neurochem. 2000 Sep;75(3):1234–1241.
71. Wataya T, Nunomura A, Smith MA, Siedlak SL, Harris PL, Shimohama S, Szweda LI, Kaminski MA, Avila J, Price DL, Cleveland DW, Sayre LM, Perry G. High molecular weight neurofilament proteins are physiological substrates of adduction by the lipid peroxidation product hydroxynonenal. J Biol Chem. 2002 Feb 15;277(7):4644–4648.
72. Perez M, Cuadros R, Smith MA, Perry G, Avila J. Phosphorylated, but not native, tau protein assembles following reaction with the lipid peroxidation product, 4-hydroxy-2-nonenal. FEBS Lett. 2000 Dec 15;486(3):270–274.
73. Perry G, Taddeo MA, Nunomura A, Zhu X, Zenteno-Savin T, Drew KL, Shimohama S, Avila J, Castellani RJ, Smith MA. Comparative biology and pathology of oxidative stress in Alzheimer and other neurodegenerative diseases: beyond damage and response. Comp Biochem Physiol C Toxicol Pharmacol. 2002 Dec;133(4):507–513.
74. Yan SD, Yan SF, Chen X, Fu J, Chen M, Kuppusamy P, Smith MA, Perry G, Godman GC, Nawroth P, et al. Non-enzymatically glycated tau in Alzheimer's disease induces neuronal oxidant stress resulting in cytokine gene expression and release of amyloid beta-peptide. Nat Med. 1995;1(7):693–699.

75. Paola D, Domenicotti C, Nitti M, Vitali A, Borghi R, Cottalasso D, Zaccheo D, Odetti P, Strocchi P, Marinari UM, Tabaton M, Pronzato MA. Oxidative stress induces increase in intracellular amyloid beta-protein production and selective activation of betaI and betaII PKCs in NT2 cells. Biochem Biophys Res Commun. 2000 Feb 16;268(2):642–646.

76. Mielke K, Herdegen T. JNK and p38 stresskinases – degenerative effectors of signal-transduction-cascades in the nervous system. Prog Neurobiol. 2000;61(1):45–60.

77. Zhu X, Lee H-G, Perry G, Raina AK, Smith MA. The MAPK pathways in Alzheimer's disease. NeuroSignals. 2002;11:270–281.

78. Zhu X, Castellani RJ, Takeda A, Nunomura A, Atwood CS, Perry G, Smith MA. Differential activation of neuronal ERK, JNK/SAPK and p38 in Alzheimer disease: the 'two hit' hypothesis. Mech Ageing Dev. 2001 Dec;123(1):39–46.

79. Zhu X, Raina AK, Perry G, Smith MA. Alzheimer's disease: the two-hit hypothesis. Lancet Neurol. 2004 Apr;3(4):219–226.

80. Zhu X, Raina AK, Lee HG, Chao M, Nunomura A, Tabaton M, Petersen RB, Perry G, Smith MA. Oxidative stress and neuronal adaptation in Alzheimer disease: the role of SAPK pathways. Antioxid Redox Signal. 2003 Oct;5(5):571–576.

81. Zhu X, Raina AK, Rottkamp CA, Aliev G, Perry G, Boux H, Smith MA. Activation and redistribution of c-Jun N-terminal kinase/stress activated protein kinase in degenerating neurons in Alzheimer's disease. J Neurochem. 2001 Jan;76(2):435–441.

82. Shoji M, Iwakami N, Takeuchi S, Waragai M, Suzuki M, Kanazawa I, Lippa CF, Ono S, Okazawa H. JNK activation is associated with intracellular beta-amyloid accumulation. Brain Res Mol Brain Res. 2000;85(1–2):221–233.

83. Pei JJ, Braak E, Braak H, Grundke-Iqbal I, Iqbal K, Winblad B, Cowburn RF. Localization of active forms of C-jun kinase (JNK) and p38 kinase in Alzheimer's disease brains at different stages of neurofibrillary degeneration. J Alzheimers Dis. 2001 Feb;3(1):41–48.

84. Suzuki YJ, Forman HJ, Sevanian A. Oxidants as stimulators of signal transduction. Free Radic Biol Med. 1997;22(1–2):269–285.

85. Zhu X, Raina AK, Perry G, Smith MA. Apoptosis in Alzheimer disease: a mathematical improbability. Curr Alzheimer Res. 2006 Sep;3(4):393–396.

86. Raina AK, Hochman A, Zhu X, Rottkamp CA, Nunomura A, Siedlak SL, Boux H, Castellani RJ, Perry G, Smith MA. Abortive apoptosis in Alzheimer's disease. Acta Neuropathol (Berl). 2001 Apr;101(4):305–310.

87. Zhu X, Wang Y, Ogawa O, Lee HG, Raina AK, Siedlak SL, Harris PL, Fujioka H, Shimohama S, Tabaton M, Atwood CS, Petersen RB, Perry G, Smith MA. Neuroprotective properties of Bcl-w in Alzheimer disease. J Neurochem. 2004 Jun;89(5):1233–1240.

88. Perry G, Nunomura A, Lucassen P, Lassmann H, Smith MA. Apoptosis and Alzheimer's disease. Science. 1998 Nov 13;282(5392):1268–1269.

89. Perry G, Zhu X, Smith MA. Do neurons have a choice in death? Am J Pathol. 2001 Jan;158(1):1–2.

90. Smith MA, Zhu X, Sun Z, Perry G. Activation of c-Jun in Alzheimer disease. J Neuropathol Exp Neurol. 2001;60:546.

91. Anderson AJ, Cummings BJ, Cotman CW. Increased immunoreactivity for Jun- and Fos-related proteins in Alzheimer's disease: association with pathology. Exp Neurol. 1994;125(2):286–295.

92. Pearson AG, Byrne UT, MacGibbon GA, Faull RL, Dragunow M. Activated c-Jun is present in neurofibrillary tangles in Alzheimer's disease brains. Neurosci Lett. 2006 May 8;398(3):246–250.

93. Smith MA, Kutty RK, Richey PL, Yan SD, Stern D, Chader GJ, Wiggert B, Petersen RB, Perry G. Heme oxygenase-1 is associated with the neurofibrillary pathology of Alzheimer's disease. Am J Pathol. 1994;145(1):42–47.

94. Renkawek K, Bosman GJ, de Jong WW. Expression of small heat-shock protein hsp 27 in reactive gliosis in Alzheimer disease and other types of dementia. Acta Neuropathol (Berl). 1994;87(5):511–519.

95. Reddy PH, McWeeney S, Park BS, Manczak M, Gutala RV, Partovi D, Jung Y, Yau V, Searles R, Mori M, Quinn J. Gene expression profiles of transcripts in amyloid precursor protein transgenic mice: up-regulation of mitochondrial metabolism and apoptotic genes is an early cellular change in Alzheimer's disease. Hum Mol Genet. 2004 Jun 15;13(12):1225–1240.

96. Manczak M, Park BS, Jung Y, Reddy PH. Differential expression of oxidative phosphorylation genes in patients with Alzheimer's disease: implications for early mitochondrial dysfunction and oxidative damage. Neuromolecular Med. 2004;5(2):147–162.

97. Dunckley T, Beach TG, Ramsey KE, Grover A, Mastroeni D, Walker DG, LaFleur BJ, Coon KD, Brown KM, Caselli R, Kukull W, Higdon R, McKeel D, Morris JC, Hulette C, Schmechel D, Reiman EM, Rogers J, Stephan DA. Gene expression correlates of neurofibrillary tangles in Alzheimer's disease. Neurobiol Aging. 2006 Oct;27(10):1359–1371.

98. Lovell MA, Smith JL, Markesbery WR. Elevated zinc transporter-6 in mild cognitive impairment, Alzheimer disease, and pick disease. J Neuropathol Exp Neurol. 2006 May;65(5):489–498.

99. Smith MA, Casadesus G, Joseph JA, Perry G. Amyloid-beta and tau serve antioxidant functions in the aging and Alzheimer brain. Free Radic Biol Med. 2002 Nov 1;33(9):1194–1199.

100. Zhu X, Rottkamp CA, Boux H, Takeda A, Perry G, Smith MA. Activation of p38 kinase links tau phosphorylation, oxidative stress, and cell cycle-related events in Alzheimer disease. J Neuropathol Exp Neurol. 2000;59(10):880–888.

101. Atzori C, Ghetti B, Piva R, Srinivasan AN, Zolo P, Delisle MB, Mirra SS, Migheli A. Activation of the JNK/p38 pathway occurs in diseases characterized by tau protein pathology and is related to tau phosphorylation but not to apoptosis. J Neuropathol Exp Neurol. 2001 Dec;60(12):1190–1197.

102. Hensley K, Floyd RA, Zheng NY, Nael R, Robinson KA, Nguyen X, Pye QN, Stewart CA, Geddes J, Markesbery WR, Patel E, Johnson GV, Bing G. p38 kinase is activated in the Alzheimer's disease brain. J Neurochem. 1999 May;72(5):2053–2058.

103. Ferrer I, Blanco R, Carmona M, Puig B. Phosphorylated mitogen-activated protein kinase (MAPK/ERK-P), protein kinase of 38 kDa (p38-P), stress-activated protein kinase (SAPK/JNK-P), and calcium/calmodulin-dependent kinase II (CaM kinase II) are differentially expressed in tau deposits in neurons and glial cells in tauopathies. J Neural Transm. 2001;108(12):1397–1415.

104. Pei JJ, Braak H, An WL, Winblad B, Cowburn RF, Iqbal K, Grundke-Iqbal I. Up-regulation of mitogen-activated protein kinases ERK1/2 and MEK1/2 is associated with the progression of neurofibrillary degeneration in Alzheimer's disease. Brain Res Mol Brain Res. 2002 Dec 30;109(1–2):45–55.

105. Hartzler AW, Zhu X, Siedlak SL, Castellani RJ, Avila J, Perry G, Smith MA. The p38 pathway is activated in Pick disease and progressive supranuclear palsy: a mechanistic link between mitogenic pathways, oxidative stress, and tau. Neurobiology of Aging. 2002;23:855–859.

106. Troy CM, Rabacchi SA, Xu Z, Maroney AC, Connors TJ, Shelanski ML, Greene LA. beta-Amyloid-induced neuronal apoptosis requires c-Jun N-terminal kinase activation. J Neurochem. 2001;77(1):157–164.

107. Morishima Y, Gotoh Y, Zieg J, Barrett T, Takano H, Flavell R, Davis RJ, Shirasaki Y, Greenberg ME. Beta-amyloid induces neuronal apoptosis via a mechanism that involves the c-Jun N-terminal kinase pathway and the induction of Fas ligand. J Neurosci. 2001;21(19):7551–7560.

108. Bozyczko-Coyne D, O'Kane TM, Wu ZL, Dobrzanski P, Murthy S, Vaught JL, Scott RW. CEP-1347/KT-7515, an inhibitor of SAPK/JNK pathway activation, promotes survival and blocks multiple events associated with Abeta-induced cortical neuron apoptosis. J Neurochem. 2001 May;77(3):849–863.

109. Wei W, Wang X, Kusiak JW. Signaling events in amyloid beta-peptide-induced neuronal death and insulin-like growth factor I protection. J Biol Chem. 2002 May 17;277(20):17649–17656.
110. Savage MJ, Lin YG, Ciallella JR, Flood DG, Scott RW. Activation of c-Jun N-terminal kinase and p38 in an Alzheimer's disease model is associated with amyloid deposition. J Neurosci. 2002 May 1;22(9):3376–3385.
111. Tamagno E, Parola M, Bardini P, Piccini A, Borghi R, Guglielmotto M, Santoro G, Davit A, Danni O, Smith MA, Perry G, Tabaton M. Beta-site APP cleaving enzyme upregulation induced by 4-hydroxynonenal is mediated by stress-activated protein kinases pathways. J Neurochem. 2005 Feb;92(3):628–636.
112. Tamagno E, Bardini P, Obbili A, Vitali A, Borghi R, Zaccheo D, Pronzato MA, Danni O, Smith MA, Perry G, Tabaton M. Oxidative stress increases expression and activity of BACE in NT2 neurons. Neurobiol Dis. 2002 Aug;10(3):279–288.
113. Kao SC, Krichevsky AM, Kosik KS, Tsai LH. BACE1 suppression by RNA interference in primary cortical neurons. J Biol Chem. 2004 Jan 16;279(3):1942–1949.
114. Zhu X, Rottkamp CA, Zevez K, Atwood CS, Perry G, Smith MA. Metal-Mediated amyloid-beta toxicity acts through the activation of JNK/SAPK pathway in Alzheimer disease. Free Radic Biol Med. 2001;31(supplement 1):S65.
115. Smith MA, Harris PL, Sayre LM, Perry G. Iron accumulation in Alzheimer disease is a source of redox-generated free radicals. Proc Natl Acad Sci U S A. 1997 Sep 2;94(18):9866–9868.
116. Velez-Pardo C, Ospina GG, Jimenez del Rio M. Abeta(25-35) peptide and iron promote apoptosis in lymphocytes by an oxidative stress mechanism: involvement of H2O2, caspase-3, NF-kappaB, p53 and c-Jun. Neurotoxicology. 2002 Sep;23(3):351–365.
117. Jang JH, Surh YJ. beta-Amyloid induces oxidative DNA damage and cell death through activation of c-Jun N terminal kinase. Ann N Y Acad Sci. 2002 Nov;973:228–236.

Nitrated Proteins in the Progression of Alzheimer's Disease: A Proteomics Comparison of Mild Cognitive Impairment and Alzheimer's Disease Brain

Rukhsana Sultana, Renã A. Sowell, and D. Allan Butterfield

Abstract Oxidative stress and nitrosative stress have been reported to play important roles in the pathogenesis of a number of diseases including neurodegenerative diseases, cancer, ischemia, etc. Reactive nitrogen species are highly reactive and unstable. One of the best ways to quantify the amount of nitrosative stress is to measure the levels of 3-nitrotyrosine level. In addition, by using proteomics selective targets of protein nitration can be identified. In this chapter we discuss the roles of proteomics-identified nitrated brain proteins to the pathology of both mild cognitive impairment and Alzheimer's disease. The identity of these nitrated proteins improves understanding of the role of nitrosative stress in the pathogenesis and progression of disease from MCI to AD. Such studies could also help in early detection and may provide therapeutic targets for early treatment that may slow disease progression.

Keywords Nitrosative stress · 3-nitrotyrosine · Proteomics · Alzheimer's disease · Mild cognitive impairment

1 Nitrosative Stress

Oxidative stress and nitrosative stress have been reported to play important roles in the pathogenesis of a number of diseases including neurodegenerative disease, cancer, ischemia, etc. [1–5]. Nitrosative stress is caused by increased levels of reactive nitrogen species (RNS). RNS are highly reactive and toxic. RNS include nitric oxide (NO), peroxynitrite, nitrogen dioxide, etc. [6, 7]. Nitric oxide synthase (NOS) produces NO, a free radical. Under physiological

---

This paper is dedicated to the life of Dr. Earl R. Stadtman (1919–2008), a good friend and accomplished scientist and mentor.

D.A. Butterfield (✉)
Department of Chemistry, University of Kentucky, Lexington, KY 40506, USA
e-mail: dabcns@uky.edu

S.C. Veasey (ed.), *Oxidative Neural Injury*,
Contemporary Clinical Neuroscience, DOI 10.1007/978-1-60327-342-8_9,
© Humana Press, a part of Springer Science+Business Media, LLC 2009

conditions NO is produced for relatively specific cellular targets [8]. However, under certain cellular conditions NO is produced by inducible NOS (iNOS) in higher amounts, which can then react with superoxide to form peroxynitrite [9].

Peroxynitrite has a half-life of less than one second and is highly reactive. NO or peroxynitrite can react with thiols to form nitrosothiols, or peroxynitrite also can react with tyrosine residues, one of the preferential sites of phosphorylation, at the meta position to form 3-nitrotyrosine (3NT) [10]. These modifications of roteins could affect the structure and thereby function of the proteins including alterations of cell signaling, catalytic activity, cytoskeletal organization, and inflammatory response [11–15]. Hence, increased levels of protein nitration can have detrimental effects on cell viability and function [16].

A number of previous studies showed that nitration of proteins may lead to inactivation of several important mammalian proteins such as antioxidant enzymes [17], structural proteins, energy-related mitochondrial proteins, and neurotransmitter-related proteins [12, 18, 19]. Several studies suggested that protein nitration is a reversible process analogous to phosphorylation and this may serve as a cellular signal [20, 21]. For example, proteins that are nitrated were reported to be more prone to proteosomal degradation than their counter-parts [18]. Since RNS themselves are unstable, one of the best ways to quantify the amount of nitrosative stress is to measure the products of RNS, e.g., levels of 3-NT [11].

2 Alzheimer's Disease and Protein Nitration

Alzheimer's disease (AD) is an age-related neurodegenerative disorder that is pathologically characterized by the presence of extracellular amyloid plaques, intracellular neurofibrillary tangles (NFT), and loss of synaptic connections within selective brain regions. Hyperphosphorylated tau protein is the main component of NFT that forms paired helical filaments and related straight filaments. Amyloid plaque has amyloid beta-peptide (Aβ) as the main compo-nent and is considered to play a causal role in the development and progression of AD [22]. Aβ peptides are generated from Aβ peptide precursor protein (APP) by the action of β- and γ-secretases. Aβ exists in many forms, such as soluble form, aggregated form, oligomeric form, protofibrils (PF), and fibrils [23, 24]. Considerable research has shown that Aβ toxicity is associated with oligomers, PF, and amyloid-derived diffusible ligands (ADDLs) [25–27].

Although the exact mechanism of AD pathogenesis is not clearly under-stood, mutation of *presenilin-1* (PS-1), *presenilin-2* (PS-2), and APP genes has been found to be associated with inherited AD [28, 29]. In addition, other genes like allele 4 of the *apolipoprotein E* (APOE) gene, *endothelial nitric oxide synthase –3* gene, and the *alpha-2-macroglobulin* gene have been associated with AD [30, 31]. Further, the amyloid cascade, excitoxicity, oxidative stress, and inflammation hypothesis have been proposed for AD mechanisms, and all these are based on the role of Aβ [32–34]. There are large number of evidences that suggest a role of oxidative stress in the pathophysiology of AD [33, 35, 36]. Oxidative stress in AD brain is manifested by decreased levels of antioxidant enzymes and also by increased protein oxidation (including protein carbonyls and 3-NT formation), lipid peroxidation, DNA oxidation, advanced glycation end products, and reactive oxygen species (ROS) formation, among other indices. The role of oxidative stress is supported by the use of vitamin E in cell culture that diminishes Aβ (1–42)-induced toxicity [37].

A number of previous studies support the role of nitrosative and oxidative injury in the pathogenesis of AD [22, 36, 38, 39]. The early markers of oxidative stress and nitrosative stress in a cell include the formation of protein carbonyls, the lipid peroxidation product, 4-hydroxy-2-nonenal (HNE), and 3-NT [12, 19, 38, 40–45]. The role of RNS in AD pathology is based on the elevated levels of nitrated proteins that were found in AD brain and cerebrospinal fluid (CSF) [12, 19, 44]. In AD brain and ventricular cerebrospinal fluid (VF) increased levels of DiTyr and 3-NT were reported [46] that probably reflect increased leakage of mitochondrial electron equivalents that may nitrate proteins. Further, immunohistochemistry showed the presence of nitrated tau in pretangles, tangles, and tau inclusions in AD brain. However, the levels of 3-NT were found to be more in pretangles of early AD brain compared to that of more advanced brain that suggest the involvement of tau nitration as an early event in AD pathogenesis [47, 48].

3 Mild Cognitive Impairment and Protein Nitration

Mild Cognitive Impairment (MCI) is characterized by loss of recent memory without dementia or significant impairment of other cognitive functions and with no loss of activities of daily living [49]. MCI is divided into two broad subtypes based on memory impairment, i.e., amnestic MCI and non-amnestic MCI. Many MCI subjects show some of the neuropathological features of AD at autopsy such as significant medical temporal lobe atrophy, while others demonstrate low CSF-β amyloid (1–42) concentrations, factors that are associated with the senile plaques common to AD. In addition, there are also genetic similarities such as mutations in allele 4 of *APOE, PS1,* and the *APP* [50, 51]

Studies from our laboratory and others have proposed the role of oxidative stress and nitrosative stress in the progression of MCI to AD [52, 53]. Like AD, in MCI patients, plasma mean levels of non-enzymatic antioxidants and activity of antioxidant enzymes appeared to be lower than in controls [53–55]. Further, studies showed increased oxidative damage in nuclear and mitochondrial DNA, as indexed by increased levels of 8-hydroxyguanosine (8-OhdG), 2,6-diamino-4-hydroxy-5-formamidopyrimidine (fapyguanine), 8-hydroxyadenine, 4,6-diamino-5-formamidopyrimidine (fapyadenine), and 5-hydroxycytosine in MCI brain [56, 57]. In addition, other markers of oxidative stress such as lipid peroxidation products isoprostane were found to be elevated in MCI plasma, urine, and CSF [58]. Our laboratory and others have shown increased levels of protein-bound and protein-free HNE [59–63] in MCI hippocampus and inferior parietal lobules compared to those of control brain [64]. Further, from our laboratory we also showed an increased level of protein-bound 3-NT levels that suggest the involvement of nitrosative stress at MCI stage [15].

4 The Proteomics Approach

4.1 Redox Proteomics

Redox proteomics focuses on the identification and quantification of oxidatively modified proteins. Several oxidative posttranslational markers have been studied by redox proteomics, e.g., protein carbonyls, HNE-bound protein, protein glutathionylation, etc. [65]. In this chapter, we focus on the application of redox proteomics to determine posttranslationally modified proteins via nitration of tyrosine residues. Redox proteomics methods most often involve the coupling of two-dimensional (2D) gel electrophoresis techniques with mass spectrometry (MS) and 2D-Western blotting analysis.

Figure 1 shows an experimental scheme of the overall approach. A detailed description is provided elsewhere [66, 67]. Briefly, the experiment is carried out as follows. Samples are split into two equal aliquots and each separated by isoelectric focusing (IEF) coupled to sodium dodecyl sulfate polyacrylamide gel electrophoresis (SDS PAGE). The IEF SDS PAGE separation is based on the physiochemical properties of isoelectric point and migration rate (proportional to molecular weight). One gel is stained with Sypro Ruby and scanned at the appropriate wavelengths for total protein detection. The other gel is used to transfer proteins onto a nitrocellulose membrane for 2D-Western blot analysis. After transfer, the 2D-Western blot is probed with anti-3NT primary antibody and nitrated proteins are visualized with a colorimetric alkaline phosphatase assay. Protein spots in gels and blots are aligned and matched utilizing powerful image analysis software (i.e., PDQuest). Nitration levels for individual proteins are calculated by normalization to total protein levels in the gel (i.e., the ratio of the spot intensity on the 2D blot to the spot intensity on the gel). This takes into account changes in protein expression levels that may influence protein nitration levels.

Statistical analysis with ANOVA or Student's *t*-tests is carried out to determine the significance of changes in nitration between age-matched control and MCI or age-matched control and AD samples, respectively. Protein spots of interest are excised, digested with trypsin, and analyzed by matrix-assisted laser desorption ionization (MALDI)-MS analysis. Mass spectra are submitted to the MASCOT database search engine for final protein identification.

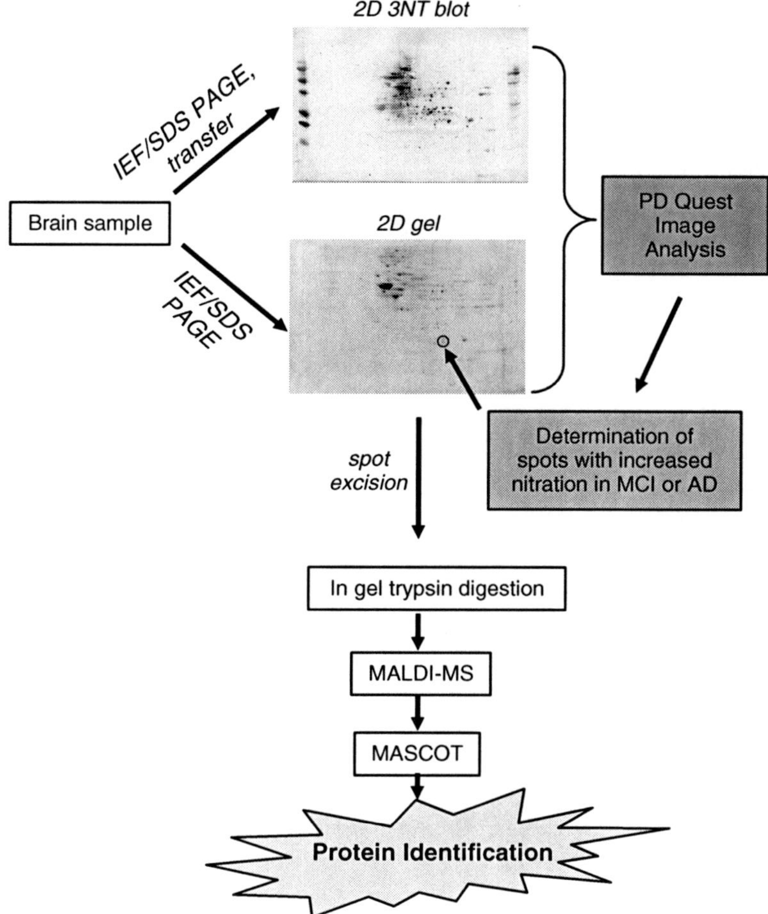

Fig. 1 Proteomics protocol utilized for the identification of nitrated proteins in MCI and AD brain

Goodness of fit of the proteins is embodied in the associated MOWSE score [$-10\log_{10} p$], where "p" represents the probability of a random identification of the protein of intent. Only proteins with MOWSE scores at or above a 95% confidence level (i.e., a score of 65 in these studies) are considered for further analysis. We note that a few proteins having a significant MOWSE score had nonsignificant increases (i.e., $p > 0.05$) in protein nitration in MCI [i.e., malate dehydrogenase, $p < 0.06$ [15]] and AD [i.e., β-actin, $p = 0.08$; lactate dehydrogenase, $p = 0.16$; γ-enolase, $p > 0.20$ [12]]. However, based on the biological functionalities of these proteins and their relationships with MCI and AD pathology, they have been included in the discussions below.

4.2 Identification of Nitrated Proteins in the MCI Brain

As mentioned above, elevated levels of nitration have been demonstrated previously in MCI inferior parietal lobule (IPL) and hippocampal brain regions [15, 52]. A comprehensive list of nitrated proteins in MCI brain, including functional categorizations, is provided in Table 1. This list includes the following proteins: α-enolase, glucose-regulated protein precursor, aldolase, malate dehydrogenase, glutathione-S-transferase Mu (GSTM3), multidrug-resistant protein (MRP3), 14-3-3-γ, peroxiredoxin 6 (PR VI), dihydropyrminidase-like protein-2 (DRP2), fascin 1, and heat shock protein 70 (HSPA8). Nitrated proteins in MCI can be grouped into several distinct biological functions and are discussed below with regard to MCI and AD pathology.

Table 1 Functional categorization of nitrated proteins identified in MCI and AD

Protein function	MCI[a]	AD
Energy dysfunction	**α-Enolase**	**α-Enolase**[b,c]
	Aldolase	γ-Enolase[b]
	Malate	Triosephosphate *isomerase*[b]
	Dehydrogenase	Glyceraldehyde-3-phosphate Dehydrogenase[c]
		Lactate dehydrogenase[b]
Mitochondrial dysfunction	–	ATP synthase α-chain[c]
		Voltage-dependent anion channel protein 1[c]
Lipid abnormalities and cholinergic dysfunction	–	Neuropolypeptide h3[c]
pH buffering and CO$_2$ transport	–	Carbonic anhydrase II[c]
Neuritic abnormalities and structural dysfunction	**DRP2**	**DRP2**[c]
	Fascin 1	β-Actin[b]
Antioxidant defense/detoxification system dysfunction	GSTM3	–
	MRP3	
	Peroxiredoxin 6	
	Glucose-regulated protein precursor	
	HSPA8	
Cell signaling dysfunction	*14-3-3-γ*	–

[a] For nitrated proteins identified in MCI, please see reference [15].
[b] For these nitrated proteins identified in AD, please see reference [12].
[c] For these nitrated proteins identified in AD, please see reference [19].
Nitrated proteins in **bold** are in common in both MCI and AD.

4.3 Energy Dysfunction

Proteins with elevated levels of nitration in MCI brain that are involved in energy metabolism are α-enolase, aldolase, and malate dehydrogenase [15]. Aldolase and α-enolase are enzymes directly involved in glycolysis. For example, aldolase catalyzes the conversion of fructose 1, 6-biphosphate into dihydroxyacetone phosphate and glyceraldehyde-3-phosphate in the second stages of glycolysis. α-Enolase is one of the subunits of the enolase enzyme that catalyzes the formation of phosphoenol pyruvate from the dehydration of 2-phosophoglycerate in the final stages of glycolysis. Malate dehydrogenase, on the other hand, catalyzes the conversion of malate to oxaloacetate in gluconeogenesis by transporting nicotinamide adenine dinucleotide (NADH) from the mitochondrion to the cytosol.

Because energy metabolism in the brain is heavily dependent on ATP generated from glycolysis any disruptions to the activities of glycolytic enzymes are detrimental to normal brain functions. Our laboratory and others have shown that oxidative modification of α-enolase in MCI and AD results in a loss of enzyme function and hence reduced amounts of ATP [12, 42, 45, 68–70]. Moreover, proper neural communication relies on the stores of ATP present in nerve terminals. Thus decreases in ATP, as a result of loss in glycolytic enzyme function, may contribute to synapse loss and dysfunction that lead to memory impairment and cognitive decline [71] as observed in amnestic MCI and AD. Other consequences of reduced ATP levels may include the following: impaired ion-motive ATPase activity with subsequent altered cell potential, loss of membrane lipid asymmetry and intercellular communication, and the induction of hypothermia which induces abnormal tau hyperphosphorylation [64, 72].

4.4 Neuritic Abnormalities and Structural Dysfunction

Maintenance of cytoskeletal structural integrity is crucial for proper neuronal transmission, especially with regard to the brain's ability to perform normal memory processes. DRP2 is involved in neuronal repair and in axonal outgrowth [73, 74] by regulating collapsin activity. Collapsin is a protein involved in dendritic elongation and axonal outgrowth. Thus, nitration of DRP2 may lead to a loss of collapsin activity and may be responsible for shortened dendritic length, which has been previously observed in AD brain [75]. Our lab has previously observed DRP2 to have increased oxidation (i.e., protein carbonyls) and decreased expression in AD brain [42, 45]. Ultimately, oxidative and nitrosative modification of DRP2 in MCI may be responsible for or enhance neuritic degeneration and synapse loss, pathological hallmarks of AD [75], resulting in cognitive deficits.

Fascin 1 is a structural protein involved in cell adhesion [76] and cell motility [77] and is a marker for dendritic functionality [78]. Nitration of fascin 1 could lead to a loss of protein function. Because fascin 1 has been shown to provide protection against oxidative insult [79], nitration and subsequently loss of function could lessen the cells natural defenses against oxidative stress. In addition, damage to fascin 1 may lead to poor neurotransmission from dendritic projections based on its role in cell adhesion. To date, this is the first association of fascin 1 with a neurodegenerative-related disorder.

4.5 Antioxidant Defense/Detoxification System Dysfunction

GSTM3, MRP3, PR VI, glucose-regulated protein precursor, and HSPA8 are involved in antioxidant defenses or detoxification processes within the cell. GSTM3 is an enzyme that functions in the detoxification of carcinogens, environmental toxins, therapeutic agents, and byproducts of oxidative stress [80, 81]. The mechanism of action of GSTM3 involves the conjugation of these toxins (e.g., xenobiotics or HNE) with glutathione. MRP3 then removes glutathione-conjugated toxic products out of the cell [82–84]. Thus, impairment of GSTM3 and MRP3 by nitration may result in impairments to the detoxification system crucial for removing ROS and toxic byproducts and maintaining low levels of oxidative stress. In addition to nitration, our lab has reported oxidative modification of GST and MRP1 by HNE in AD brain [84]. The proteomics results are consistent with reports of increased HNE levels in MCI IPL and hippocampal brain regions [66].

Peroxiredoxin VI is an efficient antioxidant enzyme that catalyzes the reduction of peroxynitrite [85] and is also involved in cell differentiation and apoptosis. Interestingly, PR VI forms a complex with GST [86], suggesting that alteration of either of these proteins can have detrimental effects on the detoxification and antioxidant systems of the brain. Changes in PR VI activity may also influence phospholipase A2 activity, a protein regulated by peptidyl prolyl *cis/trans* isomerase (Pin 1). Pin 1 is downregulated and has reduced activity in AD brain and may contribute to abnormal tau hyperphosphorylation which results in NFT formation [87, 88]. We have previously identified Pin 1 to be oxidatively modified through protein carbonylation in both MCI and AD brain [69, 89]. Nitration of GSTM3, MRP3, and PR VI may ultimately result in increased oxidative and nitrosative stress in MCI brain that contributes to disease pathology.

Glucose-regulated protein precursor is indirectly involved in energy production by regulation of glucose levels. It also belongs to the family of proteins that are molecular chaperones for the endoplasmic reticulum (ER) and thereby regulate proper protein folding of ER-associated proteins [90]. In situations where ER undergoes stress, glucose-regulated proteins (GRPs) help to protect against cell death [90]. Interestingly, inhibition of basal levels of GRP78 has been shown to increase $A\beta(1\text{--}40)$ and $A\beta(1\text{--}42)$ in cells [91]. Nitration of glucose-regulated protein precursor may influence the normal expression of GRPs under conditions of ER stress (e.g., oxidative damage). This may hinder protective mechanisms in cells, increase cell death, and lead to increased level of Aβ found in senile plaques (SP) of MCI patients.

Finally, HSPA8 is a chaperone protein belonging to the heat shock family of proteins. HSPs function by repairing misfolded proteins in response to cellular stress (i.e., oxidative damage, elevated temperature, etc.). Nitration of HSPA8 in MCI brain may result in functional impairment that may lead to elevated levels of misfolded proteins and thus increased amounts of protein aggregates [15]. Protein aggregates can cause inefficiencies in the proteasome due to "clogging," and hence these aggregates accumulate in the cells and contribute to disease pathology (e.g., Aβ peptide aggregates result in SP in AD brain). Other HSPs have been identified as oxidatively modified in AD [42] demonstrating the role of normal molecular chaperoning ability as an important pathway to reduce cellular and pathological defects observed in MCI brain.

4.6 Cell Signaling Dysfunction

14-3-3 Proteins are scaffolding proteins involved in signal transduction, protein trafficking, and metabolic processes [92, 93]. Several studies have reported 14-3-3 proteins to be increased in AD brain [94], and CSF [95] as well as in models of AD [45, 96, 97]. Nitration of 14-3-3-γ in MCI has important consequences, primarily due to the binding relationship of 14-3-3 to the proteins glycogen synthase kinase 3β (GSK 3β) and tau [98]. Conformational changes to the structure of 14-3-3 due to nitration could affect normal binding of these proteins and may enhance tau hyperphosphorylation [99]. Thus, nitration of 14-3-3-γ may indirectly contribute to NFT found in MCI and in AD brains [15].

5 Identification of Nitrated Proteins in the AD Brain

In order to fully understand the role of nitration in the progression from MCI to AD, a brief discussion of nitrated proteins that our laboratory has identified with redox proteomics techniques in AD brain is presented. Proteins with increased levels of nitration in AD brain, as listed in Table 1, are as follows: α- and γ-enolase, triosephosphate isomerase (TPI), glyceraldehyde-3-phosphate dehydrogenase (GAPDH), lactate dehydrogenase (LDH), ATP synthase α-chain, voltage-dependent anion channel 1 (VDAC), neuropolypeptide h3, carbonic anhydrase II, DRP2, and β-actin [12, 19]. In comparing the functional categories associated with these proteins to those observed in MCI, there is overlap in dysfunction of energy-related proteins and those involved in neuritic abnormalities and structural dysfunction. Because these proteins may be involved in recurring pathways in the progression of MCI to AD, the implications of nitration to proteins in these pathways as well as mitochondrial dysfunction, lipid abnormalities and cholinergic dysfunction and pH buffering and CO_2 transport are discussed.

5.1 Energy Dysfunction

Energy-related enzymes that are nitrated in AD brain are α- and γ-enolase, TPI, GAPDH, and LDH. As described above in MCI, dysfunction in energy-related enzymes can lead to decreased levels of ATP that are absolutely necessary for normal glucose metabolism in the brain and proper neuronal function. Positron emission tomography (PET) scans have shown that glucose metabolism is altered in the AD brain [100–102] and other studies have reported glucose intolerance in AD [103, 104]. These studies correlate with findings reported from our laboratory using redox proteomics of disturbances in energy-related pathways. Alterations in glucose metabolism are a pathological hallmark of AD, and as demonstrated above in these redox proteomics analyses, also in earlier disease stages such as MCI. These five enzymes are all directly or indirectly involved in the glycolysis pathway and thus are important to maintenance of proper glucose metabolism and ATP levels in the brain.

γ-Enolase is the second subunit, along with α-enolase, of the heterodimer enzyme enolase that is directly involved in glycolysis. Nitration of both enolase subunits, which are the predominant forms in the brain, can cause inefficiencies in the catalysis of 2-phosphoglycerate to phosphoenol pyruvate in glycolysis. These inefficiencies may then lead to a reduction in cellular ATP. α-Enolase has been reported as having increased levels in AD brain, oxidative modification through protein carbonylation, and a decreased activity [12, 42, 45, 70], thus providing direct evidence of its role in glucose metabolism impairment. TPI is an enzyme that provides high catalytic conversion of dihydroxyacetone phosphate into glyceraldehyde-3-phosphate in the second stages of glycolysis. In addition to nitration, our laboratory has found TPI to be oxidatively modified in AD brain [45, 70].

GAPDH catalyzes the conversion of glyceraldehyde-3-phosphate, generated from the aforementioned TPI, into 1,3-biphosphoglycerate utilizing NAD^+ as a cofactor in addition to inorganic phosphate. GAPDH has been reported to be oxidatively modified in AD brain with decreased activity [105, 106], and in other AD models [96]. Nitration and oxidative modification of GAPDH can affect its ability to serve as a NO· trap [107], resulting in increased 3NT formation in tyrosine residues of other proteins. The NADH that is generated from the GAPDH glycolytic reaction is used by LDH in its catalysis of pyruvate to lactate. Consequently, in the conversion of pyruvate to lactate, LDH also generates NAD^+ that is used by GAPDH, as previously mentioned. Thus, while nitration of LDH can influence its own activity it may also have detrimental effects on GAPDH activity (and vice versa) in AD brain.

Disturbances to enzymes in the glycolytic pathway have major consequences in MCI and in AD brain. The primary insult is a reduction in cellular ATP that is necessary for the brain to perform normal functions including neuronal communication and synapse transmission, events that are key to proper memory retrieval and storage functions [71]. ATP is also required by many cellular processes such as ATPases for maintenance of ion pumps and potential gradients and in lipid asymmetry. It is apparent that our proteomic findings of increased nitration to α- and γ-enolase, TPI, GAPDH, and LDH may contribute to glycolytic dysfunction reduced cellular ATP levels, and ultimately cognitive decline that is prevalent in AD patients.

5.2 Mitochondrial Dysfunction

ATP synthase, α-chain and VDAC are two mitochondrial-related proteins found in the inner and outer mitochondrial membranes, respectively, that we identified to have increased nitration in AD brain. Several studies have reported mitochondrial dysfunction to be an underlying cause to AD pathogenesis and neurodegeneration [108, 109]. In addition, other mitochondrial enzymes have been reported as reduced in AD brain [110, 111]. ATP synthase, α-chain is a subunit of the ATP synthesizing enzyme that is housed in the mitochondrion and generates ATP from the coupling of ADP and inorganic phosphate. ATP generation from ATP synthases is highly dependent on structural conformation such that ADP and inorganic phosphate must be close enough to each other to have a high affinity to bind. Nitration of ATP synthase, α-chain could affect the structure of the ATP synthase machinery and results in insufficient binding of ADP and inorganic phosphate. Thus, lower levels of ATP would be generated in mitochondria resulting in impairments to oxidative phosphorylation and subsequently reduced energy production. Other studies have shown ATP synthase, α-chain to be decreased in AD brain [112] and to accumulate in NFT [113]. Improper ATP synthesis can induce ROS production and leads to oxidative stress and neuronal death [45].

VDAC, also known as mitochondrial porin, is located in the mitochondrial permeability transition pore (MPTP) and is involved in the flux of metabolites, such as ATP, into and out of the mitochondrion. VDAC also plays roles in synaptic communication and apoptosis [114, 115]. Decreased expression of VDAC has been observed in various regions of AD brains [116] and deficits in learning behavior and synaptic plasticity have been reported in VDAC1-deficient mice [114]. Nitration of VDAC may alter MPTP function resulting in mitochondrial depolarization and disrupted signal transduction pathways, key elements for normal synaptic transmission and plasticity [45]. In addition, because VDAC contributes to the release of apoptotic factors such as cytochrome c [117], caspases [118], smac [119], and apoptosis-inducing factors [120], from the mitochondria, nitration of VDAC may induce apoptosis which results in cell death. Both neuron loss and synapse loss are known pathological hallmarks of AD, and apoptosis markers are elevated in AD and MCI [121].

5.3 Lipid Abnormalities and Cholinergic Dysfunction

The redox proteomics identified nitrated protein in AD that potentially is involved in lipid abnormalities and cholinergic dysfunction is neuropolypeptide h3, also known as phosphatidylethanolamine-binding protein (PEBP), hippocampal cholinergic neurostimulating peptide (HCNP), and Raf-kinase inhibitor protein (RKIP). As PEBP, nitration of neuropolypeptide h3 may influence lipid bilayer integrity. Our laboratory has shown that HNE and Aβ disrupt lipid asymmetry [122, 123] resulting in the exposure of phosphatidylserine to the outer leaflet, which induces apoptosis. Moreover, loss of phospholipid asymmetry is observed in MCI and AD brain [121] potentially coupling AD-induced ROS and RNS to cell death in MCI and AD brain.

PEBP is also the precursor of HCNP, a peptide that helps in the regulation of choline acetyltransferase and thus plays roles in signal transduction. It has been shown that a loss of choline acetyl transferase results in low levels of acetycholine, a neurotransmitter important in maintaining normal neurotransmission [124]. Cholinergic neuronal loss is associated with AD [125–127], and thus drug treatments with cholinesterase inhibitors (e.g., Aricept®) have been used to treat the symptoms in AD patients. Moreover, HCNP has been reported as decreased in AD hippocampus [128]. Nitration of HCNP could disturb choline acetyltransferase activity which has been shown to have decreased activity in AD brain [129]. This could lead to poor neurotransmission in AD which is relevant to the already known cholinergic deficits observed in AD patients.

5.4 pH Buffering and CO₂ Transport

Carbonic anhydrase II is an enzyme that catalyzes the reversible hydration of CO_2 to HCO_3^- and was identified as nitrated in AD brain. This enzyme also aids in the transport of CO_2 and HCO_3^- and in maintenance of cellular pH, electrolytic, and water balance [130]. Proper pH balance in the cell is crucial for pathways such as glycolysis and ATP synthesis and for optimal efficiency of other mitochondrial enzymes. Decreased carbonic anhydrase II activity has been previously reported in AD hippocampus [70]. Nitration of carbonic anhydrase II may lead to reduced enzymatic activity and improper pH status which would have consequences in all of the above discussed processes in AD brain. Moreover, altered pH could influence protein aggregation seen in AD.

5.5 Neuritic Abnormalities and Structural Dysfunction

As discussed above, DRP2 and β-actin are structural-related proteins. Each was identified as nitrated in AD brain. DRP2 was also identified as nitrated in MCI brain (see discussion above). Nitration of DRP2 may lead to diminished enzymatic activity and shortened dendritic length, which has been previously observed in AD brain [75]. Oxidative modification of DRP2 through protein carbonylation has been previously reported in AD brain and in AD models

[42, 45, 96]. In addition, decreased expression of DRP2 has been found both in AD brain and in Down's syndrome patients [131]. Nitrosative modification of DRP2 may be responsible for neuritic degeneration and synapse loss in AD [75], with consequent decreased inter-neuronal connection as discussed above.

β-Actin is a component of actin microfilaments found in neurons, glia, presynaptic terminals, dendritic spines, and in growth cones and plays roles in maintaining cytoskeletal structural integrity. Actin is primarily concentrated in dendritic spines and in growth cones in the brain [132] and helps in stabilizing the shape of the Golgi complex and actin filaments [133, 134]. β-Actin was found to be carbonylated in AD brain [135]. Oxidation and nitration of β-actin may influence the shape of dendrites, growth cones, and microfilaments resulting in poor neuron to neuron signal transmission, neuronal death, and ultimately, improper memory function. Both DRP2 and β-actin are important in the structural network of neurons that are necessary for normal neural communication in the brain.

6 Implication of Nitration in the Progression from MCI to AD

This chapter has discussed in some detail the effects that nitration of proteins can have in the progression of MCI to AD. Among the most profound effects are the disruptions that occur in energy-related enzymes that results in impaired glucose metabolism and reduced energy supply to the brain (via ATP). The ultimate consequences of these disturbances are neuronal death, synapse loss, poor neurotransmission, and subsequent memory loss; all of these events are characteristic of both amnestic MCI and AD. Nitration of proteins was also found to disturb neuritic and cytoskeletal structural integrity, antioxidant defenses/detoxification systems and cell signaling in MCI brain, and also mitochondrial function, cell buffering ability, lipid and cholinergic functions, as well as neuritic and cytoskeletal structural integrity in AD brain. Overall, we hypothesize that nitration of proteins can influence enzymatic activity in a manner that directly alters protein function, leading to MCI and AD pathology.

Early degradation of antioxidant and detoxification defenses in MCI can promote the induction of damaging ROS and contribute to further established AD pathology. In addition to augmented nitration of proteins in MCI, our laboratory has also identified increased carbonylation of proteins and lipid peroxidation in MCI brain [60, 62, 63, 69]. Thereby, oxidative and nitrosative stress are key elements in the earliest stages of AD (i.e., MCI) that result in worsened damage found in late-stage AD brain.

Of the many proteins that were nitrated in MCI and AD, only α-enolase and DRP2 were commonly identified as nitrated in both disease stages. Thus, these proteins may be important in the progression of MCI to AD. This implies that altered glucose metabolism, occurring from α-enolase nitration, and neuritic degeneration, resulting from DRP2 nitration, may be key components in the conversion of amnestic MCI to AD. It should be noted that α-enolase has been identified as both carbonylated and nitrated in MCI and AD in the IPL and hippocampus and as having enzymatic dysfunction [64]. Because not all individuals with MCI convert to AD and sometimes resume normal cognitive activity [136], therapeutic targets revolving around α-enolase and DRP2 pathways potentially conceivably may delay the onset of or prevent AD.

Redox proteomics studies that target MCI, an early stage in AD, are important for developing insights into the underlying causes of AD. Brain protein pathways that are revealed in common by redox proteomics analyses of MCI and AD show that energy-related pathways and structural pathways likely are involved in, and perhaps key to, disease pathogenesis. The findings presented in this chapter provide many insights into potential mechanisms of AD pathogenesis and into targets for therapeutic intervention at one of the earliest disease stage, MCI.

Acknowledgments This work was supported in part by NIH grants to D.A.B. [AG-10836 and AG-05119].

References

1. Bruijn LI, Beal MF, Becher MW, Schulz JB, Wong PC, Price DL, Cleveland DW. Elevated free nitrotyrosine levels, but not protein-bound nitrotyrosine or hydroxyl radicals, throughout amyotrophic lateral sclerosis (ALS)-like disease implicate tyrosine nitration as an aberrant in vivo property of one familial ALS-linked superoxide dismutase 1 mutant. Proc Natl Acad Sci U S A. 1997 Jul 8;94(14):7606–7611.
2. Calabrese V, Sultana R, Scapagnini G, Guagliano E, Sapienza M, Bella R, Kanski J, Pennisi G, Mancuso C, Stella AM, Butterfield DA. Nitrosative stress, cellular stress response, and thiol homeostasis in patients with Alzheimer's disease. Antioxid Redox Signal. 2006 Nov–Dec;8(11–12):1975–1986.
3. Kunz A, Park L, Abe T, Gallo EF, Anrather J, Zhou P, Iadecola C. Neurovascular protection by ischemic tolerance: role of nitric oxide and reactive oxygen species. J Neurosci. 2007 Jul 4;27(27):7083–7093.
4. Malinski T. Nitric oxide and nitroxidative stress in Alzheimer's disease. J Alzheimers Dis. 2007 May;11(2):207–218.
5. Moncada S, Bolanos JP. Nitric oxide, cell bioenergetics and neurodegeneration. J Neurochem. 2006 Jun;97(6):1676–1689.
6. Bergendi L, Benes L, Durackova Z, Ferencik M. Chemistry, physiology and pathology of free radicals. Life Sci. 1999;65(18–19):1865–1874.
7. Toader V, Xu X, Nicolescu A, Yu L, Bolton JL, Thatcher GR. Nitrosation, nitration, and autoxidation of the selective estrogen receptor modulator raloxifene by nitric oxide, peroxynitrite, and reactive nitrogen/oxygen species. Chem Res Toxicol. 2003 Oct;16(10):1264–1276.
8. Lafon-Cazal M, Culcasi M, Gaven F, Pietri S, Bockaert J. Nitric oxide, superoxide and peroxynitrite: putative mediators of NMDA-induced cell death in cerebellar granule cells. Neuropharmacology. 1993 Nov;32(11):1259–1266.
9. Kawano T, Kunz A, Abe T, Girouard H, Anrather J, Zhou P, Iadecola C. iNOS-derived NO and nox2-derived superoxide confer tolerance to excitotoxic brain injury through peroxynitrite. J Cereb Blood Flow Metab. 2007 Aug;27(8):1453–1462.

10. Souza JM, Daikhin E, Yudkoff M, Raman CS, Ischiropoulos H. Factors determining the selectivity of protein tyrosine nitration. Arch Biochem Biophys. 1999 Nov 15;371(2):169–178.

11. Butterfield DA, Stadtman ER. Protein oxidation processes in aging brain. Adv Cell Aging Gerontol; 1997. 2:161–191.

12. Castegna A, Thongboonkerd V, Klein JB, Lynn B, Markesbery WR, Butterfield DA. Proteomic identification of nitrated proteins in Alzheimer's disease brain. J Neurochem. 2003 Jun;85(6):1394–1401.

13. Koppal T, Drake J, Yatin S, Jordan B, Varadarajan S, Bettenhausen L, Butterfield DA. Peroxynitrite-induced alterations in synaptosomal membrane proteins: insight into oxidative stress in Alzheimer's disease. J Neurochem. 1999 Jan;72(1):310–317.

14. Sampson JB, Rosen H, Beckman JS. Peroxynitrite-dependent tyrosine nitration catalyzed by superoxide dismutase, myeloperoxidase, and horseradish peroxidase. Methods Enzymol. 1996;269:210–218.

15. Sultana R, Reed T, Perluigi M, Coccia R, Pierce WM, Butterfield DA. Proteomic identification of nitrated brain proteins in amnestic mild cognitive impairment: a regional study. J Cell Mol Med. 2007 Jul–Aug;11(4):839–851.

16. Sennlaub F, Courtois Y, Goureau O. Inducible nitric oxide synthase mediates retinal apoptosis in ischemic proliferative retinopathy. J Neurosci. 2002 May 15;22(10):3987–3993.

17. Ischiropoulos H, Zhu L, Chen J, Tsai M, Martin JC, Smith CD, Beckman JS. Peroxynitrite-mediated tyrosine nitration catalyzed by superoxide dismutase. Arch Biochem Biophys. 1992 Nov 1;298(2):431–437.

18. Gow AJ, Duran D, Malcolm S, Ischiropoulos H. Effects of peroxynitrite-induced protein modifications on tyrosine phosphorylation and degradation. FEBS Lett. 1996 Apr 29;385(1–2):63–66.

19. Sultana R, Poon HF, Cai J, Pierce WM, Merchant M, Klein JB, Markesbery WR, Butterfield DA. Identification of nitrated proteins in Alzheimer's disease brain using a redox proteomics approach. Neurobiol Dis. 2006 Apr;22(1):76–87.

20. Aulak KS, Koeck T, Crabb JW, Stuehr DJ. Dynamics of protein nitration in cells and mitochondria. Am J Physiol Heart Circ Physiol. 2004 Jan;286(1):H30–H38.

21. Koeck T, Fu X, Hazen SL, Crabb JW, Stuehr DJ, Aulak KS. Rapid and selective oxygen-regulated protein tyrosine denitration and nitration in mitochondria. J Biol Chem. 2004 Jun 25;279(26):27257–27262.

22. Butterfield DA. beta-Amyloid-associated free radical oxidative stress and neurotoxicity: implications for Alzheimer's disease. Chem Res Toxicol. 1997 May;10(5):495–506.

23. Caughey B, Lansbury PT. Protofibrils, pores, fibrils, and neurodegeneration: separating the responsible protein aggregates from the innocent bystanders. Annu Rev Neurosci. 2003;26:267–298.

24. Dahlgren KN, Manelli AM, Stine WB, Jr., Baker LK, Krafft GA, LaDu MJ. Oligomeric and fibrillar species of amyloid-beta peptides differentially affect neuronal viability. J Biol Chem. 2002 Aug 30;277(35):32046–32053.

25. Drake J, Link CD, Butterfield DA. Oxidative stress precedes fibrillar deposition of Alzheimer's disease amyloid beta-peptide (1–42) in a transgenic Caenorhabditis elegans model. Neurobiol Aging. 2003 May–Jun;24(3):415–420.

26. Fawzi NL, Kohlstedt KL, Okabe Y, Head-Gordon T. Protofibril assemblies of the Arctic, Dutch and Flemish mutants of the Alzheimer's A{beta}1–40 Peptide. Biophys J. 2007;44(6):2007–2016.

27. Walsh DM, Hartley DM, Condron MM, Selkoe DJ, Teplow DB. In vitro studies of amyloid beta-protein fibril assembly and toxicity provide clues to the aetiology of Flemish variant (Ala692->Gly) Alzheimer's disease. Biochem J. 2001 May 1;355(Pt 3):869–877.

28. Shen J, Kelleher RJ, 3rd. The presenilin hypothesis of Alzheimer's disease: evidence for a loss-of-function pathogenic mechanism. Proc Natl Acad Sci U S A. 2007 Jan 9;104(2):403–409.

29. Suh YH, Checler F. Amyloid precursor protein, presenilins, and alpha-synuclein: molecular pathogenesis and pharmacological applications in Alzheimer's disease. Pharmacol Rev. 2002 Sep;54(3):469–525.

30. Levy-Lahad E, Lahad A, Wijsman EM, Bird TD, Schellenberg GD. Apolipoprotein E genotypes and age of onset in early-onset familial Alzheimer's disease. Ann Neurol. 1995 Oct;38(4):678–680.

31. de la Monte SM, Lu BX, Sohn YK, Etienne D, Kraft J, Ganju N, Wands JR. Aberrant expression of nitric oxide synthase III in Alzheimer's disease: relevance to cerebral vasculopathy and neurodegeneration. Neurobiol Aging. 2000 Mar–Apr;21(2):309–319.

32. Hardy J, Selkoe DJ. The amyloid hypothesis of Alzheimer's disease: progress and problems on the road to therapeutics. Science. 2002 Jul 19;297(5580):353–356.

33. Butterfield DA, Drake J, Pocernich C, Castegna A. Evidence of oxidative damage in Alzheimer's disease brain: central role for amyloid beta-peptide. Trends Mol Med. 2001 Dec;7(12):548–554.

34. Barnham KJ, Ciccotosto GD, Tickler AK, Ali FE, Smith DG, Williamson NA, Lam YH, Carrington D, Tew D, Kocak G, Volitakis I, Separovic F, Barrow CJ, Wade JD, Masters CL, Cherny RA, Curtain CC, Bush AI, Cappai R. Neurotoxic, redox-competent Alzheimer's beta-amyloid is released from lipid membrane by methionine oxidation. J Biol Chem. 2003 Oct 31;278(44):42959–42965.

35. Lauderback CM, Hackett JM, Keller JN, Varadarajan S, Szweda L, Kindy M, Markesbery WR, Butterfield DA. Vulnerability of synaptosomes from apoE knock-out mice to structural and oxidative modifications induced by A beta(1–40): implications for Alzheimer's disease. Biochemistry. 2001 Feb 27;40(8):2548–2554.

36. Markesbery WR. Oxidative stress hypothesis in Alzheimer's disease. Free Radic Biol Med. 1997;23(1):134–147.

37. Butterfield DA, Koppal T, Subramaniam R, Yatin S. Vitamin E as an antioxidant/free radical scavenger against amyloid beta-peptide-induced oxidative stress in neocortical synaptosomal membranes and hippocampal neurons in culture: insights into Alzheimer's disease. Rev Neurosci. 1999;10(2):141–149.

38. Lovell MA, Markesbery WR. Ratio of 8-hydroxyguanine in intact DNA to free 8-hydroxyguanine is increased in Alzheimer disease ventricular cerebrospinal fluid. Arch Neurol. 2001 Mar;58(3):392–396.

39. Stadtman ER, Berlett BS. Reactive oxygen-mediated protein oxidation in aging and disease. Chem Res Toxicol. 1997 May;10(5):485–494.

40. Butterfield DA. Amyloid beta-peptide (1–42)-induced oxidative stress and neurotoxicity: implications for neurodegeneration in Alzheimer's disease brain. A review. Free Radic Res. 2002 Dec;36(12):1307–1313.

41. Butterfield DA, Lauderback CM. Lipid peroxidation and protein oxidation in Alzheimer's disease brain: potential causes and consequences involving amyloid beta-peptide-associated free radical oxidative stress. Free Radic Biol Med. 2002 Jun 1;32(11):1050–1060.

42. Castegna A, Aksenov M, Thongboonkerd V, Klein JB, Pierce WM, Booze R, Markesbery WR, Butterfield DA. Proteomic identification of oxidatively modified proteins in Alzheimer's disease brain. Part II: dihydropyrimidinase-related protein 2, alpha-enolase and heat shock cognate 71. J Neurochem. 2002 Sep;82(6):1524–1532.

43. Castegna A, Thongboonkerd V, Klein J, Lynn BC, Wang YL, Osaka H, Wada K, Butterfield DA. Proteomic analysis of brain proteins in the gracile axonal dystrophy (gad) mouse, a syndrome that emanates from dysfunctional ubiquitin carboxyl-terminal hydrolase L-1, reveals oxidation of key proteins. J Neurochem. 2004 Mar;88(6):1540–1546.

44. Smith MA, Richey Harris PL, Sayre LM, Beckman JS, Perry G. Widespread peroxynitrite-mediated damage in Alzheimer's disease. J Neurosci. 1997 Apr 15;17(8):2653–2657.

45. Sultana R, Boyd-Kimball D, Poon HF, Cai J, Pierce WM, Klein JB, Merchant M, Markesbery WR, Butterfield DA. Redox proteomics identification of oxidized proteins

in Alzheimer's disease hippocampus and cerebellum: an approach to understand pathological and biochemical alterations in AD. Neurobiol Aging. 2006 Nov;27(11):1564–1576.

46. Hensley K, Maidt ML, Yu Z, Sang H, Markesbery WR, Floyd RA. Electrochemical analysis of protein nitrotyrosine and dityrosine in the Alzheimer brain indicates region-specific accumulation. J Neurosci. 1998 Oct 15;18(20):8126–8132.

47. Horiguchi T, Uryu K, Giasson BI, Ischiropoulos H, LightFoot R, Bellmann C, Richter-Landsberg C, Lee VM, Trojanowski JQ. Nitration of tau protein is linked to neurodegeneration in tauopathies. Am J Pathol. 2003 Sep;163(3):1021–1031.

48. Zhang YJ, Xu YF, Liu YH, Yin J, Li HL, Wang Q, Wang JZ. Peroxynitrite induces Alzheimer-like tau modifications and accumulation in rat brain and its underlying mechanisms. Faseb J. 2006 Jul;20(9):1431–1442.

49. Petersen RC, Morris JC. Mild cognitive impairment as a clinical entity and treatment target. Arch Neurol. 2005 Jul;62(7):1160–1163; discussion 1167.

50. Nacmias B, Piccini C, Bagnoli S, Tedde A, Cellini E, Bracco L, Sorbi S. Brain-derived neurotrophic factor, apolipoprotein E genetic variants and cognitive performance in Alzheimer's disease. Neurosci Lett. 2004 Sep 9;367(3):379–383.

51. Almkvist O, Basun H, Backman L, Herlitz A, Lannfelt L, Small B, Viitanen M, Wahlund LO, Winblad B. Mild cognitive impairment – an early stage of Alzheimer's disease? J Neural Transm Suppl. 1998;54:21–29.

52. Butterfield DA, Reed TT, Perluigi M, De Marco C, Coccia R, Keller JN, Markesbery WR, Sultana R. Elevated levels of 3-nitrotyrosine in brain from subjects with amnestic mild cognitive impairment: implications for the role of nitration in the progression of Alzheimer's disease. Brain Res. 2007 May 7;1148:243–248.

53. Sultana R, Piroddi M, Galli F, Butterfield DA. Protein levels and activity of some antioxidant enzymes in amnestic mild cognitive impairment and control hippocampus. Neurochem Res. 2008 Dec;33(12):2540–2546.

54. Rinaldi P, Polidori MC, Metastasio A, Mariani E, Mattioli P, Cherubini A, Catani M, Cecchetti R, Senin U, Mecocci P. Plasma antioxidants are similarly depleted in mild cognitive impairment and in Alzheimer's disease. Neurobiol Aging. 2003 Nov;24(7):915–919.

55. Guidi I, Galimberti D, Lonati S, Novembrino C, Bamonti F, Tiriticco M, Fenoglio C, Venturelli E, Baron P, Bresolin N, Scarpini E. Oxidative imbalance in patients with mild cognitive impairment and Alzheimer's disease. Neurobiol Aging. 2006 Feb;27(2):262–269.

56. Wang J, Markesbery WR, Lovell MA. Increased oxidative damage in nuclear and mitochondrial DNA in mild cognitive impairment. J Neurochem. 2006 Feb;96(3):825–832.

57. Lovell MA, Markesbery WR. Oxidative DNA damage in mild cognitive impairment and late-stage Alzheimer's disease. Nucleic Acids Res. 2007;35(22):7497–7504.

58. Irizarry MC, Yao Y, Hyman BT, Growdon JH, Pratico D. Plasma F2A isoprostane levels in Alzheimer's and Parkinson's disease. Neurodegener Dis. 2007;4(6):403–405.

59. Cenini G, Sultana R, Memo M, Butterfield DA. Elevated levels of pro-apoptotic p53 and its oxidative modification by the lipid peroxidation product, HNE, in brain from subjects with amnestic mild cognitive impairment and Alzheimer's disease. J Cell Mol Med. 2007 Jun;12(3):987–994.

60. Williams TI, Lynn BC, Markesbery WR, Lovell MA. Increased levels of 4-hydroxynonenal and acrolein, neurotoxic markers of lipid peroxidation, in the brain in mild cognitive impairment and early Alzheimer's disease. Neurobiol Aging. 2006 Aug;27(8):1094–1099.

61. Markesbery WR, Lovell MA. Damage to lipids, proteins, DNA, and RNA in mild cognitive impairment. Arch Neurol. 2007 Jul;64(7):954–956.

62. Butterfield DA, Reed T, Perluigi M, De Marco C, Coccia R, Cini C, Sultana R. Elevated protein-bound levels of the lipid peroxidation product, 4-hydroxy-2-nonenal, in brain from persons with mild cognitive impairment. Neurosci Lett. 2006 Apr 24;397(3):170–173.

63. Keller JN, Schmitt FA, Scheff SW, Ding Q, Chen Q, Butterfield DA, Markesbery WR. Evidence of increased oxidative damage in subjects with mild cognitive impairment. Neurology. 2005 Apr 12;64(7):1152–1156.

64. Butterfield DA, Sultana R. Redox proteomics identification of oxidatively modified brain proteins in Alzheimer's disease and mild cognitive impairment: insights into the progression of this dementing disorder. J Alzheimers Dis. 2007 Aug;12(1):61–72.

65. Dalle-Donne I, Scaloni A, Butterfield DA. Redox proteomics: from protein modifications to cellular dysfunction and diseases. John Wiley and Sons, Hoboken, NJ. 2006.

66. Butterfield DA, Perluigi M, Sultana R. Oxidative stress in Alzheimer's disease brain: new insights from redox proteomics. Eur J Pharmacol. 2006 Sep 1;545(1):39–50.

67. Sultana R, Perluigi M, Butterfield DA. Redox proteomics identification of oxidatively modified proteins in Alzheimer's disease brain and in vivo and in vitro models of AD centered around Abeta(1–42). J Chromatogr B Analyt Technol Biomed Life Sci. 2006 Mar 20;833(1):3–11.

68. Aksenova M, Butterfield DA, Zhang SX, Underwood M, Geddes JW. Increased protein oxidation and decreased creatine kinase BB expression and activity after spinal cord contusion injury. J Neurotrauma. 2002 Apr;19(4):491–502.

69. Butterfield DA, Poon HF, St Clair D, Keller JN, Pierce WM, Klein JB, Markesbery WR. Redox proteomics identification of oxidatively modified hippocampal proteins in mild cognitive impairment: insights into the development of Alzheimer's disease. Neurobiol Dis. 2006 May;22(2):223–232.

70. Meier-Ruge W, Iwangoff P, Reichlmeier K. Neurochemical enzyme changes in Alzheimer's and Pick's disease. Arch Gerontol Geriatr. 1984 Jul;3(2):161–165.

71. Hoyer S. Memory function and brain glucose metabolism. Pharmacopsychiatry. 2003 Jun;36 Suppl 1:S62–S67.

72. Planel E, Miyasaka T, Launey T, Chui DH, Tanemura K, Sato S, Murayama O, Ishiguro K, Tatebayashi Y, Takashima A. Alterations in glucose metabolism induce hypothermia leading to tau hyperphosphorylation through differential inhibition of kinase and phosphatase activities: implications for Alzheimer's disease. J Neurosci. 2004 Mar 10;24(10):2401–2411.

73. Hamajima N, Matsuda K, Sakata S, Tamaki N, Sasaki M, Nonaka M. A novel gene family defined by human dihydropyrimidinase and three related proteins with differential tissue distribution. Gene. 1996 Nov 21;180(1–2):157–163.

74. Kato Y, Hamajima N, Inagaki H, Okamura N, Koji T, Sasaki M, Nonaka M. Post-meiotic expression of the mouse dihydropyrimidinase-related protein 3 (DRP-3) gene during spermiogenesis. Mol Reprod Dev. 1998 Sep;51(1):105–111.

75. Coleman PD, Flood DG. Neuron numbers and dendritic extent in normal aging and Alzheimer's disease. Neurobiol Aging. 1987 Nov–Dec;8(6):521–545.

76. Adams JC. Formation of stable microspikes containing actin and the 55 kDa actin bundling protein, fascin, is a consequence of cell adhesion to thrombospondin-1: implications for the anti-adhesive activities of thrombospondin-1. J Cell Sci. 1995;108:1977–1990.

77. Adams JC. Roles of fascin in cell adhesion and motility. Curr Opin Cell Biol. 2004;16:590–596.

78. Pinkus GS, Lones MA, Matsumura F, Yamashiro S, Said JW, Pinkus JL. Langerhans cell histiocytosis immunohistochemical expression of fascin, a dendritic cell marker. Am J Clin Pathol. 2002 Sep;118(3):335–343.

79. Graziewicz MA, Day BJ, Copeland WC. The mitochondrial DNA polymerase as a target of oxidative damage. Nucleic Acids Res. 2002 Jul 1;30(13):2817–2824.

80. Pearson WR, Vorachek WR, Xu SJ, Berger R, Hart I, Vannais D, Patterson D. Identification of class-mu glutathione transferase genes GSTM1-GSTM5 on human chromosome 1p13. Am J Hum Genet. 1993 Jul;53(1):220–233.

81. Tchaikovskaya T, Fraifeld V, Urphanishvili T, Andorfer JH, Davies P, Listowsky I. Glutathione S-transferase hGSTM3 and ageing-associated neurodegeneration: relationship to Alzheimer's disease. Mech Ageing Dev. 2005 Feb;126(2):309–315.

82. Joshi G, Hardas S, Sultana R, St Clair DK, Vore M, Butterfield DA. Glutathione elevation by gamma-glutamyl cysteine ethyl ester as a potential therapeutic strategy for

preventing oxidative stress in brain mediated by in vivo administration of adriamycin: Implication for chemobrain. J Neurosci Res. 2007 Feb 15;85(3):497–503.

83. Renes J, de Vries EE, Hooiveld GJ, Krikken I, Jansen PL, Muller M. Multidrug resistance protein MRP1 protects against the toxicity of the major lipid peroxidation product 4-hydroxynonenal. Biochem J. 2000 Sep 1;350 Pt 2:555–561.

84. Sultana R, Butterfield DA. Oxidatively modified GST and MRP1 in Alzheimer's disease brain: implications for accumulation of reactive lipid peroxidation products. Neurochem Res. 2004 Dec;29(12):2215–2220.

85. Peshenko IV, Shichi H. Oxidation of active center cysteine of bovine 1-Cys peroxiredoxin to the cysteine sulfenic acid form by peroxide and peroxynitrite. Free Radic Biol Med. 2001 Aug 1;31(3):292–303.

86. Ralat LA, Manevich Y, Fisher AB, Colman RF. Direct evidence for the formation of a complex between 1-cysteine peroxiredoxin and glutathione S-transferase pi with activity changes in both enzymes. Biochemistry. 2006 Jan 17;45(2):360–372.

87. Lu KP. Phosphorylation-dependent prolyl isomerization: a novel cell cycle regulatory mechanism. Prog Cell Cycle Res. 2000;4:83–96.

88. Lu PJ, Wulf G, Zhou XZ, Davies P, Lu KP. The prolyl isomerase Pin1 restores the function of Alzheimer-associated phosphorylated tau protein. Nature. 1999 Jun 24;399(6738):784–788.

89. Sultana R, Boyd-Kimball D, Poon HF, Cai J, Pierce WM, Klein JB, Markesbery WR, Zhou XZ, Lu KP, Butterfield DA. Oxidative modification and down-regulation of Pin1 in Alzheimer's disease hippocampus: a redox proteomics analysis. Neurobiol Aging. 2006 Jul;27(7):918–925.

90. Lee AS. The glucose-regulated proteins: stress induction and clinical applications. Trends Biochem Sci. 2001 Aug;26(8):504–510.

91. Hoshino T, Nakaya T, Araki W, Suzuki K, Suzuki T, Mizushima T. Endoplasmic reticulum chaperones inhibit the production of amyloid-beta peptides. Biochem J. 2007 Mar 15;402(3):581–589.

92. Dougherty MK, Morrison DK. Unlocking the code of 14-3-3. J Cell Sci. 2004 Apr 15;117(Pt 10):1875–1884.

93. Takahashi Y. The 14-3-3 proteins: gene, gene expression and function. Neurochem Res. 2003;28:1265–1273.

94. Layfield R, Fergusson J, Aitken A, Lowe J, Landon M, Mayer RJ. Neurofibrillary tangles of Alzheimer's disease brains contain 14-3-3 proteins. Neurosci Lett. 1996 May 3;209(1):57–60.

95. Burkhard PR, Sanchez JC, Landis T, Hochstrasser DF. CSF detection of the 14-3-3 protein in unselected patients with dementia. Neurology. 2001 Jun 12;56(11):1528–1533.

96. Boyd-Kimball D, Sultana R, Poon HF, Lynn BC, Casamenti F, Pepeu G, Klein JB, Butterfield DA. Proteomic identification of proteins specifically oxidized by intracerebral injection of amyloid beta-peptide (1–42) into rat brain: implications for Alzheimer's disease. Neuroscience. 2005;132(2):313–324.

97. Frautschy SA, Baird A, Cole GM. Effects of injected Alzheimer beta-amyloid cores in rat brain. Proc Natl Acad Sci U S A. 1991 Oct 1;88(19):8362–8366.

98. Agarwal-Mawal A, Qureshi HY, Cafferty PW, Yuan Z, Han D, Lin R, Paudel HK. 14-3-3 connects glycogen synthase kinase-3 beta to tau within a brain microtubule-associated tau phosphorylation complex. J Biol Chem. 2003 Apr 11;278(15):12722–12728.

99. Hashiguchi M, Sobue K, Paudel HK. 14-3-3zeta is an effector of tau protein phosphorylation. J Biol Chem. 2000 Aug 18;275(33):25247–25254.

100. Erecinska M, Silver IA. ATP and brain function. J Cereb Blood Flow Metab. 1989 Feb;9(1):2–19.

101. Hoyer S. Causes and consequences of disturbances of cerebral glucose metabolism in sporadic Alzheimer disease: therapeutic implications. Adv Exp Med Biol. 2004;541:135–152.

102. Rapoport SI. In vivo PET imaging and postmortem studies suggest potentially reversible and irreversible stages of brain metabolic failure in Alzheimer's disease. Eur Arch Psychiatry Clin Neurosci. 1999;249:46–55.
103. Mattson MP, Pedersen WA, Duan W, Culmsee C, Camandola S. Cellular and molecular mechanisms underlying perturbed energy metabolism and neuronal degeneration in Alzheimer's and Parkinson's diseases. Ann N Y Acad Sci. 1999;893:154–175.
104. Vanhanen M, Soininen H. Glucose intolerance, cognitive impairment and Alzheimer's disease. Curr Opin Neurol. 1998 Dec;11(6):673–677.
105. Kish SJ, Lopes-Cendes I, Guttman M, Furukawa Y, Pandolfo M, Rouleau GA, Ross BM, Nance M, Schut L, Ang L, DiStefano L. Brain glyceraldehyde-3-phosphate dehydrogenase activity in human trinucleotide repeat disorders. Arch Neurol. 1998 Oct;55(10):1299–1304.
106. Mazzola JL, Sirover MA. Reduction of glyceraldehyde-3-phosphate dehydrogenase activity in Alzheimer's disease and in Huntington's disease fibroblasts. J Neurochem. 2001 Jan;76(2):442–449.
107. Hara MR, Cascio MB, Sawa A. GAPDH as a sensor of NO stress. Biochim Biophys Acta. 2006 May;1762(5):502–509.
108. Blass JP, Gibson GE, Hoyer S. The role of the metabolic lesion in Alzheimer's disease. J Alzheimers Dis. 2002 Jun;4(3):225–232.
109. Bubber P, Haroutunian V, Fisch G, Blass JP, Gibson GE. Mitochondrial abnormalities in Alzheimer brain: mechanistic implications. Ann Neurol. 2005 May;57(5):695–703.
110. Bosetti F, Brizzi F, Barogi S, Mancuso M, Siciliano G, Tendi EA, Murri L, Rapoport SI, Solaini G. Cytochrome c oxidase and mitochondrial F1F0-ATPase (ATP synthase) activities in platelets and brain from patients with Alzheimer's disease. Neurobiol Aging. 2002 May–Jun;23(3):371–376.
111. Hirai K, Aliev G, Nunomura A, Fujioka H, Russell RL, Atwood CS, Johnson AB, Kress Y, Vinters HV, Tabaton M, Shimohama S, Cash AD, Siedlak SL, Harris PL, Jones PK, Petersen RB, Perry G, Smith MA. Mitochondrial abnormalities in Alzheimer's disease. J Neurosci. 2001 May 1;21(9):3017–3023.
112. Schagger H, Ohm TG. Human diseases with defects in oxidative phosphorylation. 2. F1F0 ATP-synthase defects in Alzheimer disease revealed by blue native polyacrylamide gel electrophoresis. Eur J Biochem. 1995 Feb 1;227(3):916–921.
113. Sergeant N, Wattez A, Galvan-valencia M, Ghestem A, David JP, Lemoine J, Sautiere PE, Dachary J, Mazat JP, Michalski JC, Velours J, Mena-Lopez R, Delacourte A. Association of ATP synthase alpha-chain with neurofibrillary degeneration in Alzheimer's disease. Neuroscience. 2003;117(2):293–303.
114. Weeber EJ, Levy M, Sampson MJ, Anflous K, Armstrong DL, Brown SE, Sweatt JD, Craigen WJ. The role of mitochondrial porins and the permeability transition pore in learning and synaptic plasticity. J Biol Chem. 2002 May 24;277(21):18891–18897.
115. Zaid H, Abu-Hamad S, Israelson A, Nathan I, Shoshan-Barmatz V. The voltage-dependent anion channel-1 modulates apoptotic cell death. Cell Death Differ. 2005 Jul;12(7):751–760.
116. Yoo BC, Fountoulakis M, Cairns N, Lubec G. Changes of voltage-dependent anion-selective channel proteins VDAC1 and VDAC2 brain levels in patients with Alzheimer's disease and Down syndrome. Electrophoresis. 2001 Jan;22(1):172–179.
117. Madesh M, Hajnoczky G. VDAC-dependent permeabilization of the outer mitochondrial membrane by superoxide induces rapid and massive cytochrome c release. J Cell Biol. 2001 Dec 10;155(6):1003–1015.
118. Shimizu H, Banno Y, Sumi N, Naganawa T, Kitajima Y, Nozawa Y. Activation of p38 mitogen-activated protein kinase and caspases in UVB-induced apoptosis of human keratinocyte HaCaT cells. J Invest Dermatol. 1999 May;112(5):769–774.
119. Du C, Fang, M., Li, Y., Li, L., Wang, X. Smac, a mitochondrial protein that promotes cytochrome c-dependent caspase activiation by eliminating IAP inhibition. Cell 2000;102:33–42.

120. Crompton M. The mitochondrial permeability transition pore and its role in cell death. Biochem J. 1999;341:233–249.

121. Bader Lange ML, Cenini G, Piroddi M, Mohmmad Abdul H, Sultana R, Galli F, Memo M, Butterfield DA. Loss of phospholipid asymmetry and elevated brain apoptotic protein levels in subjects with amnestic mild cognitive impairment and Alzheimer disease. Neurobiol Dis. 2008 Mar;29(3):456–464.

122. Abdul H, Butterfield D. Protection against amyloid beta-peptide(1–42)-induced loss of phospholipid asymmetry in synaptosomal membranes by tricyclodecan-9-xanthogenate (D609) and ferulic acid ethyl ester: implications for Alzheimer's disease. Biochem Biophys Acta. 2005;1741:140–148.

123. Castegna A, Lauderback CM, Mohmmad-Abdul H, Butterfield DA. Modulation of phospholipid asymmetry in synaptosomal membranes by the lipid peroxidation products, 4-hydroxynonenal and acrolein: implications for Alzheimer's disease. Brain Res. 2004 Apr 9;1004(1–2):193–197.

124. Ojika K. Hippocampal cholinergic neurostimulating peptide. Seikagaku. 1998 Sep;70(9):1175–1180.

125. Davies P, Maloney AJ. Selective loss of central cholinergic neurons in Alzheimer's disease. Lancet. 1976 Dec 25;2(8000):1403.

126. Coyle JT, Price DL, DeLong MR. Alzheimer's disease: a disorder of cortical cholinergic innervation. Science. 1983 Mar 11;219(4589):1184–1190.

127. Wevers A, Witter B, Moser N, Burghaus L, Banerjee C, Steinlein OK, Schutz U, de Vos RA, Steur EN, Lindstrom J, Schroder H. Classical Alzheimer features and cholinergic dysfunction: towards a unifying hypothesis? Acta Neurol Scand Suppl. 2000;176:42–48.

128. Maki M, Matsukawa N, Yuasa H, Otsuka Y, Yamamoto T, Akatsu H, Okamoto T, Ueda R, Ojika K. Decreased expression of hippocampal cholinergic neurostimulating peptide precursor protein mRNA in the hippocampus in Alzheimer disease. J Neuropathol Exp Neurol. 2002 Feb;61(2):176–185.

129. Davies MJ, Fu S, Wang H, Dean RT. Stable markers of oxidant damage to proteins and their application in the study of human disease. Free Radic Biol Med. 1999 Dec;27(11–12):1151–1163.

130. Sly WS, Hu PY. Human carbonic anhydrases and carbonic anhydrase deficiencies. Annu Rev Biochem. 1995;64:375–401.

131. Lubec G, Nonaka M, Krapfenbauer K, Gratzer M, Cairns N, Fountoulakis M. Expression of the dihydropyrimidinase related protein 2 (DRP-2) in Down syndrome and Alzheimer's disease brain is downregulated at the mRNA and dysregulated at the protein level. J Neural Transm Suppl. 1999;57:161–177.

132. Kaech S, Brinkhaus H, Matus A. Volatile anesthetics block actin-based motility in dendritic spines. Proc Natl Acad Sci U S A. 1999 Aug 31;96(18):10433–10437.

133. di Campli A, Valderrama F, Babia T, De Matteis MA, Luini A, Egea G. Morphological changes in the Golgi complex correlate with actin cytoskeleton rearrangements. Cell Motil Cytoskeleton. 1999;43(4):334–348.

134. Valderrama F, Luna A, Babia T, Martinez-Menarguez JA, Ballesta J, Barth H, Chaponnier C, Renau-Piqueras J, Egea G. The golgi-associated COPI-coated buds and vesicles contain beta/gamma -actin. Proc Natl Acad Sci U S A. 2000 Feb 15;97(4):1560–1565.

135. Aksenov MY, Aksenova MV, Butterfield DA, Geddes JW, Markesbery WR. Protein oxidation in the brain in Alzheimer's disease. Neuroscience. 2001;103(2):373–383.

136. Petersen RC. Mild cognitive impairment: transition between aging and Alzheimer's disease. Neurologia 2000;15(3):93–101.

Parkinson's Disease: An Overview of Pathogenesis

Pratap Chand and Irene Litvan

Abstract Parkinson's disease (PD) is clinically, pathologically, and etiologically diverse and complex combinations of genetic susceptibility and environmental risk factors like age, gender, estrogen status, race/ethnicity, exposure to pesticides, and head trauma play roles in the pathogenesis. Genetic research has identified eleven PARK loci in familial cases and genotyping technology has allowed genome-wide approaches in large populations of patients with sporadic PD. Neuropathologic studies have defined and characterized more than 30 proteins with the core filament of α-synuclein in the Lewy body inclusions of idiopathic PD. Analysis of Lewy body pathology has outlined progressive disease stages in which the intracellular deposition of α-synuclein affects medullary sites before more rostral brain regions and the late stages of involvement of cortical association neurons. In PD, there is evidence of protein misfolding, aggregation, and abnormal neuronal apoptosis. Toxin-based and recent gene mutation-based animal models of PD in mammals, flies, fish, and worms have reproduced the progressive motor and non-motor deficits observed in PD patients and shown dopamine dysfunction and delayed loss of dopaminergic neurons in the substantia nigra. Oxidative stress plays a key role in the pathogenesis of PD and the demonstration of nitrated protein within Lewy bodies and neurites provides evidence of oxidative damage in the pathogenesis of PD. Reduction in the mitochondrial enzyme complex I in PD renders neurons vulnerable to unstable oxygen free radicals and causes alpha-synuclein to aggregate into fibrils and lead to dopaminergic neuronal loss. The DJ-1 gene has antioxidant function in the cellular response to oxidative stimuli and DJ-1 is oxidatively damaged in the brains of PD patients. The occurrence of a parkinsonian syndrome in families with mutations in the DJ-1 gene also supports a role for oxidative stress in the pathogenesis of PD. Therapeutic approaches that could prevent or limit oxidative

P. Chand (✉)
Movement Disorders, Department of Neurology, Saint Louis University University
School of Medicine, St Louis, MO, USA
e-mail: pchand1@slu.edu

S.C. Veasey (ed.), *Oxidative Neural Injury*,
Contemporary Clinical Neuroscience, DOI 10.1007/978-1-60327-342-8_10,
© Humana Press, a part of Springer Science+Business Media, LLC 2009

reactions in PD represent a viable strategy for the development of newer anti-parkinsonian drugs.

Keywords Parkinson's disease · Oxidative stress · Pathogenesis

1 Introduction

Parkinson's disease is the second most common age-related neurodegenerative disease, where the prevalence increases with age to over 5% for individuals over the age of 85 years. The definition of idiopathic Parkinson's disease (PD) includes a characteristic motor phenotype (rigidity, resting tremor, bradykinesia, and postural instability), a distinctive neuropathology, and substantial loss of dopaminergic neurons from the substantia nigra associated with the presence of α-synuclein-positive inclusions in the cell body (Lewy bodies) and processes (Lewy neurites) of specific neurons of the brainstem [1]. The spectrum of PD is variable and may include nonmotor systems – olfactory dysfunction, cognition, sleep, constipation, rapid eye movement (REM), sleep behavior disorder, and depression may precede or present commonly with motor symptomatology in early PD [2]. These symptoms correlate well with the neuropathology and deposition of α-synuclein as outlined by Braak stage 1 (olfactory bulb and peripheral and central medullary autonomic neurons) and stage 2 (locus ceruleus and pontine tegmentum), which are proposed to precede the classical motor symptoms of PD (stage 3 with α-synuclein deposition in the substantia nigra) [3]. When the full spectrum of classical motor symptoms of PD is present, there is a devastating loss of dopamine neurons [4] with α-synuclein accumulation and the loss of nigral neurons is correlated with the severity of akinesia and rigidity [5]. There is also involvement of the serotoninergic, noradrenergic, and cholinergic brainstem nuclei and the olfactory, peripheral sympathetic, and myenteric nervous system [3]. Although single gene defects (e.g., in the *Parkin* gene), single environmental toxins (e.g., 1-methyl-4-phenyl-1,2,3,6-tetrahydropyridine), or single infectious agents (e.g., encephalitis lethargica) have been associated with rare forms of relatively pure parkinsonism, current theories of pathogenesis concerning idiopathic PD focus on a combination of genetic susceptibility and environmental risk factors [6–9].

2 Pathogenesis of Parkinson's Disease

2.1 *Epidemiologic, Environmental, and Demographic Issues*

Putative environmental and demographic factors that may predispose individuals to idiopathic PD include age, sex, estrogen status, race or ethnicity, exposure to pesticides, head trauma, not smoking, not drinking alcohol, and not drinking coffee. Of all the factors studied, age is the most highly linked to PD: the incidence of PD increases steeply with age in both men and women [6, 7]. A male predominance in PD has been confirmed through many observational studies [9, 10]. Epidemiologic investigations of estrogens in women [11] show that women with PD are more likely to have undergone hysterectomy (with and without unilateral oophorectomy) or bilateral oophorectomy and had more early menopause than control subjects. However, normal cyclic changes in estrogen and progesterone in premenopausal women do not correlate with changes in the signs or symptoms of parkinsonism [12]. Several studies have suggested that Caucasians are affected more often than African Americans; however, this question remains unsettled [13, 14]. An analysis of seven European populations revealed no substantial difference in the prevalence of PD across European countries [15, 16]. Other reports of variable incidence rates in different cultures may at least partly relate to the lack of uniform diagnostic criteria and differences in selected diagnostic tools [17–19].

Some studies have suggested that PD is more common in highly industrialized countries than in agricultural societies and more frequent in Europe and North America than in the Far East [20–22]. However, there remains surprising uncertainty as to whether PD prevalence rates differ across the five continents or between developed and developing countries, as suggested by recent findings from China [19, 23, 24, 25]. Environmental toxins can induce excessive production of oxygen free radicals and damaging particles that may play a major role in the neuronal degeneration leading to PD. Several studies suggest that chronic exposure to well water and living near or working with industrial chemicals or pesticide and herbicide products increases the risk for PD, especially in subjects with young onset [20, 25, 26]. In an urban United States multiethnic community, rural living, area farming, and well water drinking were associated with PD only in African Americans, whereas in Hispanics, area farming was protective and drinking unfiltered water was a risk factor [27]. In Denmark, a consistent pattern of high PD morbidity was found among occupational groups employed in agriculture and horticulture [28]. The concept that well water might accumulate chemical products more readily than free flowing water has indirectly supported an environmental hypothesis of PD. Of interest, 1-methyl-4-phenyl-1,2,3,6-tetrahydropyridine, a selective dopaminergic cell toxin associated with parkinsonism, was industrially developed as a potential herbicide with a structure resembling paraquat but was never produced commercially. It is well known that MPTP (1-methyl-4-phenyl-1,2,5,6-tetrahydropyridine), an unintended by-product when developing meperidine as a street drug, has a breakdown product (MPP+) that is toxic to dopaminergic neurons including those of the substantia nigra and leads to a severe parkinsonian syndrome. The herbicide rotenone, a common organic chemical in pesticides, is like MPP+, a potent inhibitor of mitochondrial complex I, and can increase the production of free radicals with resulting oxidative injury that in addition to dopaminergic neurons alsotargets cholinergic and gabaergic neurons [29–31]. In rats, chronic rotenone exposure leads to pathological, biochemical, and behavioral features seen in

PD and in atypical parkinsonian disorders. In Guadeloupe, the consumption of *Annona muricata* (soursop), which contains the mitochondrial toxin annonacin, is associated with the development of atypical parkinsonism resembling progressive supranuclear palsy [32]. Exposure to hydrocarbons has been linked to PD [33], and the issue of chronic manganese exposure as a risk factor for PD also remains controversial [34–37]. Neurotoxins in flour from cycad plants in Guam can (when fed to mice) lead to behavioral changes and neuronal loss much like those seen in PD. Similarly, exposure to the pesticide paraquat and the fungicide maneb has been implicated in parkinsonism in humans. Exposure to maneb or its major active element, manganese ethylene-bis-dithiocarbamate, can cause selective nigrostriatal neurodegeneration in animal and in vitro models, partially attributed to proteasomal inhibition [38]. The role of certain pathogenic factors that remain unclear includes severe head trauma in mice, which is associated with age-dependent enhanced immunoreactivity to synucleins, which could be the pathogenetic basis of PD after trauma in humans [39]; however, studies of synuclein immunoreactivity patterns in patients with PD and a history of trauma have not yet been performed. A population-based study in which trauma history was documented routinely in the medical records of a record-linkage system showed a clear association between severe head trauma and an increased risk of PD in later life [40]. Finally, based on the observations of postencephalitic parkinsonism as a sequela of epidemic encephalitis, numerous viral and bacterial exposures have been examined, but no consistent agent has been identified [41].

In addition to risk factors for PD, several protective factors have been identified. Almost all studies have confirmed a lower incidence of PD among subjects with a history of smoking [42]. Within twin pairs, the risk of developing PD was inversely associated with the dose of cigarette smoking as measured by pack and years [43]. However, the protective effect of smoking in twins was less marked when cases with PD were compared with control subjects [44]. On the other hand, once patients have developed PD there is no evidence that smoking slows disease progression or that smokers have milder disability than nonsmokers [45]. Coffee drinking has also been linked with reduced risk of PD; however, the evidence is less robust than for smoking [46, 47]. Interactions between caffeine intake and exogenous estrogen have been implicated by a study that showed a protective effect of caffeinated beverage intake in postmenopausal women without estrogen supplementation, but a higher risk in women who received hormones [48]. A pooled analysis of data from three prospective studies found that alcohol (beer) drinking reduced PD risk, even after statistical adjustment for smoking patterns [49].

3 The Role of Genetics

There have been two important approaches to genetic research studies in PD [8]. The first focuses on rare families with parkinsonism apparently following Mendelian inheritance and using the classic methodology of linkage analysis and positional cloning. In the other, an attempt is made to evaluate the PD population as a whole, using association studies and nonparametric linkage methodology and trying to define risk alleles that contribute to the sporadic form of the disease. Genetic research following the first approach has identified 11 PARK loci, each representing a genomic region linked with varying degrees of evidence, to PD-like disorders. These include dominantly inherited gene mutations in α-synuclein – PARK-1 [50], PARK-3 due to a gene mutation on chromosome 2p13, PARK-4 due to a gene mutation on chromosome 4p15 and LRRK2 – Park-8 [51, 52] as well as the recessive PD gene mutations in Parkin – Park-2 [53], PINK1 – Park-6 [54], and DJ-1 – Park-7 [55]. PARK-5 thought to be due to a mutation in the ubiquitin carboxyl-terminal hydrolase L1 (*UCHL1*) gene on chromosome 4p14-16.3 has not been substantiated in recent studies [56, 57]. These mutations cause PD in only a very small subset of families, with the exception of LRRK2, which is responsible for a significant proportion of hereditary PD (5.1–18.7%) and a smaller portion of sporadic PD (1.5–6.1%), although Ashkenazi Jews and Arabs have a higher frequency in the majority of populations studied. Nevertheless, these discoveries have been important in understanding molecular pathways involved in nigral degeneration, which includes protein aggregation, defective proteasomal degradation, mitochondrial dysfunction, and oxidative stress [58].

Identification of the *Parkin* gene mutation has furthered our understanding of the molecular mechanisms underlying PD. Impairment in the capacity of the ubiquitin–proteasome system to clear unwanted proteins has been implicated in the cell death that occurs in PD. Defects in proteasomal structure and function, as well as protein aggregates and increased levels of oxidized proteins are found in the substantia nigra of PD patients. Inhibition of proteasome activity in mesencephalic cultures induces degeneration of dopaminergic neurons coupled with the formation of proteinaceous intracellular inclusions. The gene product, the protein Parkin, is a part of the ubiquitin–proteasome system, which breaks down defective proteins in the cells. Parkin accumulates in Lewy bodies and may be important in both inherited and sporadic forms of the disease [53]. UCHL1 is an important member of the ubiquitin–proteasome system that performs "ubiquitination," a process that tags proteins for breakdown and is critical for the proper handling of misfolded proteins. The findings that LRRK2 encodes a protein from the family of mitogen-activated protein kinases and that mutations may lead to an increase of kinase activity may translate relatively quickly into novel treatment options involving kinase inhibition [59]. However, the contribution of the mechanisms of these and other genes to sporadic idiopathic PD as a whole is still poorly defined. To define this contribution is the aim of the second approach of genetic research in PD, using association studies or nonparametric linkage methodology. Unfortunately, these attempts have met with less success. Five total genome screens have been performed in populations of approximately 200 to almost 400 affected sibling pair families, searching for regions of increased allele sharing between those affected, a method that should allow mapping of genes without knowledge of the underlying genetic parameters such as mode of inheritance and disease gene frequency [60–64]. Each of these studies has provided several interesting areas of suggestive linkage, but so far none of them has led to the discovery of a PD gene. One problem seems to be that only a few of the linkage peaks overlap among studies [61, 62, 64–66], suggesting that heterogeneity between populations may be more pronounced and more important than previously thought. In addition, some of these linkage regions are still very large (up to >100 cM) and contain hundreds of genes. Moreover, the approach to identify causative variants is still poorly mapped out.

Progress in high-throughput genotyping technology has recently allowed genome-wide approaches not only in family samples (sib pairs, as described above) but also in large populations of patients with sporadic disease and control subjects using an association study design and very large panels of single nucleotide polymorphisms (SNPs). In the first very large study of this kind, Maraganore et al. [67] genotyped almost 200,000 SNPs in several hundred samples, resulting in a large number of possibly associated genetic variants. However, none of the top 13 associated SNPs could be confirmed in a follow-up study [68]. A similar study using almost 500,000 SNPs similarly did not identify unequivocal evidence for associations [69]. These results may reflect inherent limitations of the study design, such as heterogeneity of the populations studied and still insufficient coverage of the genome, rather than lack of major genetic components to the etiology of the disease.

Studying association patterns of candidate genes in large, well-characterized series of patients and matched control subjects is another commonly used approach to identify genetic contributions to complex diseases. A large-scale international collaborative study pooled data from 2,692 PD cases and 2,652 control subjects to analyze the allele length variability in the dinucleotide repeat sequence (REP1) in the α-synuclein gene promoter. They found that the α-synuclein REP1 length variability is associated with PD but did not modify age at onset [70]. In fact, increasing and repeatedly confirmed evidence suggests that genetic variations in at least two of the genes that are known to cause monogenic parkinsonian syndromes, α-synuclein and MAP tau, modify the risk for idiopathic PD and progressive supranuclear palsy, respectively [71, 72]. The underlying concept is that relatively common genetic variants (particularly SNPs) may alter the expression pattern of the gene (e.g., by changing binding sites for transcription factors) by altering splicing patterns or by influencing RNA stability or spatial distribution. As the encoded proteins seem to be crucial in the initiation of the neurodegenerative process, even small changes in protein homeostasis may push a cell across a critical threshold toward neurodegeneration. Although this variability probably still explains only a relatively small proportion of the total risk, it is very encouraging to realize that there seems to be a productive path of discovery that leads from well-defined, but rare, monogenic forms to idiopathic PD. Extreme examples of this process are rare cases of α-synuclein multiplications [73]. High concentrations of these potentially pathogenic proteins may favor rare stochastic events of initiation of the disease through the formation of pathogenic templates. After these templates have formed, other proteins may deposit on them and adopt the same conformation [74]. This theory of "permissive templating" suggests that disease propagation, once it has been initiated, may be a different process than disease initiation. Stopping the progression would require prevention of the spread of pathogenic templates between neurons. On the basis of these observations, a general framework of the molecular pathogenesis of PD, which is shared by several other common neurodegenerative diseases, is beginning to emerge: genetic variability sets the individual risk for the initiation of the disease process, which then spreads in a predetermined fashion through susceptible neuronal populations.

4 Neuropathologic Contributions

Neuropathologic studies have been essential in defining and characterizing the Lewy body inclusions of idiopathic PD. More than 30 proteins have been identified within these inclusions [75] with the core filament composed of α-synuclein [76]. These studies provide crucial information for the laboratory modeling of PD. Lewy bodies and Lewy neurites are either asymptomatic or found in pathologic conditions, grouped under the heading "Lewy body disease," which includes idiopathic PD, PD with dementia, and DLB [77]. The distribution of Lewy pathology follows three schematic profiles: brainstem, transitional (limbic Lewy body disease), and diffuse (neocortical Lewy body disease) [78], with additional widespread occurrence in olfactory and central and peripheral autonomic neurons in most cases [3, 79]. The brainstem type is generally associated with clinical PD, whereas the transitional and diffuse types are generally linked to dementia, either PD with dementia (a long history of PD followed by dementia) or DLB (dementia early in the course), usually associated with varying degrees of concurrent Alzheimer-type pathology [80]. In prospectively studied patients with PD with dementia in particular, cortical Lewy bodies appear to be the main substrate driving the progression of cognitive impairment [81, 82]. However, as discussed above, in synucleinopathies without PD, the diffuse type of Lewy body disease can occur without significant evidence of parkinsonism or dementia [83, 84]. The effects of widespread Lewy body pathology in central and peripheral autonomic regions require further clarification [85, 86]. Recent studies confirm that a proportion of the elderly population have α-synuclein aggregates without substantive clinical symptoms [83–85]. These patients may, however, have relevant nonmotor symptoms such as REM sleep behavior disorder, constipation, or other subtle features that may be reflective of this pathology [87]. There may be a significant preclinical period in which a threshold of pathology needs to be reached (both cell loss and α-synuclein accumulation) before onset of any symptoms [3]. Analysis of the cases with Lewy body pathology has led to the proposal of progressive disease stages in which the intracellular deposition of α-synuclein affects medullary sites before more rostral brain regions, with the last stages associated with the involvement of cortical association neurons [3]. It is possible that the pathologic deposition of α-synuclein does not produce profound cellular deficits, except in association with additional pathologies and/or cell loss, a finding consistent with recent clinicopathologic studies of these medullary regions [88, 89]. In the late stages, cortical deposition of α-synuclein correlates with progressive cognitive decline [83, 90].

Although there has been debate about the role of α-synuclein deposition in Lewy bodies currently, most investigators propose that α-synuclein deposition in Lewy bodies is detrimental rather than protective. The few studies of prospectively collected, autopsy-confirmed pure DLB with neocortical Lewy bodies have shown limited gross tissue atrophy and restricted cell loss [91–93]. The presence of numerous α-synuclein depositions in cases with long disease durations (e.g., PD with dementia) suggested that the neurons harboring these inclusions may remain viable within the tissue for a long time, particularly if Lewy bodies are thought to start accumulating intracellularly before the onset of symptoms. However, recent studies found that genetic mutations in α-synuclein may increase neuronal vulnerability to cellular stress in aging and PD pathogenesis [94]. In a transgenic mouse model of α-synucleinopathies, α-synuclein phosphorylated at Ser129 was abundant in α-synuclein inclusions. In vitro studies also indicate that hyperphosphorylation of Ser129 α-synuclein in pathologic inclusions may be due in part to the intrinsic properties of aggregated α-synuclein to act as substrates for kinases but not phosphatases. Further studies in transgenic mice and cultured cells suggest that cellular toxicity, including proteasomal dysfunction, increases casein kinase 2 activity, which results in elevated Ser129 α-synuclein phosphorylation [95].

In PD, there is some evidence that there is abnormal neuronal apoptosis. Three molecules are critical in the development of inherited PD – Parkin, α-synuclein, and ubiquitin, all of which interact in the normal brain. Parkin normally causes α-synuclein to bind to ubiquitin, which then triggers apoptosis causing this compound to self-destruct. In the PARK-2 variety of inherited PD, Parkin is abnormal and fails to bind α-synuclein to ubiquitin. Apoptosis does not take place and α-synuclein accumulates in Lewy bodies [96, 97].

5 The Role of Animal Models

Animal models of Parkinson's disease may be classified as toxin- or gene-based models. 6-Hydroxydopamine (6-OHDA) was the first dopaminergic toxin identified; when injected into the nigrostriatal dopaminergic neuronal populations, this catecholaminergic neurotoxin destroys nigrostriatal dopaminergic neurons and renders animals with motor behaviors similar to Parkinson's [98, 99]. This toxin does not lead to Lewy body formation. Another widely studied dopaminergic neurotoxin, 1-methyl-4-phenyl-1,2,3,6-tetrahydropyridine (MPTP), has been shown to kill dopaminergic neurons in humans, leading to severe parkinsonism [100] and has been used in animal models of Parkinson's disease. MPTP is taken up by microglia and metabolized to MPP^+ which can gain entry into dopaminergic neurons and inhibit mitochondrial complex I. Although MPTP undoubtedly kills dopaminergic neurons in humans, it remains to be proven that the mechanism by which MPTP works is similar to the disease process occurring in idiopathic PD. The use of rotenone (which, like MPTP, is a mitochondrial complex 1 inhibitor) stems from evidence that mitochondrial dysfunction may be involved in sporadic PD [98]. Similarly, paraquat and maneb represent classes of environmental agents that epidemiologic studies have implicated as risk factors for PD. These toxin-based models have provided some insight into the mechanisms of cell death of dopaminergic neurons and have been useful in identifying symptomatic therapies and elucidating the neurophysiology of PD. However, they do not necessarily reproduce proven mechanisms of PD [101].

To reproduce more closely the pathogenic mechanisms of PD in animals, investigators have taken advantage of the identification of mutations that cause the disease in rare cases of familial PD. The underlying hypothesis is that mutations causing familial PD may trigger mechanisms that also play a role in sporadic PD. Several lines of evidence support this contention. For example, mutations in Parkin and UCHL1 affect proteasomal function, which has also been implicated in sporadic PD. PINK1 and DJ-1 are mitochondrial proteins. Finally, α-synuclein accumulates in central and peripheral neurons of patients with sporadic PD. Therefore, gene-based models of Parkinson's disease have provided tremendous insight into the mechanisms by which neurons are injured in Parkinson's disease [102].

Because of its undisputed association with sporadic PD, many lines of mice were generated to overexpress either wild-type or mutant α-synuclein [103]. Several of these mice show progressive motor deficits, abnormal dopamine function, and in some cases even reproduce nonmotor deficits observed in PD patients and delayed loss of striatal dopamine and/or tyrosine-hydroxylase-positive neurons in the substantia nigra pars compacta [104–106]. The model generated by Lee et al. (transgenic mice expressing A53T α-synuclein under the prion promoter) presents a severe, rapidly progressive motor phenotype arising at approximately 10–15 months of age [105]. However, the pathologic changes are particularly severe in spinal cord, brainstem, and cortex, with death of motor neurons, but not in nigrostriatal dopaminergic cells as in PD. Nevertheless, the model is useful to determine the mechanisms by which α-synuclein accumulation leads to neuronal loss in vivo. In particular, these mice show an accumulation of truncated α-synuclein amputated from its C terminus, which may increase the misfolding of the protein, suggesting a role for proteolytic processing in the pathologic process [59]. Animal models of PD are not limited to mammals.

Dopaminergic neurons can be killed with toxins in flies, fish, or worms, and the ability to express PD-causing mutations in flies has given rise to an exciting new generation of animal models that are amenable to rapid genetic manipulations [107, 108]. The most important aspect of these models is that they allow us to decipher the pathologic processes that take place in a brain exposed to a PD-causing insult before any nigrostriatal dopaminergic neurons are lost or before the disease begins elsewhere. These models are also critical for the identification of novel therapies.

6 The Role of Oxidative Stress

The role that reactive oxygen and nitrogen species play (direct/primary or secondary phenomena) in the pathogenesis of PD has been debated for decades [109]. The first evidence in favor of a key role for oxidative stress was provided in the seminal work of Graham et al. [110], which emphasized the toxic potential of oxidative reactions involving dopamine and observations on postmortem specimens of PD brains showing an increase in markers of lipid peroxidation, a loss of antioxidant defense mechanisms including glutathione, increased iron deposition in the substantia nigra pars compacta, and increased oxidation of proteins and nucleic acids [111–113]. Recent data from genetic studies as well as toxin-induced models have reignited interest in an imbalance between prooxidant and antioxidant events in the development of neurodegeneration and other pathologic changes like inclusion body formation in PD. Oxidative stress is evident in each model early in the course of neurodegeneration.

For example, α-synuclein oxidation plays a key role in protein aggregation, including Lewy body formation. In 1996, Polymeropoulos et al. identified the first gene associated with familial parkinsonism that encodes for the protein α-synuclein [114]. Shortly after this discovery, the likely role of α-synuclein in the pathogenesis of sporadic PD was suggested by the observation that this protein is a major component of Lewy bodies, the intraneuronal inclusions that are hallmarks of PD in human brain tissue [115, 116]. Given the involvement of α-synuclein in familial and sporadic parkinsonism, it is noteworthy that evidence both in humans and experimental models strongly suggest a role of oxidative stress in α-synuclein-induced pathology. Some PD patients have a 30–40% reduction in the mitochondrial enzyme complex I making neurons vulnerable to unstable oxygen free radicals that can modify α-synuclein and cause it to aggregate into fibrils. Thus, mitochondrial dysfunction and oxidative stress may underlie an important pathway of neurodegeneration common to sporadic and familial PD and should continue to be a subject of future investigation. After staining human brain sections with antibodies against nitrated tyrosine residues of α-synuclein, Giasson et al. [117] demonstrated a widespread accumulation of nitrated protein within Lewy bodies and Lewy neurites and concluded that this accumulation "provides evidence to directly link oxidative and nitrative damage to the onset and progression" of PD. Support in favor of a relationship among α-synuclein, oxidative stress, and PD also stems from in vitro studies showing that dopamine-dependent oxidative modifications of α-synuclein could facilitate the intraneuronal accumulation of toxic "protofibrils" and thus contribute to the demise of dopaminergic neurons seen in PD [118, 119].

Perhaps no clue derived from genetic studies links parkinsonism to oxidative stress more convincingly than the observation of a parkinsonian syndrome in families with mutations in the *DJ-1* gene [55]. Although the exact function of the protein product is not known, DJ-1 has been suggested to have an antioxidant function in the cellular response to oxidative stimuli [120], and posttranslational modifications of DJ-1 at cysteine residues are important for its antioxidant properties. DJ-1 is oxidatively damaged in the brains of PD patients and, in Drosophila, inhibition of DJ-1 function through RNA interference leads to cellular accumulation of reactive oxygen species, enhanced vulnerability to oxidative stress, and degeneration of dopaminergic neurons [121–124].

Support in favor of a role of oxidative stress in the pathogenesis of PD has also been derived recently from work using toxin-induced PD models. Kaur et al. [125] administered iron to newborn mice for a 1-week period (days 10–17 postpartum) and noted that, once these animals reached adult and old age, an increase in markers of oxidative damage was paralleled by a loss of nigrostriatal function and dopaminergic cell integrity. In another mouse model of selective nigrostriatal injury caused by exposures to the herbicide paraquat, both a temporal and a causal relationship have been found between enhanced oxidative modifications in the nigrostriatal tissue and degeneration of nigral dopaminergic neurons [126, 127]. Interestingly, new experimental paradigms also implicate oxidative stress in the interaction of PD-related toxins and genes. Meulener et al. showed evidence in a Drosophila model that DJ-1 activity is specifically involved in protecting against oxidative injury caused by paraquat [128]. Wang et al. showed significant interactions between oxidative stress-inducing agents and Parkin. Addition of H_2O_2, paraquat, or iron to SY5Y neuroblastoma cells stably expressing FLAG-tagged Parkin altered the solubility of this protein, thus compromising its protective function [129].

Studies in human brains have suggested that oxidative changes (e.g., decreased glutathione levels) may be characteristic of PD. However, whether these changes underlie the neurodegenerative process of PD or are mere consequences of it (oxidative reactions could occur in tissues "after" they have been damaged) is still open to discussion [130, 131].

Uncertainty concerning the role of oxidative stress is also based on the fact that antioxidant strategies have failed to yield convincing protection against PD, as exemplified by the disappointing outcome of a clinical trial that, over 10 years ago, assessed the neuroprotective action of tocopherol in PD [132]. One could say that the ultimate proof of concept linking a specific mechanism (such as oxidative stress) to the pathogenesis of PD would be the development of therapeutic approaches that, by targeting that mechanism, result in neuroprotection. Mitochondria have been suggested to represent a significant source of reactive oxygen species (ROS) that would contribute to neuronal demise in PD.

This possibility is supported by findings with toxins, such as MPTP, that inhibit mitochondrial function and are capable of inducing PD-like pathology [112, 133]. Similarly, mitochondrial abnormalities are likely to derive from mutations in PD-associated genes, such as *PINK1* and *DJ-1* [62, 134, 135].

Besides mitochondria-mediated ROS formation, other mechanisms that could lead to the generation of oxidative species in PD have been identified. It is noteworthy that these oxidative pathways may not necessarily occur within neuronal cells as, for example, free radicals (permeable to cell membranes) could initially be formed extraneuronally as a consequence of microglial activation and inflammatory processes. In PD, there are changes in the subthalamic nucleus as glutamatergic *N*-methyl-D-aspartate (NMDA) receptors become constantly overstimulated and produce high levels of calcium ions within neurons. This in turn leads to a cascade of events that trigger oxygen free radicals and cell damage. An important role of microglia-mediated oxidative stress has been shown in a variety of experimental models of nigrostriatal degeneration, including mice treated with MPTP or paraquat and mice carrying the weaver mutation [136–138]. The relationship between neuroinflammation, oxidative stress, and PD should continue to be studied not only to clarify mechanisms of ROS generation but also to further assess the potential use of anti-inflammatory agents for neuroprotection in PD [139].

Once oxidizing species are generated, important cellular components/functions could be targeted, and their impairment could ultimately lead to neuronal degeneration. A new potential target for toxic oxidative modifications has been highlighted by studies suggesting that dysfunction of protein degradation through the proteasomal system contributes to neurodegeneration in PD [140, 141]. The issue of proteasomal impairment in PD has become somewhat controversial, because initial findings showing PD-like abnormalities in rats treated systemically with proteasomal inhibitors could not be replicated in other laboratories [142–144]. These contrasting data relate specifically to the experimental use of proteasomal inhibitors to model PD in rodents and should not be interpreted as a "fatal blow" to the hypothesis of proteasomal dysfunction in PD. In fact, the links between oxidative modifications of proteins and cellular organelles and altered protein degradation will probably be a fruitful area of future investigation into PD pathogenesis.

Many of our current uncertainties concerning the involvement of oxidative stress in PD can be resolved by careful reevaluation of the oxidative stress insight into the relationship between oxidative injury, protein aggregation, inclusion formation, and neurodegeneration and will be gained from both in vitro and in vivo studies using, for example, genetically manipulated cells in culture, invertebrate models (e.g., flies and *Caenorhabditis elegans*), transgenic mice, and animals challenged with prooxidant toxins. Mitochondrial abnormalities in PD have been supported by studies in which mitochondria (and the contained mitochondrial DNA) are inserted into culturable human cell lines depleted of their own endogenous mitochondrial DNA (so-called cybrids) [145, 146].

Examples of therapeutic approaches that could shed light upon the involvement of oxidative stress in PD are the administration of anti-inflammatory drugs and the use of coenzyme Q. The rationale for treatment with anti-inflammatory drugs stems, at least in part, from evidence of microglia-mediated oxidative injury in animal models of PD [136]. Minocycline, a second-generation semisynthetic tetracycline that possesses potent anti-inflammatory properties, has recently been recommended for Phase III clinical trials [80]. If minocycline or other anti-inflammatory drugs are found to be effective as neuroprotective agents, these clinical trials will provide much needed confirmation that prevention/limitation of oxidative reactions (within and outside dopaminergic cells) represents a viable strategy for the development of anti-parkinsonian drugs. Initial clinical findings also suggest that coenzyme Q may slow the rate of progression of idiopathic PD [90]. The mechanism of action of coenzyme Q is likely to involve an antioxidant effect derived from the interception of aberrant electrons before they react with molecular oxygen and form ROS.

In summary, although a number of new hypotheses concerning the pathogenesis of PD have been proposed over the past few years, the potential role of oxidative stress in neuronal degeneration and formation of proteinaceous inclusions (two pathologic features of PD) continues to draw a great deal of research attention. In fact, it is quite intriguing that many (if not all) of the new hypotheses (e.g., impairment of the proteasomal system) can be easily reconciled with involvement of oxidative reactions. It is also remarkable that recent findings from genetic studies further suggest that oxidative stress is implicated in the pathogenesis of familial cases of parkinsonism. Conclusive evidence linking oxidative stress to PD is still elusive. However, with the advent of new technical tools and experimental models and the integration of precious information from biochemical, pathologic and genetic studies, the involvement of oxidative damage in PD will probably cease to be just a hypothesis within the short foreseeable future. If so, our sustained efforts to unravel pathways and mechanisms of oxidative injury will also probably yield more specific and effective antioxidant strategies for neuroprotection in PD.

7 Conclusions

It is clear that PD is not a single entity; it is clinically, pathologically, and etiologically diverse. As with all disorders, the different phenotypes of PD most likely arise from complex combinations of genetics and modifiers, with many of the latter coming from the environment. A better understanding of cell biology processes, including oxidative stress, is likely to provide more insight into the pathogenesis. Animal models are certainly critical in this endeavor. Advances along these lines and those of protein misfolding and aggregation should be helpful for new approaches to therapy. Recent understanding of the nonmotor aspects of PD and degeneration outside the substantia nigra has not only increased the challenge of this work but also identifies what needs to be investigated. We have probably reached the limit of symptomatic therapy, and our patients need therapy based on the understanding of the etiology and pathogenesis of the disease.

References

1. Litvan I. Parkinsonian features: When are they Parkinson disease? JAMA 1998; 280: 1654–1655.
2. Stiasny-Kolster K, Doerr Y, Moller JC, et al. Combination of 'idiopathic' REM sleep behaviour disorder and olfactory dysfunction as possible indicator for α-synucleinopathy demonstrated by dopamine transporter FP-CIT-SPECT. Brain 2005; 128:126–137.
3. Braak H, Del Tredici K, Rub U, et al. Staging of brain pathology related to sporadic Parkinson's disease. Neurobiol Aging 2003; 24: 197–211.
4. Fearnley JM, Lees AJ. Ageing and Parkinson's disease: Substantia nigra regional selectivity. Brain 1991; 114: 2283–2301.
5. Greffard S, Verny M, Bonnet AM, et al. Motor score of the unified Parkinson disease rating scale as a good predictor of Lewy body-associated neuronal loss in the substantia nigra. Arch Neurol 2006; 63: 584–588.
6. Tanner CM, Aston DA. Epidemiology of Parkinson's disease and akinetic syndromes. Curr Opin Neurol 2000; 13: 427–430.
7. Vila M, Przedborski S. Genetic clues to the pathogenesis of Parkinson's disease. Nat Med 2004; 10(Suppl): S58–S62.
8. Litvan I, Halliday G, Hallett M et al. The etiopathogenesis of Parkinson disease and suggestions for future research. Part I. J Neuropathol Exp Neurol 2007; 66: 251–257.
9. Litvan I, Chesselet MF, Gasser T, et al. The etiopathogenesis of Parkinson disease and suggestions for future research. Part II. J Neuropathol Exp Neurol 2007; 66: 329–336.
10. Bower JH, Maraganore DM, McDonnell SK, et al. Influence of strict, intermediate, and broad diagnostic criteria on the age- and sex-specific incidence of Parkinson's disease. Mov Disord 2000; 15: 819–825.
11. Van Den Eeden SK, Tanner CM, Bernstein AL, et al. Incidence of Parkinson's disease: Variation by age, gender, and race/ethnicity. Am J Epidemiol 2003;157: 1015–1022.
12. Benedetti MD, Maraganore DM, Bower JH, et al. Hysterectomy, menopause, and estrogen use preceding Parkinson's disease: An exploratory case-control study. Mov Disord 2001; 16: 830–837.
13. Kompoliti K, Comella CL, Jaglin JA, et al. Menstrual-related changes in motoric function in women with Parkinson's disease. Neurology 2000; 55: 1572–1575.
14. Schoenberg BS, Anderson DW, Haerer AF. Prevalence of Parkinson's disease in the biracial population of Copiah County, Mississippi. Neurology 1985; 35: 841–845.
15. de Rijk MC, Launer LJ, Berger K, et al. Prevalence of Parkinson's disease in Europe: A collaborative study of population-based cohorts. Neurologic Diseases in the Elderly Research Group. Neurology 2000; 54: S21–S23.
16. Rocca WA. Dementia, Parkinson's disease, and stroke in Europe: A commentary. Neurology 2000; 54: S38–S40.
17. Schrag A, Ben-Shlomo Y, Quinn N. How valid is the clinical diagnosis of Parkinson's disease in the community? J Neurol Neurosurg Psychiatry 2002; 73: 529–534.
18. Twelves D, Perkins KS, Counsell C. Systematic review of incidence studies of Parkinson's disease. Mov Disord 2003; 18: 19–31.
19. Rocca WA. Prevalence of Parkinson's disease in China. Lancet Neurol 2005; 4: 328–329.

20. Tanner CM, Chen B, Wang W, et al. Environmental factors and Parkinson's disease: A case-control study in China. Neurology 1989; 39: 660–664.
21. Rajput AH. Environmental causation of Parkinson's disease. Arch Neurol 1993; 50: 651–652.
22. Woo J, Lau E, Ziea E, et al. Prevalence of Parkinson's disease in a Chinese population. Acta Neurol Scand 2004; 109: 228–231.
23. Ben-Shlomo Y, Whitehead AS, Smith GD. Parkinson's, Alzheimer's, and motor neuron disease. BMJ 1996; 312: 724.
24. Zhang ZX, RomanGC, Hong Z, et al. Parkinson's disease in China: Prevalence in Beijing, Xian, and Shanghai. Lancet 2005; 365: 595–597.
25. Baldereschi M, Di Carlo A, Vanni P, et al. Lifestyle-related risk factors for Parkinson's disease: A population-based study. Acta Neurol Scand 2003; 108: 239–244.
26. DiMonte D. The environment and Parkinson's disease: Is the nigrostriatal system preferentially targeted by neurotoxins? Lancet Neurol 2003; 2: 531–538.
27. Marder K, Logroscino G, Alfaro B, et al. Environmental risk factors for Parkinson's disease in an urban multiethnic community. Neurology 1998; 50: 279–281.
28. Tuchsen F, Jensen AA. Agricultural work and the risk of Parkinson's disease in Denmark, 1981–1993. Scand J Work Environ Health 2000; 26: 359–362.
29. Greenamyre JT, Sherer TB, Betarbet R, et al. Complex I and Parkinson's disease. IUBMB Life 2001; 52: 135–141.
30. Lapointe N, St-Hilaire M, Martinoli MG, et al. Rotenone induces non-specific central nervous system and systemic toxicity. FASEB J 2004; 18: 717–719.
31. Höglinger GU, Oertel WH, Hirsch EC. The rotenone model of parkinsonism – the five years inspection. J Neural Transm Suppl. 2006; 70: 269–272.
32. Lannuzel A, Höglinger GU, Verhaeghe S et al. Atypical parkinsonism in Guadeloupe: A common risk factor for two closely related phenotypes? Brain 2007; 130: 816–827.
33. Pezzoli G. Hydrocarbon exposure and Parkinson's disease. Neurology 2000; 55: 667–773.
34. Racette BA, McGee-Minnich L, Moerlein SM, et al. Welding-related parkinsonism: Clinical features, treatment, and pathophysiology. Neurology 2001; 56: 8–13
35. Levy BS, Nassetta WJ. Neurologic effects of manganese in humans: A review. Int J Occup Environ Health 2003; 9: 153–163.
36. Jankovic J. Searching for a relationship between manganese and welding and Parkinson's disease. Neurology 2005; 64: 2021–2028.
37. Zhou Y, Shie FS, Piccardo P, et al. Proteasomal inhibition induced by manganese ethylene-bis-dithiocarbamate: Relevance to Parkinson's disease. Neuroscience 2004; 128: 281–291.
38. Patel S, Sinha A, Singh MP. Identification of differentially expressed proteins in striatum of maneb-and paraquat-induced Parkinson's disease phenotype in mouse. Neurotoxicol Teratol. 2007; 29: 578–585.
39. Urye K. Age-dependent synucleins pathology following traumatic brain injury in mice. Exp Neurol 2003; 184: 214–224.
40. Bower JH, Maraganore DM, Peterson BJ, et al. Head trauma preceding PD: A case-control study. Neurology 2003; 60:1610–1615.
41. Dale RC, Church AJ, Surtees RA, et al. Encephalitis lethargica syndrome: 20 new cases and evidence of basal ganglia autoimmunity. Brain 2004; 127: 21–33.
42. Hernan MA, Takkouche B, Caamano-Isorna F, et al. A meta-analysis of coffee drinking, cigarette smoking, and the risk of Parkinson's disease. Ann Neurol 2002; 52: 276–284.
43. Tanner CM, Goldman SM, Aston DA, et al. Smoking and Parkinson's disease in twins. Neurology 2002; 58: 581–588.
44. Wirdefeldt K, Gatz M, Pawitan Y, et al. Risk and protective factors for Parkinson's disease: A study in Swedish twins. Ann Neurol 2005; 57: 27–33.
45. Alves G, Kurz M, Lie SA, et al. Cigarette smoking in PD: Influence on disease progression. Mov Disord 2004; 19: 1087–1092.

46. Benedetti MD, Bower JH, Maraganore DM, et al. Smoking, alcohol, and coffee consumption preceding Parkinson's disease: A case-control study. Neurology 2000; 55:1350–1358.
47. Ragonese P, Salemi G, Morgante L, et al. A case-control study on cigarette, alcohol, and coffee consumption preceding Parkinson's disease. Neuroepidemiology 2003; 22: 297–304.
48. Ascherio A, Chen H, Schwarzschild MA, et al. Caffeine, postmenopausal estrogen, and risk of Parkinson's disease. Neurology 2003; 60: 790–795.
49. Hernan MA, Chen H, Schwarzschild MA, et al. Alcohol consumption and the incidence of Parkinson's disease. Ann Neurol 2003; 54: 170–175.
50. Polymeropoulos MH, Lavedan C, Leroy E, et al. Mutation in the α-synuclein gene identified in families with Parkinson's disease. Science 1997; 276: 2045–2047.
51. Zimprich A, Biskup S, Leitner P, et al. Mutations in LRRK2 cause autosomal-dominant parkinsonism with pleomorphic pathology. Neuron 2004; 44: 601–607.
52. Paisan-Ruiz C, Jain S, Evans EW, et al. Cloning of the gene containing mutations that cause PARK8-linked Parkinson's disease. Neuron 2004; 44: 595–600.
53. Kitada T, Asakawa S, Hattori N, et al. Mutations in the parkin gene cause autosomal recessive juvenile parkinsonism. Nature 1998; 392: 605–608.
54. Valente EM, Abou-Sleiman PM, Caputo V, et al. Hereditary early-onset Parkinson's disease caused by mutations in PINK1. Science 2004; 304: 1158–1160.
55. Bonifati V, Rizzu P, van Baren MJ, et al. Mutations in the DJ-1 gene associated with autosomal recessive early-onset parkinsonism. Science 2003; 299: 256–259.
56. Healy DG, Abou-Sleiman PM, Casas JP et al. UCHL-1 is not a Parkinson's disease susceptibility gene. Ann Neurol 2006; 59: 627–633.
57. Hutter CM, Samii A, Factor SA et al. Lack of evidence for an association between UCHL1 S18Y and Parkinson's disease. Eur J Neurol 2008; 15: 134–139.
58. Cookson MR. The biochemistry of Parkinson's disease. Annu Rev Biochem 2005; 74: 29–52.
59. Li W, West N, Colla E, et al. Aggregation promoting C-terminal truncation of α-synuclein is a normal cellular process and is enhanced by the familial Parkinson's disease-linked mutations. Proc Natl Acad Sci U S A 2005; 102: 2162–2167.
60. DeStefano AL, Lew MF, Golbe LI, et al. PARK3 influences age at onset in Parkinson disease: A genome scan in the GenePD study. Am J Hum Genet 2002; 70:1089–1095.
61. Pankratz N, Nichols WC, Uniacke SK, et al. Genome screen to identify susceptibility genes for Parkinson disease in a sample without parkin mutations. Am J Hum Genet 2002; 71:124–135.
62. Martinez M, Brice A, Vaughan JR, et al. Genome-wide scan linkage analysis for Parkinson's disease: The European genetic study of Parkinson's disease. J Med Genet 2004; 41: 900–907.
63. Pankratz N, Nichols WC, Uniacke SK, et al. Genome-wide linkage analysis and evidence of gene-by-gene interactions in a sample of 362 multiplex Parkinson disease families. Hum Mol Genet 2003; 12: 2599–2608.
64. Scott WK, Nance MA, Watts RL, et al. Complete genomic screen in Parkinson disease: Evidence for multiple genes. JAMA 2001; 286: 2239–2244.
65. DeStefano AL, Golbe LI, Mark MH, et al. Genome-wide scan for Parkinson's disease: The GenePD Study. Neurology 2001; 57:1124–1126.
66. Hicks AA, Petursson H, Jonsson T, et al. A susceptibility gene for late-onset idiopathic Parkinson's disease. Ann Neurol 2002; 52: 549–555.
67. Maraganore DM, de Andrade M, Lesnick TG, et al. High-resolution whole-genome association study of Parkinson disease. Am J Hum Genet 2005; 77: 685–693.
68. Elbaz A, Nelson LM, Payami H, et al. Lack of replication of thirteen single-nucleotide polymorphisms implicated in Parkinson's disease: A large-scale international study. Lancet Neurol 2006; 5: 917–923.
69. Fung HC, Scholz S, Matarin M, et al. Genome-wide genotyping in Parkinson's disease and neurologically normal controls: First stage analysis and public release of data. Lancet Neurol 2006; 5: 911–916.

70. Maraganore DM, de Andrade M, Elbaz A, et al. Collaborative analysis of α-synuclein gene promoter variability and Parkinson disease. JAMA 2006; 296: 661–670.
71. Mueller JC, Fuchs J, Hofer A, et al. Multiple regions of α-synuclein are associated with Parkinson's disease. Ann Neurol 2005; 57: 535–541.
72. Baker M, Litvan I, Houlden H, et al. Association of an extended haplotype in the tau gene with progressive supranuclear palsy. Hum Mol Genet 1999; 8: 711–715.
73. Singleton AB, Farrer M, Johnson J, et al. α-Synuclein locus triplication causes Parkinson's disease. Science 2003; 302: 841.
74. Hardy J. Expression of normal sequence pathogenic proteins for neurodegenerative disease contributes to disease risk: 'Permissive templating' as a general mechanism underlying neurodegeneration. Biochem Soc Trans 2005; 33:578–581.
75. Zhang J, Goodlett DR. Proteomic approach to studying Parkinson's disease. Mol Neurobiol 2004; 29: 271–288.
76. Crowther RA, Daniel SE, Goedert M. Characterisation of isolated α-synuclein filaments from substantia nigra of Parkinson's disease brain. Neurosci Lett 2000; 292:128–130.
77. Litvan I, MacIntyre A, Goetz CG, et al. Accuracy of the clinical diagnoses of Lewy body disease, Parkinson disease, and dementia with Lewy bodies: A clinicopathologic study. Arch Neurol 1998; 55: 969–978.
78. Kosaka K. Lewy body disease with and without dementia: A clinicopathological study of 35 cases. Clin Neuropathol 1988; 7: 299–305.
79. Wakabayashi K, Takahashi H. Neuropathology of autonomic nervous system in Parkinson's disease. Eur Neurol 1997; 38: 2–7.
80. Burn DJ. Cortical Lewy body disease and Parkinson's disease dementia. Curr Opin Neurol 2006; 19: 572–579.
81. Aarsland D, Perry R, Brown A, et al. Neuropathology of dementia in Parkinson's disease: A prospective, community-based study. Ann Neurol 2005; 58: 773–776.
82. Braak H, Rüb U, Del Tredici K. Cognitive decline correlates with neuropathological stage in Parkinson's disease. J Neurol Sci 2006; 248: 255–258.
83. Parkkinen L, Kauppinen T, Pirttila T, et al. α-Synuclein pathology does not predict extrapyramidal symptoms or dementia. Ann Neurol 2005; 57: 82–91.
84. Jellinger KA. A critical reappraisal of current staging of Lewy-related pathology in human brain. Acta Neuropathol 2008; 116:1–16.
85. Bloch A, Probst A, Bissig H, et al. α-Synuclein pathology of the spinal and peripheral autonomic nervous system in neurologically unimpaired elderly subjects. Neuropathol Appl Neurobiol 2006; 32: 284–295.
86. Klos KJ, Ahlskog JE, Josephs KA, et al. α-Synuclein pathology in the spinal cords of neurologically asymptomatic aged individuals. Neurology 2006; 66:1100–1102.
87. Boeve BF, Silber MH, Ferman TJ, et al. Association of REM sleep behavior disorder and neurodegenerative disease may reflect an underlying synucleinopathy. Mov Disord 2001; 16: 622–630
88. Benarroch EE, Schmeichel AM, Sandroni P, et al. Involvement of vagal autonomic nuclei in multiple system atrophy and Lewy body disease. Neurology 2006; 66: 378–383.
89. Benarroch EE, Schmeichel AM, Low PA, et al. Involvement of medullary regions controlling sympathetic output in Lewy body disease. Brain 2005;128: 338–344.
90. Braak H, Rub U, Jansen Steur EN, et al. Cognitive status correlates with neuropathologic stage in Parkinson disease. Neurology 2005; 64: 1404–1410.
91. Lee HG, Zhu X, Takeda A, et al. Emerging evidence for the neuroprotective role of α-synuclein. Exp Neurol 2006; 200: 1–7.
92. Cordato NJ, Halliday GM, Harding AJ, et al. Regional brain atrophy in progressive supranuclear palsy and Lewy body disease. Ann Neurol 2000; 47: 718–728
93. Harding AJ, Broe GA, Halliday GM. Visual hallucinations in Lewy body disease relate to Lewy bodies in the temporal lobe. Brain 2002; 125: 391–403.

94. Jiang H, Wu YC, Nakamura M et al. Parkinson's disease genetic mutations increase cell susceptibility to stress: Mutant alpha-synuclein enhances H2O2- and Sin-1-induced cell death. Neurobiol Aging 2007; 28:1709–1717.
95. Waxman EA, Giasson BI. Specificity and regulation of casein kinase-mediated phosphorylation of alpha-synuclein. J Neuropathol Exp Neurol 2008; 67: 402–416.
96. Shimura H, Schlossmacher MG, Hattori N et al. Ubiquitination of a new form of alpha-synuclein by parkin from human brain: Implications for Parkinson's disease. Science 2001; 293: 263–269.
97. Moore DJ, West AB, Dikeman DA et al. Parkin mediates the degradation-independent ubiquitination of Hsp70. J Neurochem 2008; 105: 1806–1819.
98. Dauer W, Przedborski S. Parkinson's disease: Mechanisms and models. Neuron 2003; 39: 889–909.
99. Chesselet MF, Delfs JM. Basal ganglia and movement disorders: An update. Trends Neurosci 1996; 19: 417–422.
100. Langston JW, Ballard P, Tetrud JW, Irwin I. Chronic Parkinsonism in humans due to a product of meperidine-analog synthesis. Science 1983; 219: 979–980.
101. Meissner W, Hill MP, Tison F et al. Neuroprotective strategies for Parkinson's disease: Conceptual limits of animal models and clinical trials. Trends Pharmacol Sci 2004; 25: 249–253.
102. Chesselet MF, Fernagut PO, Fleming S. Parkinson's disease models: From toxins to genes. Drug Disc Today Dis Models 2005; 2: 299–303.
103. Fernagut PO, Chesselet MF. α-Synuclein and transgenic mouse models. Neurobiol Dis 2004; 17: 123–130.
104. Fleming SM, Fernagut PO, Chesselet MF. Genetic mouse models of parkinsonism: Strengths and limitations. NeuroRx 2005; 2: 495–503.
105. Lee MK, Stirling W, Xu Y, et al. Human α-synuclein-harboring familial Parkinson's disease-linked Ala-53 → Thr mutation causes neurodegenerative disease with α-synuclein aggregation in transgenic mice. Proc Natl Acad Sci U S A 2002; 99: 8968–8973.
106. Thiruchelvam MJ, Powers JM, Cory-Slechta DA et al. Risk factors for dopaminergic neuron loss in human α-synuclein transgenic mice. Eur J Neurosci 2004; 19: 845–854.
107. Sang T-K, Jackson G. Drosophilia models of neurodegenerative disease. NeuroRx 2005; 2: 438–446.
108. Nass R, Blakely RD. The Caenorhabditis elegans dopaminergic system: Opportunities for insights into dopamine transport and neurodegeneration. Annu Rev Pharmacol Toxicol 2003; 43: 521–544.
109. Calne DB. The free radical hypothesis in idiopathic parkinsonism: Evidence against it. Ann Neurol 1992; 32: 799–803.
110. Graham DG. Oxidative pathways for catecholamines in the genesis of neuromelanin and cytotoxic quinones. Mol Pharmacol 1978; 14: 633–643.
111. Dexter DT, Carter CJ, Wells FR, et al. Basal lipid peroxidation in substantia nigra is increased in Parkinson's disease. J Neurochem 1989; 52: 381–389.
112. Sian J, Dexter DT, Lees AJ, Daniel S, Jenner P, Marsden CD. Glutathione-related enzymes in brain in Parkinson's disease. Ann Neurol 1994; 36: 356–361.
113. Youdim MB, Ben-Shachar D, Riederer P. The possible role of iron in the etiopathology of Parkinson's disease. Mov Disord 1993; 8: 1–12.
114. Polymeropoulos MH, Higgins JJ, Golbe LI, et al. Mapping of a gene for Parkinson's disease to chromosome 4q21-q23. Science 1996; 274: 1197–1199.
115. Spillantini MG, Crowther RA, Jakes R et al. α-Synuclein in filamentous inclusions of Lewy bodies from Parkinson's disease and dementia with Lewy bodies. Proc Natl Acad Sci U S A 1998; 95: 6469–6473.
116. Lee VM, Trojanowski JQ. Mechanisms of Parkinson's disease linked to pathological α-synuclein: New targets for drug discovery. Neuron 2006; 52: 33–38.
117. Giasson BI, Duda JE, Murray IV, et al. Oxidative damage linked to neurodegeneration by selective α-synuclein nitration in synucleinopathy lesions. Science 2000; 290: 985–989.

118. Conway KA, Rochet JC, Bieganski RM, Lansbury PT Jr. Kinetic stabilization of the α-synuclein protofibril by a dopamine-α-synuclein adduct. Science 2001; 294: 1346–1349.
119. Ischiropoulos H. Oxidative modifications of α-synuclein. Ann N Y Acad Sci 2003; 991: 93–100.
120. Mitsumoto A, Nakagawa Y. DJ-1 is an indicator for endogenous reactive oxygen species elicited by endotoxin. Free Radic Res 2001; 35: 885–893.
121. Canet-Aviles RM, Wilson MA, Miller DW, et al. The Parkinson's disease protein DJ-1 is neuroprotective due to cysteine–sulfinic acid-driven mitochondrial localization. Proc Natl Acad Sci U S A 2004; 101: 9103–9108.
122. Meulener MC, Xu K, Thomson L et al. Mutational analysis of DJ-1 in Drosophila implicates functional inactivation by oxidative damage and aging. Proc Natl Acad Sci U S A 2006; 103: 12517–12522.
123. Choi J, Sullards MC, Olzmann JA, et al. Oxidative damage of DJ-1 is linked to sporadic Parkinson and Alzheimer diseases. J Biol Chem 2006; 281: 10816–10824.
124. Yang Y, Gehrke S, Haque ME, et al. Inactivation of Drosophila DJ-1 leads to impairments of oxidative stress response and phosphatidylinositol 3-kinase/Akt signaling. Proc Natl Acad Sci U S A 2005; 102: 13670–13675.
125. Kaur D, Peng J, Chinta SJ, et al. Increased murine neonatal iron intake results in Parkinson-like neurodegeneration with age. Neurobiol Aging 2007; 28: 907–913.
126. McCormack AL, Atienza JG, Johnston LC et al. Role of oxidative stress in paraquat-induced dopaminergic cell degeneration. J Neurochem 2005; 93:1030–1037.
127. Peng J, Stevenson FF, Doctrow SR et al. Superoxide dismutase/catalase mimetics are neuroprotective against selective paraquat-mediated dopaminergic neuron death in the substantial nigra: Implications for Parkinson disease. J Biol Chem 2005; 280: 29194–29198.
128. Meulener M, Whitworth AJ, Armstrong-Gold CE, et al. Drosophila DJ-1 mutants are selectively sensitive to environmental toxins associated with Parkinson's disease. Curr Biol 2005; 15: 1572–1577.
129. Wang C, Ko HS, Thomas B, et al. Stress-induced alterations in parkin solubility promote parkin aggregation and compromise parkin's protective function. Hum Mol Genet 2005; 14: 3885–3897.
130. Smith MT, Sandy MS, Di Monte D. Free radicals, lipid peroxidation, and Parkinson's disease. Lancet 1987; 1: 38.
131. Di Monte DA, Chan P, Sandy MS. Glutathione in Parkinson's disease: A link between oxidative stress and mitochondrial damage? Ann Neurol 1992; 32 (Suppl): S111–S115.
132. The Parkinson Study Group. Effects of tocopherol and deprenyl on the progression of disability in early Parkinson's disease. N Engl J Med 1993; 286: 176–183.
133. Betarbet R, Sherer TB, MacKenzie G et al. Chronic systemic pesticide exposure reproduces features of Parkinson's disease. Nat Neurosci 2000; 3: 1301–1306.
134. Ved R, Saha S, Westlund B, et al. Similar patterns of mitochondrial vulnerability and rescue induced by genetic modification of α-synuclein, parkin, and DJ-1 in Caenorhabditis elegans. J Biol Chem 2005; 280: 42655–42668.
135. Yang Y, Gehrke S, Imai Y, et al. Mitochondrial pathology and muscle and dopaminergic neuron degeneration caused by inactivation of Drosophila Pink1 is rescued by Parkin. Proc Natl Acad Sci U S A 2006; 103:10793–10798.
136. Wu DC, Teismann P, Tieu K, et al. NADPH oxidase mediates oxidative stress in the 1-methyl-4-phenyl-1,2,3,6-tetrahydropyridine model of Parkinson's disease. Proc Natl Acad Sci U S A 2003;100: 6145–6150.
137. Purisai MG, McCormack AL, Cumine S et al. Microglial activation as a priming event leading to paraquat-induced dopaminergic cell degeneration. Neurobiol Dis 2007; 25: 392–400.
138. Peng J, Xie L, Stevenson FF, Melov S et al. Nigrostriatal dopaminergic neurodegeneration in the Weaver mouse is mediated via neuroinflammation and alleviated by minocycline administration. J Neurosci 2006; 26: 11644–11651.

139. The NINDS NET-PD Investigators. A randomized, double-blind, utility clinical trial of creatine and minocycline in early Parkinson disease. Neurology 2006; 66: 664–671.
140. Snyder H, Wolozin B. Pathological proteins in Parkinson's disease: Focus on the proteasome. J Mol Neurosci 2004; 24: 25–42.
141. Moore DJ, West AB, Dawson VL et al. Molecular pathophysiology of Parkinson's disease. Annu Rev Neurosci 2005; 28:57–87.
142. McNaught KS, Perl DP, Brownell AL et al. Systemic exposure to proteasome inhibitors causes a progressive model of Parkinson's disease. Ann Neurol 2004; 56: 149–162.
142. Bove J, Zhou C, Jackson-Lewis V, et al. Proteasome inhibition and Parkinson's disease modeling. Ann Neurol 2006; 60: 260–264.
144. Manning-Bog AB, Reaney SH, Chou VP, et al. Lack of nigrostriatal pathology in a rat model of proteasome inhibition. Ann Neurol 2006; 60: 256–260.
145. Ghosh SS, Swerdlow RH, Miller SW, Sheeman B, Parker WD Jr, Davis RE. Use of cytoplasmic hybrid cell lines for elucidating the role of mitochondrial dysfunction in Alzheimer's disease and Parkinson's disease. Ann N Y Acad Sci 1999; 893: 176–191.
146. Trimmer PA, Borland MK, Keeney PM, Bennet JP Jr, Parker WD Jr. Parkinson's disease transgenic mitochondrial cybrids generate Lewy inclusion bodies. J Neurochem 2004; 88: 800–812.

Protein Oxidation Triggers the Unfolded Protein Response and Neuronal Injury in Chemically Induced Parkinson Disease

Alison I. Bernstein and Karen L. O'Malley

Abstract The recent identification of genetic mutations linked to Parkinson's disease (PD), such as α-synuclein, parkin, and LRRK2, has highlighted the role of aberrant protein handling and degradation in this disorder. Moreover, a growing body of data suggests that environmental toxins that mimic PD also exhibit faulty protein handling, providing a mechanistic link between toxicity and the identified PD mutations. In particular, toxin-mediated cell stress and/or some PD mutations can trigger unfolded protein response, a cell-protective mechanism intended for surviving short-term cellular perturbations. If this process cannot overcome the insult, it is thought that apoptosis is rapidly activated. Although the toxicity of several parkinsonian mimetics is thought to stem from the production of reactive oxygen species, whether oxidative stress and other forms of cell stress are subsequent or parallel events is not well established. Emerging data collected using molecular, biochemical, and cellular techniques suggest that oxidative stress precedes the appearance of unfolded protein response which, in turn, precedes apoptosis. Knowledge of the signaling pathways utilized by parkinsonian mimetics as well as their temporal induction may aid in designing more effective interventions in models of PD and ultimately to treat PD in humans.

Keywords 6-OHDA · MPTP · UPR · ER stress · Reactive oxygen species · Apoptosis

1 Introduction

Parkinson's disease (PD) is a progressive neurological disorder characterized by loss of dopaminergic neurons in the substantia nigra (SN) and the formation of cytoplasmic inclusions in the nervous system called Lewy bodies. The latter are

K.L. O'Malley (✉)
Washington University in St. Louis, St. Louis, MO, USA
e-mail: omalleyk@pcg.wustl.edu

S.C. Veasey (ed.), *Oxidative Neural Injury*,
Contemporary Clinical Neuroscience, DOI 10.1007/978-1-60327-342-8_11,
© Humana Press, a part of Springer Science+Business Media, LLC 2009

spherical eosinophilic cytoplasmic protein aggregates composed of a number of proteins, including α-synuclein, parkin, ubiquitin, and neurofilaments [1, 2]. Loss of neurons in the SN leads to depletion of dopamine (DA) in the nigrostriatal pathway. This striatal DA deficiency produces the symptoms of PD, which include resting tremor, rigidity, bradykinesia, and postural instability. Current treatments for PD are symptomatic; replenishment of striatal DA by administration of the DA precursor, levodopa (L-DOPA), improves most of the motor symptoms of PD, at least initially. The etiology of PD is not well understood and drugs are not yet available that stop or slow degeneration of dopaminergic neurons.

Although the molecular mechanisms underlying the etiology of PD remain unclear, oxidative stress and impaired protein degradation have been implicated in the pathogenesis of the disease based on postmortem studies, toxin models, and the identification of genetic mutations linked to the disorder [1, 3, 4]. There is extensive evidence that oxidative stress plays a central role in PD [1, 3]. There are several biochemical defects indicative of oxidative stress in the SN of PD patients, including oxidative damage to DNA and proteins, mitochondrial dysfunction, the depletion of reduced glutathione content, and increased iron levels [4]. Oxidized α-synuclein has been shown to have an increased propensity to misfold and aggregate and lead to neurodegeneration [5]. Oxidative modification of other proteins, including proteins linked to familial PD, such as DJ-1, UCH-L1, and parkin has also been identified in PD brains [6–8]. Thus, oxidative stress may lead to oxidative protein damage and impaired proteolysis, which, in turn, trigger cellular responses such as endoplasmic reticulum (ER) stress, upregulation of the unfolded protein response (UPR), and eventual apoptosis.

2 ER Stress, UPR, and the Ubiquitin–Proteasome System

In response to cell stress conditions that interfere with protein folding, highly specific signaling pathways of the UPR are activated to cope with the accumulation of unfolded and misfolded proteins in the ER [9, 10]. Subsequently, these proteins are retro-translocated to the cytoplasm where they are ubiquitinated and degraded by the proteasome in a process known as ER-associated degradation (ERAD). There are three main pathways by which the UPR regulates ERAD and other adaptive mechanisms. These include three gatekeeper transmembrane signaling proteins such as dsRNA-activated protein kinase (PKR)-like ER kinase (PERK), inositol-requiring protein 1α (IRE1α), and activating transcription factor 6 (ATF6). Under non-stress conditions, the binding immunoglobulin protein (BiP), an ER molecular chaperone, binds to the luminal domains of IRE1, PERK, and ATF6 preventing them from activating UPR. In the presence of unfolded and misfolded proteins, BiP binds to the misfolded proteins leading to the release of one or more of the "gatekeepers."

This results in the activation of IRE1 and PERK and the transport of ATF6 to the Golgi. In the Golgi, ATF6 is cleaved and then translocates to the nucleus where it too activates transcription of UPR target genes, including X-box binding protein 1 (Xbp1). IRE1 cleaves Xbp1 mRNA, producing a novel transcript encoding a transcription factor for UPR target genes. Activation of PERK results in the phosphorylation of eukaryotic translation initiation factor 2 (eIF2α), which attenuates protein synthesis to help relieve ER stress.

If the overload of misfolded proteins is not reduced, prolonged activation of the UPR may lead to cell death. Two pathways of cell death are mediated by IRE1 and PERK. Activated IRE1 can bind to c-Jun-N-terminal inhibitory kinase, which eventually activates mitochondrial/apoptotic protease activating factor-1 (Apaf-1)-dependent apoptosis. Prolonged UPR activation also leads to expression of activating transcription factor 3 (ATF3) through the PERK–eIF2α pathway. ATF3 induces expression of activating transcription factor 4 (ATF4), which, in turn, induces expression of the stress-induced transcription factor C/EBP homologous protein (CHOP), which can activate caspase-3 through unknown intermediates.

Recent evidence suggests that dysfunction of the ubiquitin–proteasome system (UPS), one of the major pathways of protein degradation, may also be an important factor in PD [11]. Proteins are targeted for degradation by the proteasome by a series of enzymes. First, ubiquitin-activating enzymes (E1) activate ubiquitin in an ATP-dependent manner. Next, activated ubiquitin is transferred to ubiquitin-conjugating enzymes (E2). Last, ubiquitin is ligated to the target protein by ubiquitin ligases (E3), which are responsible for the specificity of substrate recognition. Additional activated ubiquitins are then added to form a polyubiquitin chain. Polyubiquitinated proteins are usually degraded by the proteasome complex. The UPS is responsible for the normal turnover of many cellular proteins as well as the clearance of misfolded, mutant, and oxidatively damaged proteins.

3 ER Stress, the Unfolded Protein Response, and the Ubiquitin–Proteasome System in PD

One of the first clues suggesting that ER stress and altered protein degradation may play a role in the etiology of PD came from the presence of Lewy bodies in the surviving dopaminergic neurons of the SN. Lewy bodies are cytoplasmic inclusions containing a variety of proteins, including ubiquitin, neurofilament, proteasomal components, and oxidatively damaged α-synuclein [4, 5]. Whether these aggregates are the result or cause of ER stress or dysregulation of the UPS system is unclear: the inclusions could result from the accumulation of aberrant proteins if the underlying cause of the disease is loss of proteasomal function. Conversely, PD-linked mutations could result in aggregated proteins which, in turn, impair the UPS system.

The identification of mutations associated with familial cases of PD in α-synuclein, parkin, and UCH-L1 underscore the importance of the ubiquitin proteasomal system (UPS) and misfolded proteins in the etiology of PD [2]. As described, α-synuclein is not only a major component of Lewy bodies, but has also been linked to familial PD [12, 13]. Mutations in parkin have also been linked to autosomal recessive juvenile parkinsonism (AR-JP) [14]. Parkin has been identified as an E3 ubiquitin-protein ligase that acts in conjunction with E2 ubiquitin-conjugating enzymes to ubiquitinate target proteins [15]. Mutations associated with AR-JP disrupt the E3 ligase activity of parkin, thus preventing the targeting of specific proteins to the proteasome. Although parkin null mice do not develop a loss of dopaminergic neurons in the SN and have normal brain morphology, identification of parkin substrates has provided some clues as to how parkin mutations may lead to PD [16]. For example, cell division control-related protein (CDCrel-1), which is ubiquitinated by parkin, is involved in the regulation of synaptic vesicle release [17]. Parkin mutations could, therefore, affect the regulation of DA and possibly contribute to parkinsonism. Another parkin substrate, parkin-associated endothelial-like receptor (Pael-R) is a putative G-protein-coupled transmembrane protein that, when overexpressed in cells, becomes misfolded, aggregates and triggers UPR and cell death [18]. Parkin ubiquitinates insoluble Pael-R and targets it to the proteasome for degradation. Pael-R is present in high levels in dopaminergic neurons and insoluble Pael-R is known to accumulate in the brains of AR-JP patients [19]. A third parkin substrate, synphilin-1, is an α-synuclein-associated protein [20]. Despite the possible association between these substrates and PD, none of the parkin substrates have yet been linked to PD pathogenesis.

Mutations in the gene for ubiquitin carboxy-terminal hydrolase L1 (UCH-L1) have been associated with autosomal dominant PD [21]. UCH-L1 is part of a family of de-ubiquitinating enzymes that hydrolyze polyubiquitin chains to monomeric ubiquitin [22]. Similar to genetic animal models of parkin mutations, UCH-L1 mutations in mice do not produce parkinsonian symptoms [23]. However, mutations in the UCH-L1 gene decrease its catalytic activity, suggesting that loss of this activity could lead to impaired clearance of proteins by the UPS. Therefore, loss of UCH-L1 activity in PD may lead to the accumulation of specific proteins that may be toxic or may lead to a general defect in the clearance of abnormal proteins.

Most cases of PD are sporadic but the identification of mutations linked to familial PD has provided clues to the etiology of sporadic PD. The presence of elevated levels of oxidatively damaged proteins, increased protein aggregation, and impaired proteolysis in the SN of PD patients are consistent with the idea that oxidative damage and impaired proteolysis lead to the accumulation of intracellular proteins in both familial and sporadic cases of PD [24, 25]. Additionally, impairment of proteasomal activity and reduced expression of proteasomal subunits have been found in the SN of postmortem tissue from PD patients [25]. Taken together, these findings from familial and sporadic PD patients suggest that dysfunction of the UPS and protein folding machinery may underlie the loss of dopaminergic neurons in PD.

4 Neurotoxin Models of PD

The neurotoxins 6-hydroxydopamine (6-OHDA) and 1-methyl 4-phenyl 1,2,3,6-tetrahydropyridine (MPTP) have been shown to mimic many of the behavioral, pharmacological, and pathological symptoms of PD and hence, have been widely used to create animal models of this disorder [26, 27]. For example, stereotactic injection of 6-OHDA leads to degeneration of dopaminergic neurons in the SN and depletion of striatal DA. 6-OHDA also induces cell death in primary mesencephalic cultures and the mouse dopaminergic cell line, MN9D, through the generation of hydrogen peroxide and derived hydroxyl radicals. Several studies have confirmed that 6-OHDA produces oxidative stress both in vivo and in vitro and that antioxidants are protective against 6-OHDA toxicity [1, 28, 29].

MPTP, a by-product in the production of synthetic meperidine, mimics both the motor deficits and the pathology characteristic of PD in humans [30, 31]. Administration of MPTP to non-human primates and rodents also produces PD-like symptoms and selective dopaminergic loss [32]. MPTP is converted to its active metabolite, 1-methyl-4-phebylpyridinium (MPP^+), in glia and MPP^+ is taken up into dopaminergic neurons by the dopamine transporter (DAT). In in vitro systems, direct application of MPP^+ is able to induce cell death specifically in dopaminergic neurons. MPP^+ induces cell death through the production of ROS, inhibition of mitochondrial complex I, ATP depletion, and superoxide production [1]. Inasmuch as these processes generate free radical species, lead to oxidative stress and contribute to ER stress/UPR, these findings underscore the connection between oxidative stress, mitochondrial dysfunction, and impaired protein degradation in PD.

5 Apoptosis in Toxin Models of PD

Previous results from this lab have demonstrated that both 6-OHDA and MPP^+ rapidly induce ROS production, activate cell stress pathways in dopaminergic cells and trigger cell death in a protein synthesis-dependent manner [28, 29, 33, 34]. Cell death triggered by 6-OHDA in the dopaminergic cell line, MN9D, and in primary dopaminergic neurons has many characteristics of canonical apoptosis: cytochrome c release from mitochondria, a collapse in mitochondrial membrane potential, caspase activation, membrane blebbing, chromatin condensation, and DNA fragmentation [1, 27]. Studies in vivo reveal that, following intrastriatal injection of 6-OHDA in rats, caspase-3-like proteases are activated [35]. In contrast to canonical mitochondrial-mediated apoptosis, however, overexpression of the anti-apoptotic protein Bcl-2 is not protective against 6-OHDA, nor is knocking out the pro-apoptotic protein Bax [36, 37]. Finally, the "extrinsic" death receptor pathway does not seem to be involved either [6, 34]. Therefore, 6-OHDA induces a non-canonical apoptotic pathway.

In contrast to 6-OHDA-induced death, MPP$^+$ toxicity is not blocked by caspase inhibitors in MN9D cells or primary mesencephalic cultures [28, 29]. Moreover, treatment with MPP$^+$ does not induce annexin-V staining or loss of mitochondrial membrane potential in primary mesencephalic cultures [28]. In contrast, MPTP treatment in mice induces both caspase-3 activation and PARP cleavage in the SN if the toxin is given in small doses over 5 days (chronic model) but not if given in small doses within a single day (acute model) [1]. Thus, depending upon the mode of administration of MPTP, apoptosis or necrosis occurs.

6 ROS in Toxin Models

We have previously demonstrated that both 6-OHDA and MPP$^+$ lead to an increase in ROS levels in primary cultures and MN9D cells, as detected with fluorescent ROS sensitive dyes [28, 29, 38]. Cell death induced by 6-OHDA is clearly dependent on this increase in ROS since pretreatment of cells with antioxidants (NAC, MnTBAP, or C3 carboxyfullerene) prevent both increased ROS levels and cell death [29, 39, 40]. However, the role of ROS in initiating cell death following MPP$^+$ treatment is less clear. Specifically, despite the induction of ROS production resulting from DA oxidation by MPP$^+$, pretreatment of cells with antioxidants does not appear to protect cells against MPP$^+$ [29, 39]. In contrast, pretreatment with the C3 carboxyfullerene, provides partial rescue against MPP$^+$ in primary mesencephalic cultures [28]. Therefore, cell death induced by 6-OHDA is completely dependent on the production of ROS, while cell death induced by MPP$^+$ is only partially mediated by increased ROS levels.

ROS is known to cause damage to protein, lipids and DNA, but which of these insults is the trigger for cell death in response to 6-OHDA or MPP$^+$ is currently unknown. Our lab and others have found evidence for oxidative protein damage. For example, within 30 min, 6-OHDA-generated ROS rapidly oxidize various intracellular proteins, including ERp57 and peroxiredoxin [40–43]. Protein oxidation can be prevented by treatment with anti-oxidants and, at least in the dissociated culture model, appears to lie upstream of UPR induction. Thus protein oxidation is a good candidate for triggering ER stress and UPR [39]. Evidence for DNA damage is less clear. DNA is particularly sensitive to oxidation, resulting in oxidized base lesions such as 7,8-dihydro-8-oxoguanine, 8-hydroxy-2-deoxyguanosine, and 5-hydroxyuracil, as well as strand breaks and DNA/protein cross-linking [44]. Attempts to measure DNA damage by staining for mutagenic base modifications in 6-OHDA-treated cell lines or primary dopaminergic neurons have been inconclusive in our hands (Holtz, Kim-Han, and O'Malley, unpublished observations), although earlier studies demonstrated the appearance of single strand breaks detected via agarose DNA fragmentation assays or TUNEL staining detected 18–24 h after 6-OHDA treatment [45–47]. As base oxidation can lead to DNA strand breaks, these data imply that DNA oxidation may occur in a parallel yet delayed fashion versus the more rapid temporal order of UPR events seen in our model system [44]. However, without better tools with which to probe DNA oxidation, it would be premature to completely rule it out as an upstream factor contributing to 6-OHDA-mediated cell death. Besides damaging proteins and DNA, ROS also damage biological membranes with deleterious consequences to receptors, transporters, and channels, as well as signal transduction and cytoskeletal proteins [48]. Lipid peroxidation generates highly reactive aldehydes including 4-hydroxynonenal (HNE). HNE is increased in cortical and brain stem neurons in PD providing support for the notion that lipid peroxidation occurs in this disorder [49]. In contrast, attempts to measure hallmarks of lipid peroxidation in toxin-treated cell lines or primary dopaminergic neurons using various immunological and biochemical strategies have lacked sufficient sensitivity to unequivocally determine temporal induction, if any, in this paradigm (Holtz and O'Malley, unpublished observation). Taken together, current findings support an important initiating role for oxidative protein damage and subsequent ER stress in 6-OHDA- and MPP$^+$-mediated cell death.

7 ER Stress and UPR in Toxin Models

Since 6-OHDA and MPP$^+$ appear to induce a non-canonical apoptosis, we and other labs used genomic and bioinformatics approaches to identify biological processes occurring in response to these toxins [34, 50–52]. Ryu and colleagues employed serial analysis of gene expression in PC12 cells treated with 6-OHDA and saw an increase in the expression of transcripts associated with ER stress and found that 6-OHDA induced a rapid activation of the IRE1 and PERK pathways. In addition to 6-OHDA, they also looked at ER stress and UPR markers following treatment with MPP$^+$, which also led to activation of the PERK and IRE1 activation in PC12 cells. Finally, they determined that loss of PERK increased the sensitivity of sympathetic neurons to 6-OHDA. Further analysis revealed that the upregulated transcripts were associated with many cellular processes including the response to oxidative stress, transcription factors, energetics and metabolism, apoptosis (both pro- and anti-apoptotic genes), and cell cycle regulation. They concluded that neuronal degeneration may result from the combinatorial action of multiple pathways and that stressed neurons may simultaneously trigger pro-apoptotic and protective mechanisms.

Our lab used a microarray approach to assess changes in gene expression following 6-OHDA and MPP$^+$ treatment in MN9D cells. The most highly induced transcript in both conditions was CHOP; other genes involved in UPR were also upregulated by both toxins, including ATF3 and ATF4. 6-OHDA also induced a number of transcripts that were unaffected by MPP$^+$, including molecular chaperones and genes involved in protein folding, trafficking, and the ubiquitin–proteasome pathway, indicating that the two toxins promote distinct yet overlapping pathways (Fig. 1). The microarray results were confirmed by RT-PCR, Western blotting, and immunostaining. Together, these results indicated that 6-OHDA induces all three branches (IRE1, ATF6, and PERK) of UPR, while MPP$^+$ induces only the PERK pathway. Subsequently, we sought to determine whether this upregulation of

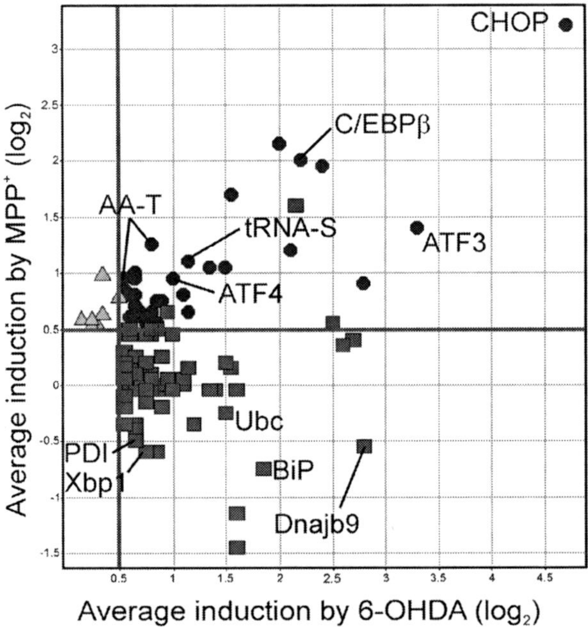

Fig. 1 Microarray analysis reveals both common and distinct transcriptional changes induced by 6-OHDA and MPP$^+$. Total RNA from MN9D cells treated with 6-OHDA or MPP$^+$ in addition to untreated control was used for Affymetrix MG-U74Av2 GeneChip array probe hybridization. Data were analyzed by Affymetrix Microarray Suite version 5 as well as Spotfire Decision Site for Functional Genomics. Transcriptional changes were defined as described in the text. *Graph* shows known genes induced by 6-OHDA or MPP$^+$ treatment plotted as average-fold induction on the *x*-axis and *y*-axis, respectively, with a scale of log$_2$. Several genes of interest involved in UPR have been labeled. Independent of their position on the plot, genes were grouped according to those induced by 6-OHDA but not induced by MPP$^+$ (*squares*), those induced by MPP$^+$ but not induced by 6-OHDA (*triangles*), or those induced by both 6-OHDA and MPP$^+$ (*circles*). Reproduced from [34] with permission from American Society for Biochemistry and Molecular Biology, Inc.

UPR by 6-OHDA is dependent on the production of ROS since oxidative stress is thought to play a key role in toxin models of PD and in PD itself. We found that 6-OHDA rapidly induced the production of ROS and this increase in ROS could be blocked by treatment with antioxidants. 6-OHDA also induced oxidative damage to proteins, increased proteasome activity and levels of polyubiquitinated proteins in an ROS-dependent manner, suggesting that 6-OHDA-induced ER stress arises as a result of oxidative stress. We also demonstrated that the upregulation of UPR markers, cytochrome c release, and eventual cell death induced by 6-OHDA could be blocked by antioxidants. These experiments allowed us to assemble a timeline in which 6-OHDA-induced oxidative stress leads to ER stress and the upregulation of UPR; prolonged UPR activation then leads to apoptosis (Fig. 2).

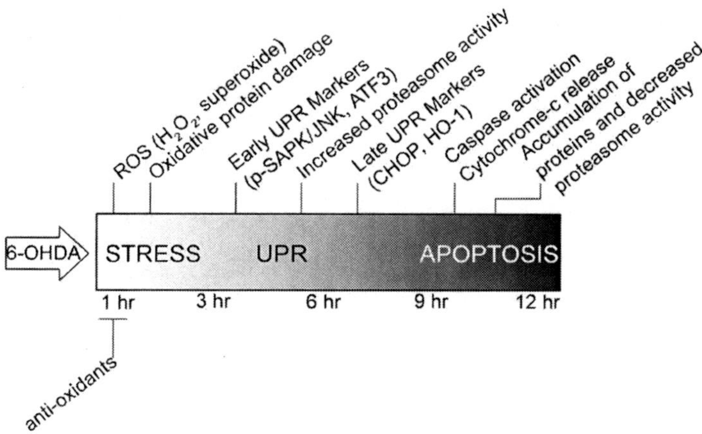

Fig. 2 Timeline of 6-OHDA-induced events. 6-OHDA exposure leads to the rapid induction of ROS and subsequent oxidative damage to proteins. Oxidized proteins are sensed by ER mechanisms leading to upregulation of the UPR and proteasome activity. Activation of UPR-associated events can be generally defined as occurring early (~3 h) or late (~6–9 h). After 8 h, release of cytochrome c from the mitochondria is observed which corresponds to the appearance of activated caspases. A late event associated with failure of cellular protein clearance and declining proteasome activity is the accumulation of polyubiquitinated (polyUb) proteins. Reproduced from [40] with permission from Blackwell Publishing Company

8 Possible Mechanisms of ER Stress-Mediated Activation of Apoptosis

Prolonged ER stress and UPR upregulation can lead to apoptosis through both mitochondrial-dependent and -independent pathways. In rodents, ER stress induces cleavage of caspase-12, which can either bypass the mitochondria and directly cleave caspase-9 or potentially interact with other pro-apoptotic proteins such as Bap31, leading to mitochondrial fission and loss of cytochrome c [53, 54]. Other evidence suggests that loss of calcium from the ER and concomitant uptake by mitochondria can trigger the collapse of the mitochondrial membrane potential and lead to apoptosis [55, 56]. The exact mechanism by which calcium is transferred from one organelle to another is not known but appears to require various pro-apoptotic proteins such as Bax, Bak, and Bid, as well as the inositol-1,4,5-trisphosphate (IP3) receptor [55, 57–59]. We have demonstrated that Bax is not the mediator in 6-OHDA- or MPP$^+$-mediated cell death in primary mesencephalic cultures since deletion of Bax does not protect dopaminergic neurons against either toxin [37]. In some systems, the structurally related pro-apoptotic proteins, Bak or Bok, can substitute for Bax. In response to 6-OHDA in MN9D cells, an ROS-dependent oligomerization of Bak, but not Bok, can be detected, suggesting that Bak serves as the upstream apoptotic trigger [40]. A third possible mechanism of ER stress-induced apoptosis is via pathways involving the oxidation and reduction of disulfide bonds, which result in the accumulation of ROS, reduced mitochondrial activity, and ensuing cell death [60–62]. It is also possible that a BH3-only protein serves as

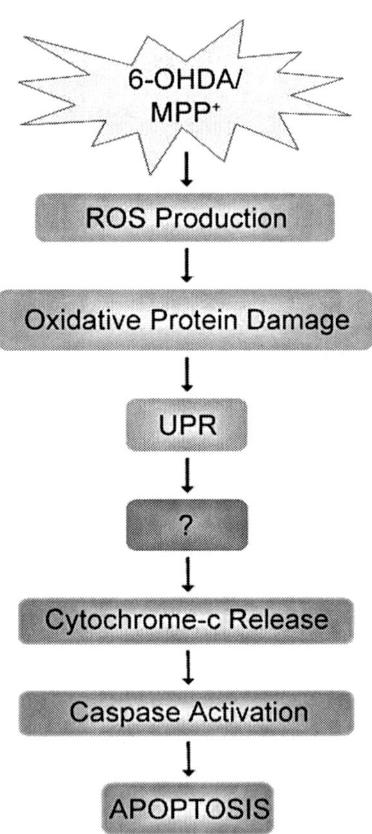

Fig. 3 Model of ER stress-induced apoptosis by parkinsonian mimetics. In this model, 6-OHDA and MPP$^+$ induce a rapid increase in ROS and oxidative protein damage, which leads to protein misfolding and impaired clearance of damaged proteins. The resulting accumulation of misfolded proteins causes ER stress and subsequent upregulation of UPR. Activation of UPR triggers a signal from the ER to the mitochondria to initiate the mitochondrial induction of apoptosis

an ER stress-induced trigger of apoptosis. BH3-only proteins are pro-apoptotic members of the Bcl-2 protein family that are known to initiate apoptosis in response to many toxins [63]. PUMA/bbc3 (p53-upregulated mediator or apoptosis/bcl-2 binding component 3) is one such BH3-only protein that is known to be upregulated in response to ER stress and to cause cytochrome c release, caspase activation, and subsequent apoptosis [64–71]. Which of these pathways predominates following a particular insult depends on both the cell type and the cell stressor.

9 Conclusions

Together, these experiments demonstrate that ER stress and UPR are induced by the parkinsonian mimetics, 6-OHDA and MPP$^+$ (Fig. 3). UPR is temporally upstream of apoptotic and cell death markers, suggesting that it may be functionally upstream of cell death as well. However, the question of whether UPR and apoptosis lie in a sequential or parallel pathway remains to be determined. It is currently unclear whether UPR represents a protective or pro-apoptotic mechanism, or perhaps a protective UPR initially followed by a pro-apoptotic mechanism in response to overwhelming injury. Current work in our lab is focused on answering this question.

References

1. Blum D, Torch S, Lambeng N, et al. Molecular pathways involved in the neurotoxicity of 6-OHDA, dopamine and MPTP: contribution to the apoptotic theory in Parkinson's disease. Prog Neurobiol 2001;65(2):135–72.
2. Dauer W, Przedborski S. Parkinson's disease: mechanisms and models. Neuron 2003; 39(6):889–909.
3. Jenner P. Oxidative stress in Parkinson's disease. Ann Neurol 2003;53 Suppl 3:S26–36; discussion S-8.
4. McNaught KS, Olanow CW, Halliwell B, Isacson O, Jenner P. Failure of the ubiquitin-proteasome system in Parkinson's disease. Nat Rev Neurosci 2001;2(8):589–94.
5. Giasson BI, Duda JE, Murray IV, et al. Oxidative damage linked to neurodegeneration by selective alpha-synuclein nitration in synucleinopathy lesions. Science 2000;290(5493): 985–9.
6. Choi J, Levey AI, Weintraub ST, et al. Oxidative modifications and down-regulation of ubiquitin carboxyl-terminal hydrolase L1 associated with idiopathic Parkinson's and Alzheimer's diseases. J Biol Chem 2004;279(13):13256–64.
7. Choi J, Sullards MC, Olzmann JA, et al. Oxidative damage of DJ-1 is linked to sporadic Parkinson and Alzheimer diseases. J Biol Chem 2006;281(16):10816–24.
8. LaVoie MJ, Ostaszewski BL, Weihofen A, Schlossmacher MG, Selkoe DJ. Dopamine covalently modifies and functionally inactivates parkin. Nat Med 2005;11(11):1214–21.
9. Forman MS, Lee VM, Trojanowski JQ. 'Unfolding' pathways in neurodegenerative disease. Trends Neurosci 2003;26(8):407–10.
10. Zhang K, Kaufman RJ. Signaling the unfolded protein response from the endoplasmic reticulum. J Biol Chem 2004;279(25):25935–8.

11. Betarbet R, Sherer TB, Greenamyre JT. Ubiquitin-proteasome system and Parkinson's diseases. Exp Neurol 2005;191 Suppl 1:S17–27.
12. Polymeropoulos MH, Higgins JJ, Golbe LI, et al. Mapping of a gene for Parkinson's disease to chromosome 4q21-q23. Science 1996;274(5290):1197–9.
13. Polymeropoulos MH, Lavedan C, Leroy E, et al. Mutation in the alpha-synuclein gene identified in families with Parkinson's disease. Science 1997;276(5321):2045–7.
14. Kitada T, Asakawa S, Hattori N, et al. Mutations in the parkin gene cause autosomal recessive juvenile parkinsonism. Nature 1998;392(6676):605–8.
15. Imai Y, Soda M, Takahashi R. Parkin suppresses unfolded protein stress-induced cell death through Its E3 ubiquitin-protein ligase activity. J Biol Chem 2000;275(46):35661–4.
16. Goldberg MS, Fleming SM, Palacino JJ, et al. Parkin-deficient mice exhibit nigrostriatal deficits but not loss of dopaminergic neurons. J Biol Chem 2003;278(44):43628–35.
17. Zhang Y, Gao J, Chung KK, Huang H, Dawson VL, Dawson TM. Parkin functions as an E2-dependent ubiquitin- protein ligase and promotes the degradation of the synaptic vesicle-associated protein, CDCrel-1. Proc Natl Acad Sci U S A 2000;97(24):13354–9.
18. Imai Y, Soda M, Inoue H, Hattori N, Mizuno Y, Takahashi R. An unfolded putative transmembrane polypeptide, which can lead to endoplasmic reticulum stress, is a substrate of Parkin. Cell 2001;105(7):891–902.
19. Dawson TM, Dawson VL. Rare genetic mutations shed light on the pathogenesis of Parkinson disease. J Clin Invest 2003;111(2):145–51.
20. Chung KK, Zhang Y, Lim KL, et al. Parkin ubiquitinates the alpha-synuclein-interacting protein, synphilin-1: implications for Lewy-body formation in Parkinson disease. Nat Med 2001;7(10):1144–50.
21. Leroy E, Boyer R, Auburger G, et al. The ubiquitin pathway in Parkinson's disease. Nature 1998;395(6701):451–2.
22. Pickart CM. Ubiquitin in chains. Trends Biochem Sci 2000;25(11):544–8.
23. Saigoh K, Wang YL, Suh JG, et al. Intragenic deletion in the gene encoding ubiquitin carboxy-terminal hydrolase in gad mice. Nat Genet 1999;23(1):47–51.
24. McNaught KS, Jenner P. Proteasomal function is impaired in substantia nigra in Parkinson's disease. Neurosci Lett 2001;297(3):191–4.
25. McNaught KS, Belizaire R, Isacson O, Jenner P, Olanow CW. Altered proteasomal function in sporadic Parkinson's disease. Exp Neurol 2003;179(1):38–46.
26. Beal MF. Experimental models of Parkinson's disease. Nat Rev Neurosci 2001;2(5):325–34.
27. Speciale SG. MPTP: insights into parkinsonian neurodegeneration. Neurotoxicol Teratol 2002;24(5):607–20.
28. Lotharius J, Dugan LL, O'Malley KL. Distinct mechanisms underlie neurotoxin-mediated cell death in cultured dopaminergic neurons. J Neurosci 1999;19(4):1284–93.
29. Choi WS, Yoon SY, Oh TH, Choi EJ, O'Malley KL, Oh YJ. Two distinct mechanisms are involved in 6-hydroxydopamine- and MPP+-induced dopaminergic neuronal cell death: role of caspases, ROS, and JNK. J Neurosci Res 1999;57(1):86–94.
30. Davis GC, Williams AC, Markey SP, et al. Chronic parkinsonism secondary to intravenous injection of meperidine analogues. Psychiatry Res 1979;1(3):249–54.
31. Langston JW, Ballard P, Tetrud JW, Irwin I. Chronic parkinsonism in humans due to a product of meperidine-analog synthesis. Science 1983;219(4587):979–80.
32. Kopin IJ, Schoenberg DG. MPTP in animal models of Parkinson's disease. Mt Sinai J Med 1988;55(1):43–9.
33. Coelln RV, Kugler S, Bahr M, Weller M, Dichgans J, Schulz JB. Rescue from death but not from functional impairment: caspase inhibition protects dopaminergic cells against 6-hydroxydopamine-induced apoptosis but not against the loss of their terminals. J Neurochem 2001;77(1):263–73.
34. Holtz WA, O'Malley KL. Parkinsonian mimetics induce aspects of unfolded protein response in death of dopaminergic neurons. J Biol Chem 2003;278(21):19367–77.

35. Jeon BS, Kholodilov NG, Oo TF, et al. Activation of caspase-3 in developmental models of programmed cell death in neurons of the substantia nigra. J Neurochem 1999;73(1):322–33.
36. Oh YJ, Wong SC, Moffat M, O'Malley KL. Overexpression of Bcl-2 attenuates MPP+, but not 6-ODHA, induced cell death in a dopaminergic neuronal cell line. Neurobiol Dis 1995;2(3):157–67.
37. O'Malley KL, Liu J, Lotharius J, Holtz W. Targeted expression of BCL-2 attenuates MPP+ but not 6-OHDA induced cell death in dopaminergic neurons. Neurobiol Dis 2003;14(1):43–51.
38. Lotharius J, O'Malley KL. The parkinsonism-inducing drug 1-methyl-4-phenylpyridinium triggers intracellular dopamine oxidation. A novel mechanism of toxicity. J Biol Chem 2000;275(49):38581–8.
39. Han BS, Hong HS, Choi WS, Markelonis GJ, Oh TH, Oh YJ. Caspase-dependent and -independent cell death pathways in primary cultures of mesencephalic dopaminergic neurons after neurotoxin treatment. J Neurosci 2003;23(12):5069–78.
40. Holtz WA, Turetzky JM, Jong YJ, O'Malley KL. Oxidative stress-triggered unfolded protein response is upstream of intrinsic cell death evoked by parkinsonian mimetics. J Neurochem 2006;99(1):54–69.
41. Kim-Han JS, O'Malley KL. Cell stress induced by the parkinsonian mimetic, 6-hydroxydopamine, is concurrent with oxidation of the chaperone, ERp57, and aggresome formation. Antioxid Redox Signa 2007;9(12):2255–64.
42. Lee YM, Park SH, Shin DI, et al. Oxidative modification of peroxiredoxin is associated with drug-induced apoptotic signaling in experimental models of Parkinson disease. J Biol Chem 2008;283(15):9986–98.
43. Saito Y, Nishio K, Ogawa Y, et al. Molecular mechanisms of 6-hydroxydopamine-induced cytotoxicity in PC12 cells: involvement of hydrogen peroxide-dependent and -independent action. Free Radic Biol Med 2007;42(5):675–85.
44. Martin LJ. DNA damage and repair: relevance to mechanisms of neurodegeneration. J Neuropathol Exp Neurol 2008;67(5):377–87.
45. Takai N, Nakanishi H, Tanabe K, et al. Involvement of caspase-like proteinases in apoptosis of neuronal PC12 cells and primary cultured microglia induced by 6-hydroxydopamine. J Neurosci Res 1998;54(2):214–22.
46. Li H, Ding JH, Hu G. Group I mGluR ligands fail to affect 6-hydroxydopamine-induced death and glutamate release of PC12 cells. Acta Pharmacol Sin 2003;24(7):641–5.
47. Jiang H, Ren Y, Zhao J, Feng J. Parkin protects human dopaminergic neuroblastoma cells against dopamine-induced apoptosis. Hum Mol Genet 2004;13(16):1745–54.
48. Stark G. Functional consequences of oxidative membrane damage. J Membr Biol 2005;205(1):1–16.
49. Zarkovic K. 4-hydroxynonenal and neurodegenerative diseases. Mol Aspects Med 2003;24(4–5):293–303.
50. Ryu EJ, Harding HP, Angelastro JM, Vitolo OV, Ron D, Greene LA. Endoplasmic reticulum stress and the unfolded protein response in cellular models of Parkinson's disease. J Neurosci 2002;22(24):10690–8.
51. Ryu EJ, Angelastro JM, Greene LA. Analysis of gene expression changes in a cellular model of Parkinson disease. Neurobiol Dis 2005;18(1):54–74.
52. Holtz WA, Turetzky JM, O'Malley KL. Microarray expression profiling identifies early signaling transcripts associated with 6-OHDA-induced dopaminergic cell death. Antioxid Redox Signal 2005;7(5–6):639–48.
53. Nakagawa T, Yuan J. Cross-talk between two cysteine protease families. Activation of caspase-12 by calpain in apoptosis. J Cell Biol 2000;150(4):887–94.
54. Nakagawa T, Zhu H, Morishima N, et al. Caspase-12 mediates endoplasmic-reticulum-specific apoptosis and cytotoxicity by amyloid-beta. Nature 2000;403(6765):98–103.

55. Nutt LK, Pataer A, Pahler J, et al. Bax and Bak promote apoptosis by modulating endoplasmic reticular and mitochondrial Ca2+ stores. J Biol Chem 2002;277(11):9219–25.
56. Scorrano L, Oakes SA, Opferman JT, et al. BAX and BAK regulation of endoplasmic reticulum Ca2+: a control point for apoptosis. Science 2003;300(5616):135–9.
57. Boehning D, Patterson RL, Sedaghat L, Glebova NO, Kurosaki T, Snyder SH. Cytochrome c binds to inositol (1,4,5) trisphosphate receptors, amplifying calcium-dependent apoptosis. Nat Cell Biol 2003;5(12):1051–61.
58. Darios F, Lambeng N, Troadec JD, Michel PP, Ruberg M. Ceramide increases mitochondrial free calcium levels via caspase 8 and Bid: role in initiation of cell death. J Neurochem 2003;84(4):643–54.
59. Oakes SA, Scorrano L, Opferman JT, et al. Proapoptotic BAX and BAK regulate the type 1 inositol trisphosphate receptor and calcium leak from the endoplasmic reticulum. Proc Natl Acad Sci U S A 2005;102(1):105–10.
60. Tu BP, Weissman JS. Oxidative protein folding in eukaryotes: mechanisms and consequences. J Cell Biol 2004;164(3):341–6.
61. Harding HP, Zhang Y, Zeng H, et al. An integrated stress response regulates amino acid metabolism and resistance to oxidative stress. Mol Cell 2003;11(3):619–33.
62. Haynes CM, Titus EA, Cooper AA. Degradation of misfolded proteins prevents ER-derived oxidative stress and cell death. Mol Cell 2004;15(5):767–76.
63. Huang DC, Strasser A. BH3-Only proteins-essential initiators of apoptotic cell death. Cell 2000;103(6):839–42.
64. Yu J, Zhang L, Hwang PM, Rago C, Kinzler KW, Vogelstein B. Identification and classification of p53-regulated genes. Proc Natl Acad Sci U S A 1999;96(25):14517–22.
65. Nakano K, Vousden KH. PUMA, a novel proapoptotic gene, is induced by p53. Mol Cell 2001;7(3):683–94.
66. Sun YF, Yu LY, Saarma M, Timmusk T, Arumae U. Neuron-specific Bcl-2 homology 3 domain-only splice variant of Bak is anti-apoptotic in neurons, but pro-apoptotic in non-neuronal cells. J Biol Chem 2001;276(19):16240–7.
67. Han J, Flemington C, Houghton AB, et al. Expression of bbc3, a pro-apoptotic BH3-only gene, is regulated by diverse cell death and survival signals. Proc Natl Acad Sci U S A 2001;98(20):11318–23.
68. Reimertz C, Kogel D, Rami A, Chittenden T, Prehn JH. Gene expression during ER stress-induced apoptosis in neurons: induction of the BH3-only protein Bbc3/PUMA and activation of the mitochondrial apoptosis pathway. J Cell Biol 2003;162(4):587–97.
69. Villunger A, Michalak EM, Coultas L, et al. p53- and drug-induced apoptotic responses mediated by BH3-only proteins puma and noxa. Science 2003;302(5647):1036–8.
70. Jeffers JR, Parganas E, Lee Y, et al. Puma is an essential mediator of p53-dependent and -independent apoptotic pathways. Cancer Cell 2003;4(4):321–8.
71. Melino G, Bernassola F, Ranalli M, et al. p73 Induces apoptosis via PUMA transactivation and Bax mitochondrial translocation. J Biol Chem 2004;279(9):8076–83.

Treating Oxidative Neural Injury: Methionine Sulfoxide Reductase Therapy for Parkinson's Disease

Ramez Wassef, Stefan H. Heinemann, and Toshinori Hoshi

Abstract Parkinson's disease is a common neurodegenerative disease that is characterized by loss of dopaminergic neurons in the substantia nigra and impaired motor function. The disease is multifactorial but oxidative injury is associated with the pathology and contributes to neuronal injury. Fibrillations of α-synuclein are present in pathological lesions in this disease. Oxidation of the sulfur moieties of methionine residues on α-synuclein can contribute to α-synuclein fibrillation. An anti-oxidant enzyme methionine sulfoxide reductase can reverse the methionine oxidation on α-synuclein and act as a sink scavenging reactive oxygen species. Thus, boosting methionine sulfoxide reductase activity may prevent oxidative injury in dopaminergic neurons that contributes to neurodegeneration and impaired neuronal function in Parkinson's disease. One promising approach to augmenting methionine sulfoxide reductase activity in neurons is to provide a naturally occurring substrate for methionine sulfoxide reductase A, S-methyl-L-cysteine. Recent work in our lab with this compound supports the promise of this substance in preventing or delaying motor dysfunction in multiple model systems of Parkinson's disease.

Keywords Parkinson's · α-synuclein · methionine oxidation · methionine sulfoxide reductase · S-methyl-Lcysteine

1 Introduction

1.1 Parkinson's Disease

More than 600,000 individuals in the United States suffer from Parkinson's disease (PD), a devastating neurodegenerative disorder [1]. Several risk factors for the disease, including oxidative stress, neurotoxins, pesticides, and metal ions, have been identified but the fundamental cause of the disease is not yet known

T. Hoshi (✉)
Department of Physiology, University of Pennsylvania, Philadelphia, PA 19104, USA
e-mail: hoshi@hoshi.org

S.C. Veasey (ed.), *Oxidative Neural Injury*,
Contemporary Clinical Neuroscience, DOI 10.1007/978-1-60327-342-8_12,

[2, 3]. The progressive disease is associated with loss of dopaminergic neurons in the substantia nigra and characterized by four main symptoms: bradykinesia, postural instability, cogwheel rigidity, and resting tremors [2]. These symptoms of parkinsonism may arise with any disorder or drug that decreases the bioavailability of dopamine in the central nervous system, the end result of which is reduced input to the cortical motor system [4]. However, the presence of characteristic eosinophilic cytoplasmic inclusions in surviving nerve cells and neuronal processes serves as a hallmark feature of PD. The inclusions in nerve cells and neuronal processes in PD are called Lewy bodies (LBs) and Lewy neuritis (LNs), respectively, for they were first described in the substantia nigra by the German neurologist Frederic Lewy in the early 20th century. In addition to the substantia nigra, Lewy inclusions were also found in the olfactory bulb, the anterior olfactory nucleus, the dorsal motor nucleus of the vagus, and the intermediate reticular zone in the lower brain stem, probably contributing to some of the non-motor symptoms of PD, such as olfactory dysfunction and sleep disturbances [5].

1.2 *α-Synuclein*

One of the main components of LBs and LNs is the protein α-synuclein [6]. The function of this protein is yet to be established but its presynaptic localization, in the cytoplasmic and/or membrane compartments, suggests a role in vesicular transport and neurotransmitter release [7]. Structurally, human α-synuclein is composed of 140 amino acids with a large amino-terminal segment of 67 amino acids followed by a hydrophobic domain essential for the aggregation of the protein [8], and a C-terminus domain rich in glutamic acid, aspartic acid, and proline. Under native conditions, α-synuclein is largely a disordered soluble protein, but it forms fibrillar aggregates with ubiquitin and tubulin within LBs and LNs [6]. In humans, multiplications and point mutations of the α-synuclein gene (A30P, A53T) are associated with rare familial forms of PD [9–11]. In addition, in patients with idiopathic PD, the mRNA levels of α-synuclein are elevated in the substantia nigra dopaminergic neurons [12] and aggregation of α-synuclein is clearly observed [13]. The observations such as those summarized above led to the idea that aggregation of α-synuclein may be a critical event in the pathogenesis of PD [14]. In particular, aggregation of α-synuclein has been hypothesized to follow a nucleation polymerization pattern [14]. According to this postulate, different risk factors implicated in PD, such as oxidative stress, neurotoxins, pesticides, metal ions, and multiplications/point mutations of select genes, enhance the formation of soluble oligomers of α-synuclein, which is then followed by protofibrils and finally fibril formation [15]. The direct role of α-synuclein aggregation/fibrillation in motor impairment and neurodegeneration in the substantia nigra neurons in PD has been extensively studied in many animal models but thus far failed to yield unequivocal results [16, 17]. Motor impairment by overexpression of α-synuclein was observed in some studies [16] but absent in others [17]. One study reported motor impairment caused by α-synuclein overexpression but the pathology was associated with denervation of the neuromuscular junctions, yet sparing the substantia nigra [18]. Furthermore, the neuronal inclusions did not show any fibrillar structures [18].

The mutation A53T in human α-synuclein is associated with a familial form of PD [53] and a C-terminal truncation of this mutant protein facilitates aggregation in vitro compared with the wild-type protein [19]. Despite the greater aggregation propensity of the C-terminal truncated A53T α-synuclein protein, its overexpression failed to induce LB-positive pathology [19]. Interestingly, the presence of motor dysfunction accompanied by nigral neuronal loss in the absence of fibrillar inclusions was reported in some transgenic mouse models as well as in rats overexpressing α-synuclein, suggesting that the neuronal degeneration may be mediated by the soluble oligomers of α-synuclein [20, 21].

2 α-Synuclein, Methionine Oxidation and Methionine Sulfoxide Reductase

Several risk factors associated with PD have been identified, including select metal ions and oxidative stress [22, 23]. While α-synuclein lacks cysteine and tryptophan, which are readily oxidized in many other proteins, it does contain four methionine residues that have been suggested to be subject to oxidative modification under physiological conditions [24]. Oxidation of the sulfur moieties of these methionine residues plays a role in fibrillation of α-synuclein in a pH-dependent manner [24–26]. At low pH, oxidation of the methionine residues *promotes* fibril formation by favoring a less extended conformation [25]. In contrast, at physiological pH, the oxidation *inhibits* fibrillation of α-synuclein by stabilizing soluble oligomers of the protein [25]. The degree of fibrillation of methionine-oxidized α-synuclein may also depend on the availability of metal ions, which generally promote fibrillation. However, Hg^{2+}, Ca^{2+}, and Cu^{2+} interestingly inhibit fibrillation [27]. These in vitro studies clearly indicate a role of methionine oxidation in α-synuclein fibrillation, either by enhancing or inhibiting fibrillation depending on the experimental conditions used.

Many physiological oxidants, such as superoxide ($O_2^{\bullet-}$), $HO^{\bullet}$, H_2O_2, chloramines, and peroxynitrites, readily oxidize methionine [28], producing two optical enantiomers of methionine sulfoxide: methionine-R-sulfoxide (Met-R-O) and methionine-S-sulfoxide (Met-S-O) by the addition of an oxygen atom to the sulfur atom of methionine (Fig. 1) [28]. Unlike other oxidized amino acids, the conversion to Met-S/R-O is reversible and enzymatically catalyzed; Met-S-O and Met-R-O are catalytically reduced back to methionine by the methionine sulfoxide reductases (MSRs) in a stereo-specific manner. Methionine sulfoxide reductase A (MSRA) typically coded by one gene reduces Met-S-O while methionine sulfoxide reductase B (MSRB1, 2, and 3) coded by three genes

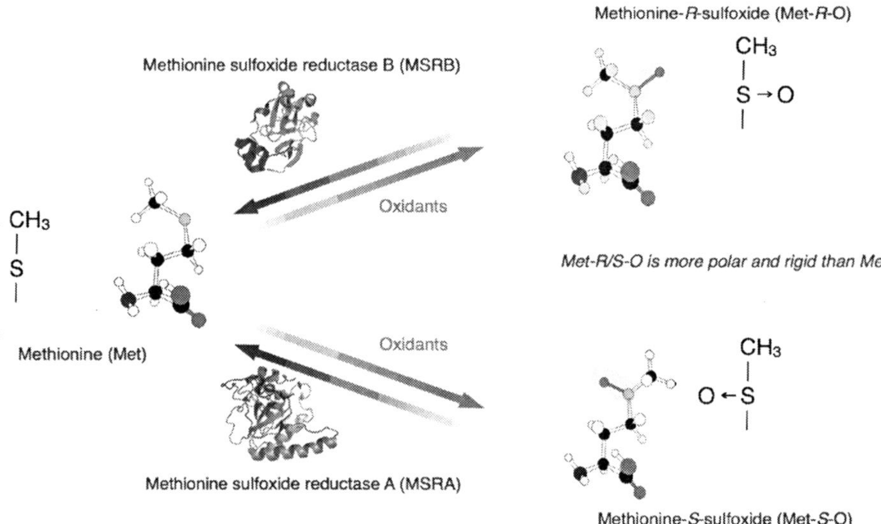

Fig. 1 Reversible oxidation of methionine enabled by methionine sulfoxide reductases. Methionine is oxidized by a variety of reactive molecules by the addition of an oxygen atom to the sulfur atom to form two optical enantiomers of methionine sulfoxide, Met-R-O and Met-S-O. Methionine sulfoxide reductase A (MSRA) reduces Met-S-O and methionine sulfoxide reductase B (MSRB) reduces Met-R-O back to Met. The structures of MSRA (1FVG; [75]) and MSRB (1L1D; [76]) were rendered using MacPyMol (http://pymol.source-forge.net/)

reduces Met-R-O back to methionine [29]. The unique enzymatic reduction of Met-S/R-O suggests the physiological importance of reversible oxidation of methionine involving the MSRs. Accumulating results from a variety of model cell systems and organisms illustrate the general importance of MSRA in protection against oxidative damage, hence a role in many disease states where oxidative stress is implicated. For example, in *E. coli*, where the msrA gene was first isolated [30], a genetic disruption of the gene renders the cells more prone to oxidative damage caused by reactive oxygen species (ROS) and reactive nitrogen intermediates [31]. Human lens epithelial cells with low levels of MSRA have decreased viability when challenged with ROS [32]. MSRA deficient mice were reported to have increased sensitivity to oxidative stress, an abnormal walking pattern, and a shortened lifespan [33]. Conversely, over-expression of MSRA confers greater oxidative stress resistance in human T-lymphocytes, PC12 cells, WI-38 human fibroblasts, and cardiac myocytes [34–37]. In the fruit fly *Drosophila melanogaster*, overexpression of MSRA preferentially in the central nervous system using a bipartite expression strategy [38] extends the lifespan by about 70% [39]. This lifespan extension is accompanied by a marked delay in the age-associated decrease in physical activity and also by a greater protection against the oxidative stress caused by feeding paraquat, a pesticide that produces the oxidant $O_2^{\bullet -}$ and causes a variety of oxidative modifications including severe lipid peroxidation [40]. It is worth noting that paraquat induces phenotypes reminiscent of PD symptoms in other laboratory animals [41] and is implicated as a risk factor for PD [3].

The results of many studies, such as those mentioned above, collectively suggest that disruptions of MSRs, especially MSRA that is coded by only one gene in mammals [29], compromise the oxidative stress resistance, while over-expression of MSRA often confers greater oxidative stress resistance. Two mechanisms of the MSRA action to enhance oxidative stress protection have been proposed: the repair hypothesis and the sink hypothesis [42, 43]. The repair hypothesis postulates that oxidation of select methionine residues in proteins interfere with their functions and that MSRA restores their normal functionality. One of the supporting observations for this idea is that, in calmodulin, oxidation of C-terminal methionine residues impairs its function, and reduction by MSRA restores the function [44]. In contrast, according to the "sink hypothesis" of the MSRA action, oxidation of select surface-exposed methionine residues has no impact on the protein function but such oxidation scavenges potentially damaging reactive compounds, thus sparing other more critical targets from oxidative damage [43]. The role of MSRA is then to regenerate the scavenging sites. Both of the two mechanisms described above are most probably important in a subcellular-, cellular-, and tissue-specific manner. In fact, the msrA transcript is alternatively spliced to produce multiple variant proteins, which are targeted to specific subcellular loci, such as the cytoplasm and mitochondria [45, 46].

Within the mouse brain, both MSRA and MSRBs are expressed in a region-specific manner with notably high levels of RNA expression in the hippocampus and the cerebellum (http://www.brain-map.org). In contrast, the expression in the substantia nigra is not particularly high, suggesting that the protection mediated by the MSRA system is not at a saturating or maximal level. This less-than-maximal expression of MSRs may contribute to the vulnerability of the substantia nigra neurons to oxidative damage, contributing to the PD pathogenesis [47]. While no firm link has been established between PD and a disruption of the MSRA system, the MSRA activity was suggested to be lower in the brain of Alzheimer's disease patients [48].

3 The Fruit Fly Model of PD

3.1 Advantages of the Model

Mechanistic understandings of neurodegenerative diseases, including PD, are greatly facilitated by the availability of robust animal models of the disorders. Recently, the fruit fly *Drosophila melanogaster* has emerged as an easy-to-use yet potentially powerful model system of PD [49, 50]. The relatively short lifespan, typically up to 2–3 months in many laboratories, combined with some relatively "high-level" behavioral patterns makes *Drosophila* a suitable model to decipher the molecular and cellular mechanisms of PD. Multiple genes implicated in PD have been manipulated in *Drosophila* to study the molecular pathways involved in PD [49]. *Drosophila* may be particularly well suited to study physiological and pathophysiological roles of α-synuclein in PD as the animal does not have an α-synuclein gene [51]. Taking advantage of this, Feany and Bender overexpressed human α-synuclein as well as α-synuclein with A53T and A30P mutations in *Drosophila (52)*. These two mutations of α-synuclein are associated with autosomal dominant forms of PD [53, 54]. Pan-neuronal overexpression of the wild-type and mutant α-synuclein resulted in premature age-dependent deterioration of the locomotor behavior as measured using the geotactic response. After tapping or vortexing, the young animals climb rapidly to the top of a vial and this reactive climbing ability [55] deteriorates markedly with age (Fig. 2A), essentially recapitulating the age-dependent motor decline observed in most other species, including human. Overexpression of human α-synuclein in the *Drosophila* nervous system accelerates the age-dependent loss and the time course is more accelerated when A30P α-synuclein associated with a familial form of PD is expressed [52] (also see Fig. 2A). As found in PD patients in which a loss of dopaminergic neurons is observed resulting in impairment of motor functions [56], the accelerated decline in the locomotor ability caused by α-synuclein in *Drosophila* is also accompanied by a marked loss of dopaminergic neurons in multiple areas of the *Drosophila* brain as suggested by a decrease in immunostaining for the enzyme tyrosine hydroxylase involved in dopamine synthesis [57]. Thus, the α-synuclein *Drosophila* model recapitulates many features of PD albeit the phenotypes may not be fully penetrant under every laboratory condition [58].

3.2 Overexpression of MSRA Alleviates the Parkinson's Disease Phenotype

Overexpression of MSRA in the *Drosophila* nervous system confers greater oxidative stress resistance, extends lifespan and delays age-dependent decline of locomotor behavior compared to control animals [39]. This finding in combination with the previous suggestions that oxidative stress may be an important risk factor in PD raises the question whether MSRA could have any protective role in PD. The ease of genetic manipulations in *Drosophila* enabled the development of double transgenic flies expressing both α-synuclein and MSRA [57]. Overexpression of MSRA markedly mitigates the deleterious effects of α-synuclein overexpression on the dopaminergic neuronal loss and climbing ability (Fig. 2A). Similar obliteration of the PD-like phenotypes by MSRA overexpression was also observed in the flies overexpressing the mutant A30P α-synuclein associated with a familial form of PD [57].

In addition to the well-known movement symptoms such as bradykinesia, rigidity, and resting tremors, PD patients may display many other symptoms,

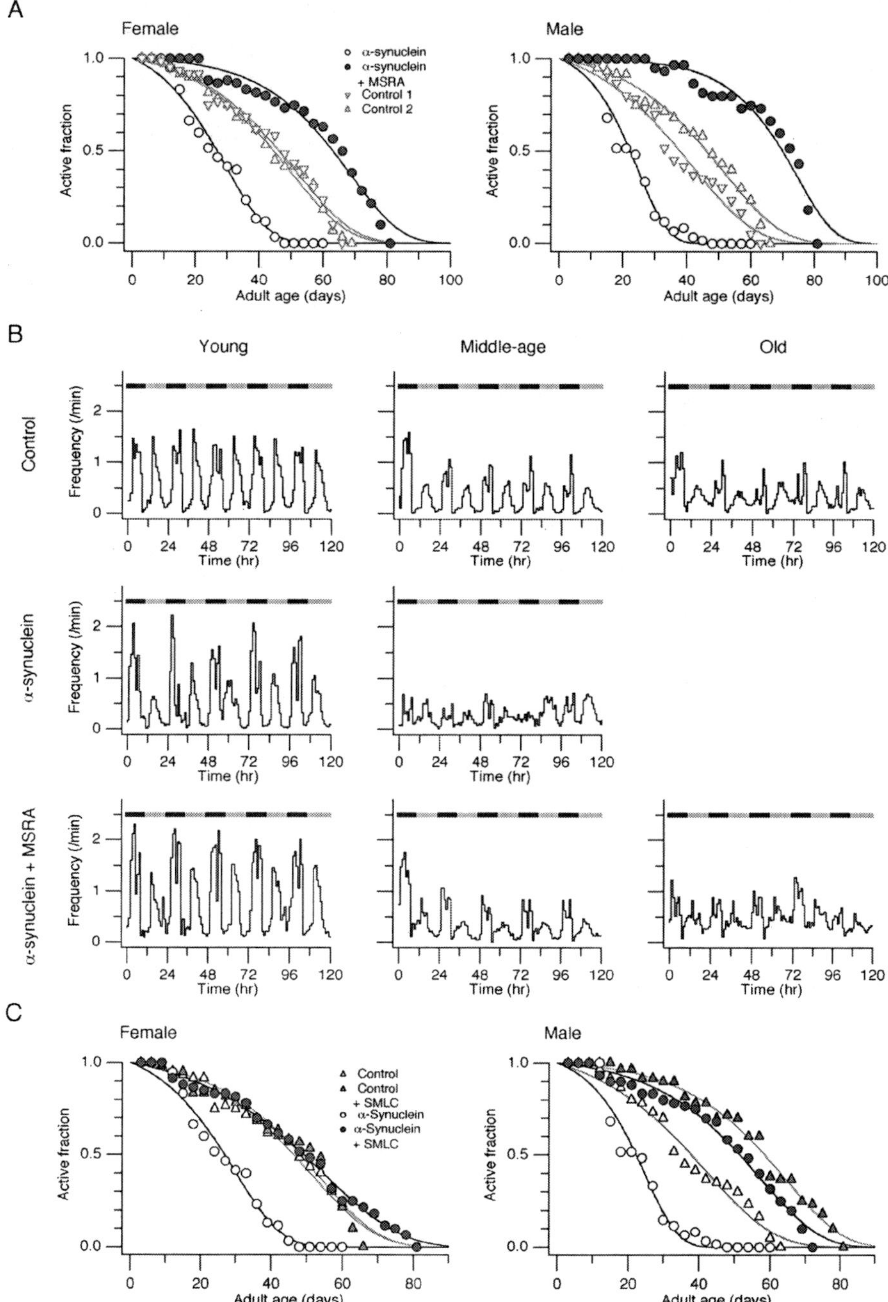

Fig. 2 MSRA and SMLC alleviate the defects in locomotor behavior and circadian rhythm caused by overexpression of α-synuclein in the nervous system. (**A**) Fraction of "active" animals capable of climbing up a laboratory vial as a function of age. Overexpression of

including sleep disturbances, and similar non-motor features are also found in the *Drosophila* overexpressing human α-synuclein [57]. Young *Drosophila melanogaster* exhibits clear "rest/activity" patterns with a peak high-activity phase occurring every 12 h under a laboratory 12 h light/dark condition [59, 60]. The periodic motor activity patterns are considered to represent the underlying circadian clock in *Drosophila* [59] and the activity pattern becomes much less apparent with age [61, 62]. The time course of the progressive decline in the circadian rhythmicity is dramatically accelerated by α-synuclein and restored by MSRA overexpression [57] (Fig. 2B). The improvement is dramatic such that the rest/activity cycles of the animals expressing MSRA and α-synuclein at 60 days is similar to that of the animals expressing α-synuclein at 30 days [57].

The finding that overexpression of MSRA by genetic means alleviates the adverse phenotypes caused by overexpression of human α-synuclein in *Drosophila* suggests that manipulations of the effectiveness of the MSRA catalytic antioxidant system may hold promise as a new therapeutic strategy for PD. One way to increase the MSRA activity is via gene therapy using viral vectors; however, the use of viral vectors may pose unacceptable risk. The long-term effectiveness of cell transplantation therapy is yet uncertain [63, 64]. Clearly safe inexpensive pharmacotherapies are preferred. As postulated by the "sink" hypothesis of the MSRA action, one way to enhance the overall efficacy of the MSRA system is to increase the number of methionine-like sites available to scavenge ROS that can be catalytically regenerated by MSRA. Each cyclic oxidation–reduction mediated by MSRA eliminates one equivalent of ROS [29], thus protecting other vital cellular components from oxidative damage. One obvious strategy to enhance oxidative stress protection is to introduce more methionine, possibly as a dietary supplement. The methionine supplementation is, however, associated with many adverse effects: an increase in the plasma levels of cholesterol and lipoproteins [65, 66]. High dietary intake of methionine increases the levels of homocysteine and may therefore contribute to cardiovascular diseases [67, 68]. While methionine itself may not be a suitable substrate to enhance the catalytic antioxidant mechanism, other Met-*S*-O-like MSRA substrates are available [69]. For example, *S*-methyl-L-cysteine (SMLC) is a methionine analog found abundantly in garlic, turnip, and cabbage, and,

---

Fig. 2 (continued) α-synuclein in the nervous system markedly accelerates the time course of the age-dependent decline (*open circles*) compared with the control groups (*triangles*) and overexpression of MSRA eliminates the adverse effect (*filled circles*). (**B**) Spontaneous movements of the control animals, the animals overexpressing both MSRA and α-synuclein animals at different ages (young: 10 days, middle-age: 30 days, and old: 60 days). The animal activity was measured using an infrared light-based instrument. The *horizontal dark segments* indicate dark periods. (**C**) Dietary supplementation with SMLC also alleviates the accelerated decline in the climbing ability caused by α-synuclein. The results are adapted from Wassef et al. [57]

when it is oxidized, it is a substrate for MSRA [57]. Unlike methionine, which produces adverse health effects, dietary supplementation of SMLC lowers serum cholesterol levels [70]. Furthermore, in rodents, SMLC prevents hepatic cancer and delays diabetic deterioration [71, 72]. According to the catalytic antioxidant "sink hypothesis" of the MSRA action, dietary supplementation of SMLC should bolster the overall efficacy of the endogenous MSRA system, essentially mimicking the effect of MSRA overexpression. This prediction was verified in the *Drosophila* overexpressing α-synuclein [57]. When SMLC is mixed with food, flies overexpressing α-synuclein show a marked improvement in the reactive climbing ability and maintained a lifespan that resembled that of control animals (Fig. 2C). Furthermore, SMLC protects against the loss of dopaminergic neurons caused by α-synuclein overexpression [57]. Besides the beneficial effects of SMLC on *Drosophila* overexpressing human α-synuclein, SMLC also extends the lifespan of male wild-type flies (Fig. 3). In cultured cells, the protective effect of SMLC requires the presence of MSRA, further illustrating the importance of the catalytic antioxidant mechanism involving MSRA [57]. The available evidence suggests that SMLC is well tolerated in animals and humans [70] and may have potential as a viable measure against PD.

In addition to SMLC, other compounds that are possible MSRA substrates when oxidized exist. Some of these compounds are already in clinical use. For example, sulindac (Clinoril®) is a non-steroidal anti-inflammatory drug used for the treatment of osteoarthritis. It has a preventive and therapeutic activity against colon cancer and familial adenomatous polyposis although it is not an FDA-approved indication [73]. Another compound is sulforaphane, which is found in cruciferous vegetables and has anticancer activity [74]. It is not known whether these compounds have any significant effect in the α-synuclein *Drosophila* model of PD or in PD patients; however, these additional compounds are also certainly worthy of investigation.

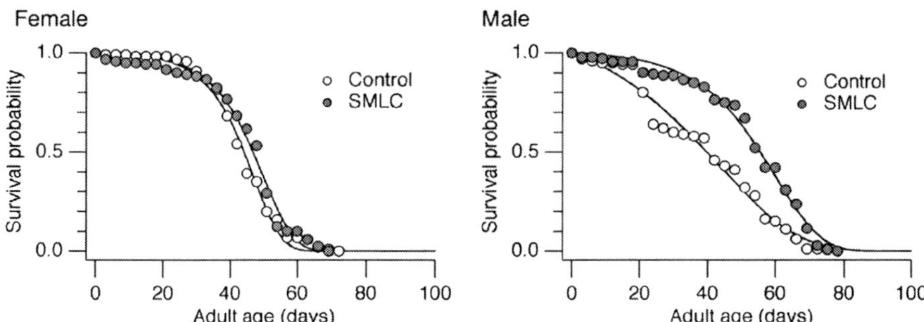

Fig. 3 Supplementation of mashed banana food with SMLC (3 mM) starting at eclosion extends the lifespan of wild-type flies. The effect on the median lifespan is statistically significant in male flies ($p = 0.04$; resampling). Control animals remained on the mashed banana food. The lifespan trial was conducted and the results were analyzed as in Wassef et al. [57]

4 Conclusions

Evidence from a variety of model systems, cultured cells, *Drosophila melanogaster*, and mice, highlighted the importance of reversible oxidation of methionine involving MSRs as an essential component of oxidative damage protection. While oxidative stress itself is unlikely to be the fundamental cause of disease states including PD, oxidative damage is undoubtedly an important contributing factor, thus suggesting that the MSR system could represent a new therapeutic strategy for PD as well as for other diseases in which oxidative stress is implicated. The overall effectiveness of the MSR catalytic antioxidant mechanism can be increased by increasing the catalytic activity and/or by increasing the number of suitable scavenging substrates site. SMLC, a substrate for MSRA, may hold a promise as a way to bolster the endogenous MSRA system thereby conferring greater oxidative stress resistance.

Acknowledgments The authors' laboratories are supported in part by grants from NIH and IZKF Jena.

References

1. Huse DM, Schulman K, Orsini L, Castelli-Haley J, Kennedy S, Lenhart G. Burden of illness in Parkinson's disease. Mov Disord. 2005 Nov;20(11):1449–1454.
2. de Lau LM, Breteler MM. Epidemiology of Parkinson's disease. Lancet Neurol. 2006 Jun;5(6):525–535.
3. Gorell JM, Johnson CC, Rybicki BA, Peterson EL, Richardson RJ. The risk of Parkinson's disease with exposure to pesticides, farming, well water, and rural living. Neurology. 1998 May;50(5):1346–1350.
4. Alexander GE, Crutcher MD. Functional architecture of basal ganglia circuits: neural substrates of parallel processing. Trends Neurosci. 1990 Jul;13(7):266–271.
5. Hawkes CH, Del Tredici K, Braak H. Parkinson's disease: a dual-hit hypothesis. Neuropathol Appl Neurobiol. 2007 Dec;33(6):599–614.
6. Spillantini MG, Schmidt ML, Lee VM, Trojanowski JQ, Jakes R, Goedert M. α-Synuclein in Lewy bodies. Nature. 1997 Aug 28;388(6645):839–840.
7. Cooper AA, Gitler AD, Cashikar A, Haynes CM, Hill KJ, Bhullar B, Liu K, Xu K, Strathearn KE, Liu F, Cao S, Caldwell KA, Caldwell GA, Marsischky G, Kolodner RD, Labaer J, Rochet JC, Bonini NM, Lindquist S. α-synuclein blocks ER-Golgi traffic and Rab1 rescues neuron loss in Parkinson's models. Science. 2006 Jul 21;313(5785):324–328.
8. Giasson BI, Murray IV, Trojanowski JQ, Lee VM. A hydrophobic stretch of 12 amino acid residues in the middle of α-synuclein is essential for filament assembly. J Biol Chem. 2001 Jan 26;276(4):2380–2386.
9. Polymeropoulos MH, Lavedan C, Leroy E, Ide SE, Dehejia A, Dutra A, Pike B, Root H, Rubenstein J, Boyer R, Stenroos ES, Chandrasekharappa S, Athanassiadou A, Papapetropoulos T, Johnson WG, Lazzarini AM, Duvoisin RC, Di Iorio G, Golbe LI, Nussbaum RL. Mutation in the α-synuclein gene identified in families with Parkinson's disease. Science. 1997 Jun 27;276(5321):2045–2047.
10. Singleton AB, Farrer M, Johnson J, Singleton A, Hague S, Kachergus J, Hulihan M, Peuralinna T, Dutra A, Nussbaum R, Lincoln S, Crawley A, Hanson M, Maraganore D, Adler C, Cookson MR, Muenter M, Baptista M, Miller D, Blancato J, Hardy J,

Gwinn-Hardy K. α-Synuclein locus triplication causes Parkinson's disease. Science. 2003 Oct 31;302(5646):841.

11. Chartier-Harlin MC, Kachergus J, Roumier C, Mouroux V, Douay X, Lincoln S, Levecque C, Larvor L, Andrieux J, Hulihan M, Waucquier N, Defebvre L, Amouyel P, Farrer M, Destee A. α-Synuclein locus duplication as a cause of familial Parkinson's disease. Lancet. 2004 Sep 25–Oct 1;364(9440):1167–1169.

12. Grundemann J, Schlaudraff F, Haeckel O, Liss B. Elevated α-synuclein mRNA levels in individual UV-laser-microdissected dopaminergic substantia nigra neurons in idiopathic Parkinson's disease. Nucleic Acids Res. 2008;36(7):e38.

13. Braak E, Braak H. Silver staining method for demonstrating Lewy bodies in Parkinson's disease and argyrophilic oligodendrocytes in multiple system atrophy. J Neurosci Methods. 1999 Feb 1;87(1):111–115.

14. Paleologou KE, Irvine GB, El-Agnaf OM. α-synuclein aggregation in neurodegenerative diseases and its inhibition as a potential therapeutic strategy. Biochem Soc Trans. 2005 Nov;33(Pt 5):1106–1110.

15. Wood SJ, Wypych J, Steavenson S, Louis JC, Citron M, Biere AL. α-synuclein fibrillogenesis is nucleation-dependent. Implications for the pathogenesis of Parkinson's disease. J Biol Chem. 1999 Jul 9;274(28):19509–19512.

16. Kahle PJ, Neumann M, Ozmen L, Muller V, Jacobsen H, Schindzielorz A, Okochi M, Leimer U, van Der Putten H, Probst A, Kremmer E, Kretzschmar HA, Haass C. Subcellular localization of wild-type and Parkinson's disease-associated mutant α-synuclein in human and transgenic mouse brain. J Neurosci. 2000 Sep 1;20(17):6365–6373.

17. Masliah E, Rockenstein E, Veinbergs I, Mallory M, Hashimoto M, Takeda A, Sagara Y, Sisk A, Mucke L. Dopaminergic loss and inclusion body formation in α-synuclein mice: implications for neurodegenerative disorders. Science. 2000 Feb 18;287(5456):1265–1269.

18. van der Putten H, Wiederhold KH, Probst A, Barbieri S, Mistl C, Danner S, Kauffmann S, Hofele K, Spooren WP, Ruegg MA, Lin S, Caroni P, Sommer B, Tolnay M, Bilbe G. Neuropathology in mice expressing human α-synuclein. J Neurosci. 2000 Aug 15;20(16):6021–6029.

19. Wakamatsu M, Ishii A, Iwata S, Sakagami J, Ukai Y, Ono M, Kanbe D, Muramatsu S, Kobayashi K, Iwatsubo T, Yoshimoto M. Selective loss of nigral dopamine neurons induced by overexpression of truncated human α-synuclein in mice. Neurobiol Aging. 2008 Apr;29(4):574–585.

20. Lo Bianco C, Ridet JL, Schneider BL, Deglon N, Aebischer P. α-synucleinopathy and selective dopaminergic neuron loss in a rat lentiviral-based model of Parkinson's disease. Proc Natl Acad Sci U S A. 2002 Aug 6;99(16):10813–10818.

21. Nuber S, Petrasch-Parwez E, Winner B, Winkler J, von Horsten S, Schmidt T, Boy J, Kuhn M, Nguyen HP, Teismann P, Schulz JB, Neumann M, Pichler BJ, Reischl G, Holzmann C, Schmitt I, Bornemann A, Kuhn W, Zimmermann F, Servadio A, Riess O. Neurodegeneration and motor dysfunction in a conditional model of Parkinson's disease. J Neurosci. 2008 Mar 5;28(10):2471–2484.

22. Hashimoto M, Hsu LJ, Xia Y, Takeda A, Sisk A, Sundsmo M, Masliah E. Oxidative stress induces amyloid-like aggregate formation of NACP/α-synuclein *in vitro*. Neuroreport. 1999 Mar 17;10(4):717–721.

23. Uversky VN, Li J, Fink AL. Pesticides directly accelerate the rate of α-synuclein fibril formation: a possible factor in Parkinson's disease. FEBS Lett. 2001 Jul 6;500(3):105–108.

24. Glaser CB, Yamin G, Uversky VN, Fink AL. Methionine oxidation, α-synuclein and Parkinson's disease. Biochim Biophys Acta. 2005 Jan 17;1703(2):157–169.

25. Uversky VN, Yamin G, Souillac PO, Goers J, Glaser CB, Fink AL. Methionine oxidation inhibits fibrillation of human α-synuclein *in vitro*. FEBS Lett. 2002 Apr 24;517(1–3):239–244.

26. Norris EH, Giasson BI, Hodara R, Xu S, Trojanowski JQ, Ischiropoulos H, Lee VM. Reversible inhibition of α-synuclein fibrillization by dopaminochrome-mediated conformational alterations. J Biol Chem. 2005 Jun 3;280(22):21212–21219.

27. Yamin G, Glaser CB, Uversky VN, Fink AL. Certain metals trigger fibrillation of methionine-oxidized alpha-synuclein. J Biol Chem. 2003 Jul 25;278(30):27630–27635.
28. Vogt W. Oxidation of methionyl residues in proteins: tools, targets, and reversal. Free Radic Biol Med. 1995 Jan;18(1):93–105.
29. Weissbach H, Resnick L, Brot N. Methionine sulfoxide reductases: history and cellular role in protecting against oxidative damage. Biochim Biophys Acta. 2005 Jan 17;1703(2):203–212.
30. Rahman MA, Nelson H, Weissbach H, Brot N. Cloning, sequencing, and expression of the *Escherichia coli* peptide methionine sulfoxide reductase gene. J Biol Chem. 1992 Aug 5;267(22):15549–15551.
31. St John G, Brot N, Ruan J, Erdjument-Bromage H, Tempst P, Weissbach H, Nathan C. Peptide methionine sulfoxide reductase from *Escherichia coli* and Mycobacterium tuberculosis protects bacteria against oxidative damage from reactive nitrogen intermediates. Proc Natl Acad Sci U S A. 2001 Aug 14;98(17):9901–9906.
32. Marchetti MA, Lee W, Cowell TL, Wells TM, Weissbach H, Kantorow M. Silencing of the methionine sulfoxide reductase A gene results in loss of mitochondrial membrane potential and increased ROS production in human lens cells. Exp Eye Res. 2006 Nov;83(5):1281–1286.
33. Moskovitz J, Bar-Noy S, Williams WM, Requena J, Berlett BS, Stadtman ER. Methionine sulfoxide reductase (MsrA) is a regulator of antioxidant defense and lifespan in mammals. Proc Natl Acad Sci U S A. 2001 Nov 6;98(23):12920–12925.
34. Picot CR, Petropoulos I, Perichon M, Moreau M, Nizard C, Friguet B. Overexpression of MsrA protects WI-38 SV40 human fibroblasts against H_2O_2-mediated oxidative stress. Free Radic Biol Med. 2005 Nov 15;39(10):1332–1341.
35. Yermolaieva O, Xu R, Schinstock C, Brot N, Weissbach H, Heinemann SH, Hoshi T. Methionine sulfoxide reductase A protects neuronal cells against brief hypoxia/reoxygenation. Proc Natl Acad Sci U S A. 2004 Feb 3;101(5):1159–1164.
36. Moskovitz J, Flescher E, Berlett BS, Azare J, Poston JM, Stadtman ER. Overexpression of peptide-methionine sulfoxide reductase in *Saccharomyces cerevisiae* and human T cells provides them with high resistance to oxidative stress. Proc Natl Acad Sci U S A. 1998 Nov 24;95(24):14071–14075.
37. Prentice HM, Moench IA, Rickaway ZT, Dougherty CJ, Webster KA, Weissbach H. MsrA protects cardiac myocytes against hypoxia/reoxygenation induced cell death. Biochem Biophys Res Commun. 2008 Feb 15;366(3):775–778.
38. Brand AH, Manoukian AS, Perrimon N. Ectopic expression in *Drosophila*. Methods Cell Biol. 1994;44:635–654.
39. Ruan H, Tang XD, Chen ML, Joiner ML, Sun G, Brot N, Weissbach H, Heinemann SH, Iverson L, Wu CF, Hoshi T. High-quality life extension by the enzyme peptide methionine sulfoxide reductase. Proc Natl Acad Sci U S A. 2002 Mar 5;99(5):2748–2753.
40. Raj HG, Sharma RK, Garg BS, Parmar VS, Jain SC, Goel S, Tyagi YK, Singh A, Olsen CE, Wengel J. Mechanism of biochemical action of substituted 4-methylbenzopyran-2-ones. Part 3: A novel mechanism for the inhibition of biological membrane lipid peroxidation by dioxygenated 4-methylcoumarins mediated by the formation of a stable ADP-Fe-inhibitor mixed ligand complex. Bioorg Med Chem. 1998 Nov;6(11):2205–2212.
41. Ossowska K, Wardas J, Smialowska M, Kuter K, Lenda T, Wieronska JM, Zieba B, Nowak P, Dabrowska J, Bortel A, Kwiecinski A, Wolfarth S. A slowly developing dysfunction of dopaminergic nigrostriatal neurons induced by long-term paraquat administration in rats: an animal model of preclinical stages of Parkinson's disease? Eur J Neurosci. 2005 Sep;22(6):1294–1304.
42. Hoshi T, Heinemann S. Regulation of cell function by methionine oxidation and reduction. J Physiol. 2001 Feb 15;531(Pt 1):1–11.
43. Levine RL, Mosoni L, Berlett BS, Stadtman ER. Methionine residues as endogenous antioxidants in proteins. Proc Natl Acad Sci U S A. 1996 Dec 24;93(26):15036–15040.

44. Yao Y, Yin D, Jas GS, Kuczer K, Williams TD, Schoneich C, Squier TC. Oxidative modification of a carboxyl-terminal vicinal methionine in calmodulin by hydrogen peroxide inhibits calmodulin-dependent activation of the plasma membrane Ca-ATPase. Biochemistry. 1996 Feb 27;35(8):2767–2787.

45. Hansel A, Kuschel L, Hehl S, Lemke C, Agricola H-J, Hoshi T, Heinemann SH. Mitochondrial targeting of the human peptide methionine sulfoxide reductase (MSRA), an enzyme involved in the repair of oxidized proteins. FASEB J. 2002;16:911–913.

46. Haenold R, Wassef R, Hansel A, Heinemann SH, Hoshi T. Identification of a new functional splice variant of the enzyme methionine sulphoxide reductase A (MSRA) expressed in rat vascular smooth muscle cells. Free Radic Res. 2007 Sep 28:1–13.

47. Chiueh CC, Andoh T, Lai AR, Lai E, Krishna G. Neuroprotective strategies in Parkinson's disease: protection against progressive nigral damage induced by free radicals. Neurotox Res. 2000;2(2–3):293–310.

48. Gabbita SP, Aksenov MY, Lovell MA, Markesbery WR. Decrease in peptide methionine sulfoxide reductase in Alzheimer's disease brain. J Neurochem. 1999;73(4):1660–1666.

49. Muqit MM, Feany MB. Modelling neurodegenerative diseases in Drosophila: a fruitful approach? Nat Rev Neurosci. 2002 Mar;3(3):237–243.

50. Whitworth AJ, Wes PD, Pallanck LJ. *Drosophila* models pioneer a new approach to drug discovery for Parkinson's disease. Drug Discov Today. 2006 Feb;11(3–4):119–126.

51. Coulom H, Birman S. Chronic exposure to rotenone models sporadic Parkinson's disease in *Drosophila melanogaster*. J Neurosci. 2004 Dec 1;24(48):10993–10998.

52. Feany MB, Bender WW. A *Drosophila* model of Parkinson's disease. Nature. 2000 Mar 23;404(6776):394–398.

53. Polymeropoulos MH. Autosomal dominant Parkinson's disease and α-synuclein. Ann Neurol. 1998 Sep;44(3 Suppl 1):S63–64.

54. Kruger R, Kuhn W, Muller T, Woitalla D, Graeber M, Kosel S, Przuntek H, Epplen JT, Schols L, Riess O. Ala30Pro mutation in the gene encoding α-synuclein in Parkinson's disease. Nat Genet. 1998 Feb;18(2):106–108.

55. Le Bourg E, Lints FA. Hypergravity and aging in *Drosophila melanogaster*. 4. Climbing activity. Gerontology. 1992;38(1–2):59–64.

56. Braak H, Del Tredici K, Rub U, de Vos RA, Jansen Steur EN, Braak E. Staging of brain pathology related to sporadic Parkinson's disease. Neurobiol Aging. 2003 Mar–Apr;24(2):197–211.

57. Wassef R, Haenold R, Hansel A, Brot N, Heinemann SH, Hoshi T. Methionine sulfoxide reductase A and a dietary supplement *S*-methyl-*L*-cysteine prevent Parkinson's-like symptoms. J Neurosci. 2007 Nov 21;27(47):12808–12816.

58. Pesah Y, Burgess H, Middlebrooks B, Ronningen K, Prosser J, Tirunagaru V, Zysk J, Mardon G. Whole-mount analysis reveals normal numbers of dopaminergic neurons following misexpression of α-synuclein in *Drosophila*. Genesis. 2005 Apr;41(4):154–159.

59. Williams JA, Sehgal A. Molecular components of the circadian system in *Drosophila*. Annu Rev Physiol. 2001;63:729–755.

60. Helfrich-Forster C. Differential control of morning and evening components in the activity rhythm of *Drosophila melanogaster*– sex-specific differences suggest a different quality of activity. J Biol Rhythms. 2000 Apr;15(2):135–154.

61. Krishnan B, Dryer SE, Hardin PE. Circadian rhythms in olfactory responses of *Drosophila melanogaster*. Nature. 1999 Jul 22;400(6742):375–378.

62. Fernandez JR, Grant MD, Tulli NM, Karkowski LM, McClearn GE. Differences in locomotor activity across the lifespan of *Drosophila melanogaster*. Exp Gerontol. 1999 Aug;34(5):621–631.

63. Kordower JH, Chu Y, Hauser RA, Freeman TB, Olanow CW. Lewy body-like pathology in long-term embryonic nigral transplants in Parkinson's disease. Nat Med. 2008 Apr 6.

64. Li JY, Englund E, Holton JL, Soulet D, Hagell P, Lees AJ, Lashley T, Quinn NP, Rehncrona S, Bjorklund A, Widner H, Revesz T, Lindvall O, Brundin P. Lewy bodies

in grafted neurons in subjects with Parkinson's disease suggest host-to-graft disease propagation. Nat Med. 2008 May;14(5):501–503.

65. Sugiyama K, Kumazawa A, Zhou H, Saeki S. Dietary methionine level affects linoleic acid metabolism through phosphatidylethanolamine N-methylation in rats. Lipids. 1998 Mar;33(3):235–242.

66. Sugiyama K, Yamakawa A, Kumazawa A, Saeki S. Methionine content of dietary proteins affects the molecular species composition of plasma phosphatidylcholine in rats fed a cholesterol-free diet. J Nutr. 1997 Apr;127(4):600–607.

67. Hirche F, Schroder A, Knoth B, Stangl GI, Eder K. Methionine-induced elevation of plasma homocysteine concentration is associated with an increase of plasma cholesterol in adult rats. Ann Nutr Metab. 2006;50(2):139–146.

68. Hirche F, Schroder A, Knoth B, Stangl GI, Eder K. Effect of dietary methionine on plasma and liver cholesterol concentrations in rats and expression of hepatic genes involved in cholesterol metabolism. Br J Nutr. 2006 May;95(5):879–888.

69. Moskovitz J, Weissbach H, Brot N. Cloning the expression of a mammalian gene involved in the reduction of methionine sulfoxide residues in proteins. Proc Natl Acad Sci U S A. 1996 Mar 5;93(5):2095–2099.

70. Yeh YY, Liu L. Cholesterol-lowering effect of garlic extracts and organosulfur compounds: human and animal studies. J Nutr. 2001 Mar;131(3s):989S–993S.

71. Takada N, Yano Y, Wanibuchi H, Otani S, Fukushima S. S-methylcysteine and cysteine are inhibitors of induction of glutathione S-transferase placental form-positive foci during initiation and promotion phases of rat hepatocarcinogenesis. Jpn J Cancer Res. 1997 May;88(5):435–442.

72. Hsu CC, Yen HF, Yin MC, Tsai CM, Hsieh CH. Five cysteine-containing compounds delay diabetic deterioration in Balb/cA mice. J Nutr. 2004 Dec;134(12):3245–3249.

73. Agarwal B, Rao CV, Bhendwal S, Ramey WR, Shirin H, Reddy BS, Holt PR. Lovastatin augments sulindac-induced apoptosis in colon cancer cells and potentiates chemopreventive effects of sulindac. Gastroenterology. 1999 Oct;117(4):838–847.

74. Juge N, Mithen RF, Traka M. Molecular basis for chemoprevention by sulforaphane: a comprehensive review. Cell Mol Life Sci. 2007 May;64(9):1105–1127.

75. Lowther WT, Brot N, Weissbach H, Matthews BW. Structure and mechanism of peptide methionine sulfoxide reductase, an "anti-oxidation" enzyme. Biochemistry. 2000;39(44):13307–13312.

76. Lowther WT, Weissbach H, Etienne F, Brot N, Matthews BW. The mirrored methionine sulfoxide reductases of *Neisseria gonorrhoeae* pilB. Nat Struct Biol. 2002 May;9(5):348–352.